Lecture Notes in Civil Engineering

Volume 679

Lecture Notes in Civil Engineering (LNCE) publishes the latest developments in Civil Engineering—quickly, informally and in top quality. Though original research reported in proceedings and post-proceedings represents the core of LNCE, edited volumes of exceptionally high quality and interest may also be considered for publication. Volumes published in LNCE embrace all aspects and subfields of, as well as new challenges in, Civil Engineering. Topics in the series include:

- Construction and Structural Mechanics
- Building Materials
- Concrete, Steel and Timber Structures
- Geotechnical Engineering
- Earthquake Engineering
- Coastal Engineering
- Ocean and Offshore Engineering; Ships and Floating Structures
- Hydraulics, Hydrology and Water Resources Engineering
- Environmental Engineering and Sustainability
- Structural Health and Monitoring
- Surveying and Geographical Information Systems
- Indoor Environments
- Transportation and Traffic
- Risk Analysis
- Safety and Security

To submit a proposal or request further information, please contact the appropriate Springer Editor:

- Pierpaolo Riva at pierpaolo.riva@springer.com (Europe and Americas);
- Swati Meherishi at swati.meherishi@springer.com (Asia—except China, Australia, and New Zealand);
- Wayne Hu at wayne.hu@springer.com (China).

All books in the series now indexed by Scopus and EI Compendex database!

Limao Zhang · Yue Pan · Penghui Lin ·
Mirosław J. Skibniewski

Intelligent Construction in Tunnels

Springer

Limao Zhang
School of Civil and Hydraulic Engineering
Huazhong University of Science and Technology (HUST)
Wuhan, Hubei, China

Penghui Lin
School of Civil and Environmental Engineering
Nanyang Technological University
Singapore, Singapore

Yue Pan
Department of Civil Engineering
Shanghai Jiao Tong University
Shanghai, China

Mirosław J. Skibniewski
Department of Civil and Environmental Engineering
University of Maryland
College Park, MD, USA

ISSN 2366-2557 ISSN 2366-2565 (electronic)
Lecture Notes in Civil Engineering
ISBN 978-981-96-8784-8 ISBN 978-981-96-8785-5 (eBook)
https://doi.org/10.1007/978-981-96-8785-5

This Springer imprint is published by the registered company Springer Nature Singapore Pte Ltd.
The registered company address is: 152 Beach Road, #21-01/04 Gateway East, Singapore 189721, Singapore

Preface

Traditional tunnel construction works heavily depend on the extensive expertise of seasoned professionals, with projects often characterized by complex logistics and inherent safety risks. In response, intelligent approaches in tunnel construction are ushering in a new era of operational excellence and risk mitigation. Advanced sensor networks, real-time data analytics, and sophisticated machine learning algorithms are seamlessly integrated to monitor underground conditions, evaluate safety risks, and optimize the construction. These state-of-the-art techniques complement traditional engineering expertise by converting complex datasets into actionable insights, thereby predicting unforeseen conditions and automating the construction process. The application of intelligent approaches in tunnel construction not only reduces operational uncertainties but also significantly boosts efficiency, safety, and sustainability in tunnel construction, paving the way for more resilient and competitively viable tunnel projects globally.

This book underlines the latest applications of intelligent approaches in the domain of tunnel construction. Relying on the representative techniques, we emphasize 14 hot research topics of intelligent approaches to tunnel construction, including "Advance Geological Forecast" (Chapter Two), "Geological Features Characterization" (Chapter Three), "Geological Condition Detection" (Chapter Four), "Tunnel Face Reliability Modeling" (Chapter Five), "Tunneling Equipment Selection" (Chapter Six), "TBM Performance Prediction" (Chapter Seven), "Advanced TBM Attitude Control" (Chapter Eight), "Synchronous Excavation and Segment Assembly" (Chapter Nine), "TBM Fault Detection" (Chapter Ten), "Tunnel Line Alignment" (Chapter Eleven), "Safety Risk Assessment" (Chapter Twelve), "Risk Mitigation Planning" (Chapter Thirteen), and "Digital Twin Platform Development" (Chapter Fourteen). Different intelligent approaches have been proposed to serve different functions that are beneficial to tunnel construction in terms of automation, safety, reliability, and digitalization. These applications are evident to affirm the remarkable value of intelligent approaches in revolutionizing the tunnel construction industry, leading to a more reliable, efficient, automated, sustainable, and safer construction process of tunnels.

This book expects to (1) explore prevalent intelligent construction approaches and seek their possible applications in tunneling works; (2) display techniques and workflows in a detailed manner on how to apply intelligent approaches in tunnel construction; (3) present evident practices to underline the advantages of intelligent approaches in providing automatic, efficient, and reliable solutions by learning from multiple data resources; (4) enhance the understanding of the current development and future advances in intelligent construction approaches and promote their applications in tunnel construction practices.

This book is an interesting reference for scientists, engineers, and graduate students in various disciplines, including but not limited to civil engineering, intelligent construction, computer engineering, system engineering, mathematics, etc.

Wuhan, China — Limao Zhang
Shanghai, China — Yue Pan
Singapore — Penghui Lin
Maryland, USA — Mirosław J. Skibniewski
February 2025

Acknowledgments

It is our sincere desire to thank the anonymous peer reviewers for this work. We would like to extend our sincere gratitude to our esteemed colleagues for their invaluable contributions to this scholarly endeavor. Our profound appreciation is extended to Dr. Kai Guo for his erudite guidance in theoretical framework development, Dr. Xianlei Fu for his methodological insights that significantly strengthened our research design, and Dr. Weiyi Chen for her critical analysis of statistical interpretations. Furthermore, we acknowledge the technical support provided by Mr. Yongsheng Li in data visualization, Mr. Zhang Xia for his meticulous assistance in literature curation and citation management, Ms. Jiaxin Huang for her meticulous stewardship of bibliographic resources and cross-disciplinary citation synthesis, and Ms. Lulu Wang for her exceptional organizational coordination throughout the book preparation process. Their collective expertise and professional dedication were instrumental in enhancing the academic rigor and presentational quality of this work.

Contents

Intelligent Proactive Decisions

Introduction to Intelligent Construction

1 What Is Intelligent Construction

As one of the biggest industries in the world, the construction industry remains notably under-digitized, with project management primarily relying on the expertise and experience of professionals. This traditional approach is further compounded by a global shortage of skilled construction workers, exacerbating challenges such as project delays, cost overruns, suboptimal quality, and complex management tasks [1]. This challenge is especially pertinent in specialized areas such as tunneling engineering, where the intricate nature of underground construction demands exceptional precision, safety protocols, and a highly skilled workforce. To tackle these issues, the industry increasingly incorporates innovative digital technologies, including building information modeling (BIM), artificial intelligence (AI), the Internet of Things (IoT), and robotics. The essential task, therefore, is to develop a broad spectrum of intelligent solutions that can drive the construction industry's digital transformation [2]. In particular, tunneling engineering stands to benefit substantially from the integration of these technologies, enhancing both operational efficiency and safety. This technological shift has given rise to the emerging field of "intelligent construction," which aims to revolutionize traditional construction practices by leveraging advanced digital tools to optimize project delivery, particularly in high-risk and technically demanding areas such as tunneling [3].

At present, intelligent construction is gaining substantial attention within both academic and industry communities. Though no universally accepted definition exists, intelligent construction is broadly understood as a system that integrates software, hardware, sensors, cloud computing, and computational tools aimed at facilitating automation, data-driven technologies, and advanced digital applications. Central to intelligent construction are its capabilities for autonomous perception, learning, analysis, decision-making, and execution, which collectively enable dynamic adaptation to evolving environmental conditions. By adopting digital technologies, intelligent construction reduces time and labor while improving quality,

L. Zhang et al., *Intelligent Construction in Tunnels*, Lecture Notes in Civil Engineering 679, https://doi.org/10.1007/978-981-96-8785-5_1

handling complexity, and managing uncertainty throughout construction processes. This shift is expected to drive the construction industry toward greater servitization, networking, customization, and intelligence, with a significant emphasis on human-centric, resilient, and sustainable practices in areas like tunneling. These advancements align with the principle of Industry 5.0, which prioritizes societal value over purely economic gains and stresses the need for better collaboration between advanced machinery and human expertise [4].

Intelligent construction, with a focus on enhancing automation, productivity, safety, and sustainability, is gradually gaining traction in many countries. Despite significant infrastructure gaps in technology integration, regions like North America, Europe, and Asia-Pacific are seizing opportunities in the smart construction market. For instance, the United Kingdom introduced its Vision for Construction 2025 in 2013, aiming to invest in smart construction and digital design to modernize the industry. Japan's i-Construction initiative, launched in 2015, seeks to enhance project planning and logistics through advanced technologies, promoting digital transformation in construction processes. Similarly, the U.S. proposed the "Infrastructure For Rebuilding America" strategy in 2017, utilizing cutting-edge technology to drive socio-economic growth and connectivity. In 2020, Chinese government bodies issued the "Guiding Opinions on Promoting the Coordinated Development of Intelligent Construction and Building Industrialization" to integrate intelligent construction concepts throughout the industry, facilitating its industrialization. As a result, the global intelligent construction market is expected to grow significantly, potentially reaching multi-billion USD by 2028. This trend presents a crucial opportunity for countries to lead in the construction sector, particularly in specialized areas like tunneling, where complex underground infrastructure projects stand to gain significantly from the application of digital innovations.

Motivated by the popularity of intelligent construction, several review papers have emerged to shed light on this rapidly expanding field [5]. However, these existing reviews suffer from some limitations. Firstly, many adopt narrow perspectives, focusing on specific application topics and risking the neglect of critical areas for project improvement. This lack of a comprehensive overview can create a unilateral understanding of intelligent construction, making it challenging for readers to explore new research avenues. Therefore, a holistic analysis is needed to inspire further theoretical and methodological advancements. Secondly, while the role of AI in enhancing accuracy, automation, and efficiency is frequently emphasized, there is a lack of focus on other innovative techniques that are transforming the industry, such as advancements in robotics, sensor networks, and real-time data analytics. This oversight is particularly evident in specialized areas like tunneling engineering, where advanced technologies are revolutionizing operations but are often excluded from broader discussions on intelligent construction. Furthermore, the construction industry remains one of the lowest-ranked sectors in terms of digitalization according to the McKinsey digital globalization index. This late onset of intelligent development has resulted in underdeveloped theoretical foundations and technological frameworks, especially in niche sectors like tunneling.

To bridge these gaps, a comprehensive and systematic investigation is necessary to create a robust body of knowledge on intelligent construction. This includes a detailed summary of implementation strategies, prominent research topics, prevailing challenges, and future research directions. Such an investigation is crucial for guiding research, policy development, and practical applications, particularly as the industry prepares for the next phase of digital and intelligent transformation under Industry 5.0. Specifically, for tunneling engineering, this book aims to provide critical insights into the technological innovations and methodologies driving the evolution of underground construction. By synthesizing this knowledge, researchers will be positioned to chart a strategic development pathway that integrates digital technologies with value-optimizing strategies, thus advancing the realization of intelligent construction in the context of ongoing digitalization.

2 Key Supporting Technologies for Intelligent Construction

The development of intelligent construction cannot be separated from the support of various technologies. As a practical approach in the construction industry, intelligent construction incorporates a wide range of professional technologies, which are then localized and adapted to specific project requirements and real-world scenarios. At present, the main technologies of intelligent construction can be summarized as "4S" enabling technologies, namely Smart sensing and devices, Signal transmission and communications technology, Software with computational intelligence, and System integration for improved decision-making. These technologies play a pivotal role in realizing construction modeling, data transmission, intelligent decision-making, and overall project implementation, ensuring the full integration of intelligence throughout the construction lifecycle. In tunneling engineering, for example, these technologies offer vital solutions for managing the complexities of underground projects, including safety monitoring, real-time performance tracking, and efficient resource management. Figure 1 illustrates these key "4S" technologies of intelligent construction, with a detailed presentation of their primary functions and relationships provided below. The advantages and disadvantages of the relevant technologies are shown in Table 1.

(1) Smart sensing and devices

Smart sensing and devices are crucial data acquisition tools in intelligent construction, enabling the perception of engineering environments and data collection. These advanced systems surpass basic sensors, effectively addressing the growing demands for automation and efficiency in modern construction management [6]. A variety of engineering sensing devices have been developed to ensure fast, accurate data acquisition with minimal noise across diverse environments. Common smart sensors include various types of sensors, cameras, and point cloud radar, all designed for effective information collection and exchange. In the construction industry, these devices can monitor personnel, equipment, and environmental conditions throughout

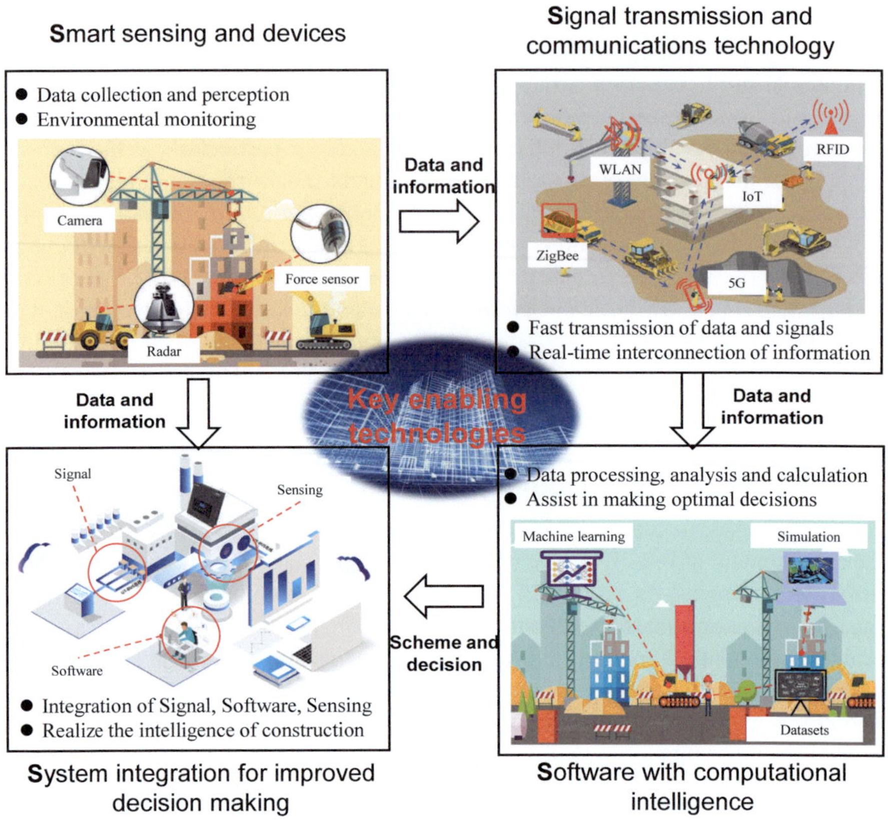

Fig. 1 Key "4S" technologies of intelligent construction

the entire design, construction, and maintenance lifecycle. For instance, sensors track the operational status of machinery and the safety of buildings, while cameras capture construction progress and personnel activity. Point cloud radar can assess interactions between structures and their environments, particularly regarding ground settlement. By utilizing smart sensing technologies, real-time tracking of construction progress and status becomes possible, providing a foundation for data-driven management and control. However, the widespread adoption of these devices faces challenges, primarily concerning reliability. The harsh conditions of construction sites impose strict requirements for waterproofing, dustproofing, and shock resistance. Additionally, managing the acquisition and processing of large volumes of data within limited structural confines poses a significant hurdle for smart sensing technologies.

(2) Signal transmission and communications technology

Signal transmission and communications technology facilitate efficient information exchange in intelligent construction, enabling real-time interaction among

Table 1 The advantages and disadvantages of the relevant technologies

Key enabling technologies	AI-based technologies	Advantages and efficacy	Limitations and challenges	References
Smart sensing and devices	Smart sensors	• Various types, suitable for different demand scenarios • Small size, no additional burden on the building structure • Strong anti-interference	• The layout scheme needs to be designed according to the site requirements • Weak data processing ability • As accuracy increases, the cost increases dramatically	[7]
	Depth cameras	• Good visualization • Long service cycle, can be a long time uninterrupted operation	• Weak anti-interference • Mainly collect pictures and video information • Do not have perspective functions	[8]
	Point cloud radar	• Wide monitoring range • High monitoring accuracy	• High costs • Slow information collection speed	[9]
Signal transmission and communications technology	5G	• Higher network speed • Low latency and high reliability • Low power mass connection	• High power consumption • Base stations need to be laid out on a large scale	[10]
	IoT	• High communication efficiency • High integration • High security	• Not suitable for cross-regional applications • Consistent communication protocols are required	[11]
	Bluetooth	• Low power, easy to operate battery powered equipment • Cheap and can be applied to low-cost devices	• Limited transmission distance • Poor penetration • Weak anti-interference	[12]

(continued)

Table 1 (continued)

Key enabling technologies	AI-based technologies	Advantages and efficacy	Limitations and challenges	References
Software with computational intelligence	Simulation and modeling	• Good visualization • High precision, in line with the real scene	• High computational cost • The boundary conditions are difficult to determine	[13]
	Machine learning	• High data processing efficiency • Strong interpretability	• Not suitable for large amounts of data • Poor portability	[14]
	Deep learning	• Suitable for large amounts of data • Good portability	• Poor interpretability • High computational cost	[15]
	Reinforcement learning	• The model can be continuously enhanced • No manual supervision is required	• High training cost • Poor convergence • Complex environment design	[16]
	Multi-objective optimization	• Low training cost • Good convergence	• The result is locally optimal • Difficult to continuously optimize	[17]
System integration for improved decision making	Knowledge system	• Good stability • Strong reasoning ability	• Limited transferability to other domains • Design lots of rules	[18]
	Cyber–physical systems	• Good computation ability • Good communication ability Good control ability	• Requires a lot of computing resources • Mainly used in the verification of scientific principles	[19]
	Digital twins	• Good visualization • Real-time and fidelity	• High cost of construction • Mainly used in engineering projects	[20]

various engineering entities. Complementing smart sensing devices, these technologies ensure the seamless sharing of collected data among on-site and off-site participants throughout the construction process. In the Internet of Everything era, information flow extends beyond human interactions to include machine-to-machine and human-machine collaborations, especially in construction monitoring and safety management. Currently, common communication technologies include 5G networks and WLAN (Wireless Local Area Network), along with various IoT protocols (e.g., MQTT, CoAP) and wireless protocols (e.g., Wi-Fi, Bluetooth, RFID, ZigBee, Sigfox, LoRaWAN). In practice, these technologies connect construction site units; for example, designers can access real-time site data while excavator drivers receive updates on tower crane status. This connectivity enhances interaction and feedback among components of intelligent construction, significantly improving crisis management and operational efficiency. Compared to traditional construction, these technologies streamline the entire construction process, reducing information transmission costs. The main barrier to signal transmission and communications technology is the inefficient transmission of information in extreme operating environments. Signal transmission at high altitudes, extremely cold, and closed construction usually causes data loss or signal disconnection, which is easy to adversely affect the efficiency and safety of construction.

(3) Software with computational intelligence

Software with computational intelligence is a tool for information processing in intelligent construction. Software with computational intelligence plays a vital role in the "engineering brain" in intelligent construction, and thus it can assist construction personnel in making optimal decisions. By applying various data application methods and various commercial software, it can effectively make up for the errors caused by traditional manual decision-making and calculation. Through various data collected by Smart sensing and devices and transmitted by Signal transmission and communications technology, software with computational intelligence is responsible for implementing data analysis, processing, and application in an intelligent way. With increasing demands for intelligence, numerous data analysis approaches, including simulation and modeling, machine learning, deep learning, and reinforcement learning, have been integrated into existing software, enhancing accuracy and automation. Common simulation methods include the finite element, finite difference, and finite volume methods, with representative software such as ANSYS and ABAQUS. Machine learning algorithms like support vector machines and random forests handle low-dimensional data and are often packaged in software like Python's SMT and Matlab's Dace. Deep learning, which manages high-dimensional data, uses algorithms such as convolutional and recurrent neural networks, implemented through libraries like Keras and TensorFlow. Reinforcement learning techniques, such as deep Q-networks and proximal policy optimization, continuously improve models based on real-world experiences. Additionally, mature standard datasets, like the site security and subway leakage datasets, exist for method validation. Designers can also create custom software to address specific engineering needs. However, a significant barrier in the construction industry is the "Information Island" phenomenon, where the abundance of confidential data hampers the updating and promotion of intelligent algorithms. Overall, integrating computational intelligence into software is crucial for enhancing intelligent decision-making in construction.

(4) System integration for improved decision making

In intelligent construction, smart sensing and devices, signal transmission and communications technology, software with computational intelligence, and other technologies are usually integrated into a set of systems, where specific functions are realized through corresponding integrated systems. The integrated system is believed to be the core terminal of intelligent construction, which undertakes all tasks of the whole construction process from perception and transmission to decision-making and operation. Integrated systems, such as BIM, cyber-physical systems, and digital twins, can orchestrate elements and units in the construction industry, which ensure the standardization, digitization, and intellectualization of intelligent construction. Compared with traditional construction, system integration can effectively improve construction efficiency, increase coordination between entities, and ensure the effective implementation of various technologies in intelligent construction. Besides, techniques, like data perception, information transmission, intelligent optimization, and other work, can be applied to develop sorts of integrated systems, which finally provide feedback to the engineering entity to improve the existing construction decision-making. In other words, the effective integration of various technologies of intelligent construction can promote a high level of digitalization, intelligence, standardization, and collaboration in the whole construction cycle. The main barrier to system integration mainly comes from the separation of design, construction, and maintenance in the construction industry. Only by coordinating the whole life cycle of the building can the maximum advantages of system integration in intelligent construction be brought into play.

3 Intelligent Construction Practices Over the World

Intelligent construction has emerged as a transformative force within the construction industry, particularly in tunneling engineering, where its application is revolutionizing traditional practices. Intelligent construction inherits the characteristics of intelligence, digitalization, and sustainability. Through the integration of cutting-edge technologies, including the Internet of Things (IoT), big data analytics, 5G connectivity, machine learning, and artificial intelligence (AI), intelligent construction is effectively addressing persistent challenges such as high pollution, energy inefficiency, and labor-intensive workflows. This evolution is steering the sector towards a new industrial revolution, characterized by enhanced efficiency, reduced environmental impact, and improved operational sustainability.

Building upon this foundation, the growing significance of intelligent construction is further illustrated by the semantic word cloud presented in Fig. 2, derived from the literature review. The cloud visualizes key terms and concepts that are central to the ongoing digital transformation in the field. Larger words in the cloud indicate a higher frequency of occurrence, thereby emphasizing emerging research hotspots. Notably, key terms like "smart city," "machine learning," "deep learning," and "artificial intelligence" are prominent, signaling heightened interest in these technologies within the context of intelligent construction. These terms underscore the increasing role of digital tools and AI-driven solutions in transforming traditional tunneling practices. For instance, the prominence of "Smart City" and "IoT" highlights the critical function these technologies play in facilitating broader digitalization and interconnected systems. In tunneling projects, IoT devices, such as sensors and monitoring systems, are used to collect real-time data on ground conditions, structural integrity, and equipment performance. This data, transmitted via high-speed 5G networks, fosters seamless communication between on-site teams and remote experts. Furthermore, advancements in "Machine Learning," "Deep Learning," and "Artificial Intelligence" are reshaping operational practices. These technologies enable machines to analyze extensive datasets, optimize tunneling processes, predict potential risks, and enhance decision-making. For instance, AI algorithms can predict potential ground collapses or equipment failures, allowing for proactive mitigation measures. Machine learning models can improve the prediction of TBM operations. The integration of big data analytics further enhances these efforts by enabling engineers to process complex datasets efficiently, identify trends, optimize resource allocation, and refine project planning strategies.

As reviewed, the adoption of intelligent construction in tunneling offers numerous advantages that significantly enhance project performance across several key dimensions. Safety is notably improved through the deployment of real-time monitoring systems and predictive analytics, which facilitate the early detection of potential hazards, such as ground movements or gas leaks. This enables proactive interventions, mitigating risks before they escalate. The integration of AI-driven technologies, such as automated tunnel boring machines (TBMs), further enhances operational efficiency by optimizing resource utilization, reducing construction timelines, and minimizing material waste, ultimately accelerating project completion. Sustainability is advanced through the use of smart ventilation systems, energy-efficient machinery, and waste-reduction strategies, which optimize energy consumption and minimize

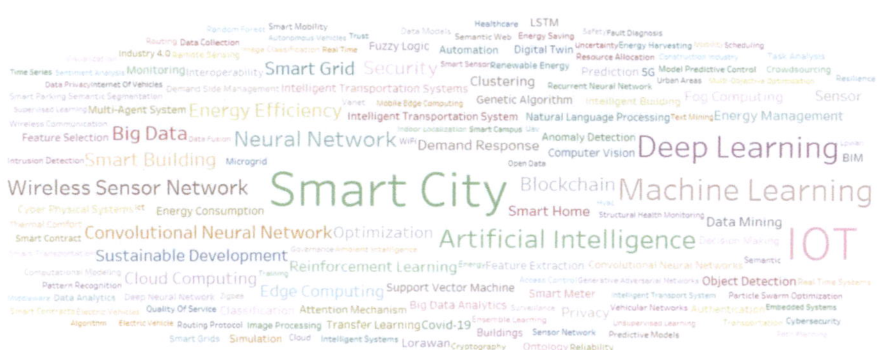

Fig. 2 Word cloud of keywords in the selected papers

environmental impact, contributing to the development of more sustainable infrastructure. Moreover, the implementation of predictive maintenance protocols and optimized resource allocation reduces operational downtime, resulting in substantial cost savings throughout the lifecycle of tunneling projects. Finally, the integration of IoT, AI, and big data analytics empowers data-driven decision-making, which not only enhances project outcomes but also ensures adherence to rigorous safety and quality standards, facilitating a more streamlined and intelligent construction process.

The distribution of selected publications on intelligent construction across various countries and regions is illustrated in Fig. 3, highlighting the global contributions to this rapidly evolving field. It is evident that countries/regions in Asia, North America, and Australia are at the forefront of this research, with China, the United States, the United Kingdom, Spain, and South Korea ranking as the top five contributors. Notably, China leads with 2661 publications, accounting for 29.82% of the total 8924 papers published on intelligent construction. This dominance can be attributed to a combination of government-backed initiatives, robust policies, and significant investments in advancing construction technologies. For example, in tunneling engineering, intelligent construction has proven to be a game-changer, offering substantial benefits in terms of safety, efficiency, and sustainability. Furthermore, the Chinese government has played a pivotal role in this transformation. The "Strategic Research about China Construction 2035" project, launched by the Chinese Academy of Engineering in 2019, emphasizes intelligent construction as a cornerstone of future development, exploring innovative ways to modernize the construction industry with Chinese characteristics. The Ministry of Housing and Urban-Rural Development

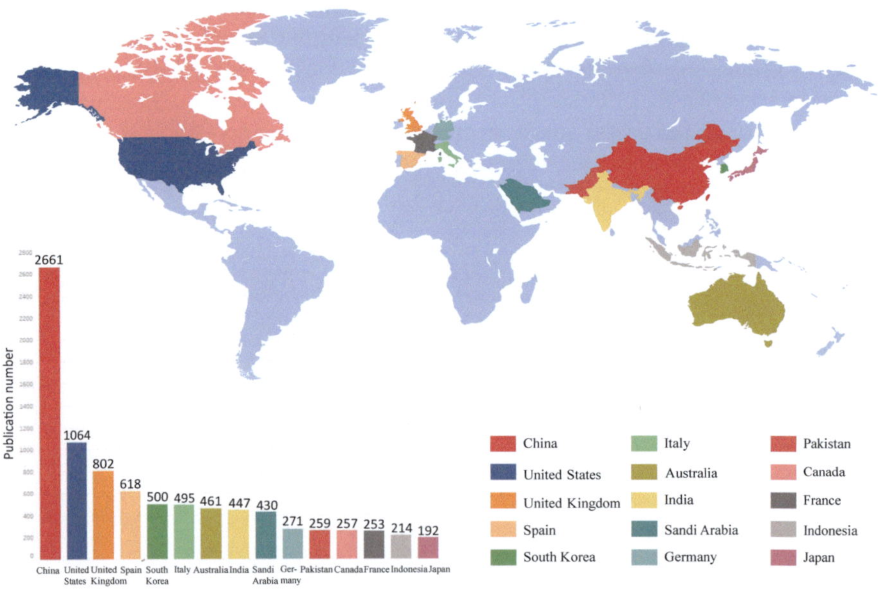

Fig. 3 Number of publications distributed in countries/regions

further reinforces this shift through the "14th Five-Year Plan," which aims to elevate industrialization, digitalization, and the intelligence of construction, setting China on a path to become a global leader in the field.

The favorable environment in these leading countries—characterized by robust policy support, comprehensive personnel training systems, advanced technical extension mechanisms, and sophisticated assessment frameworks—has been a critical catalyst for the rapid advancement of intelligent construction, particularly in tunneling and other large-scale infrastructure projects. These efforts not only drive innovation but also contribute to enhanced safety, reduced costs, and a more sustainable construction approach, fostering a competitive edge in global infrastructure development.

4 Intelligent Construction in Tunneling Works

Tunnel construction is a complex and high-risk endeavor. Traditional tunnel construction techniques have historically faced numerous challenges including geological uncertainty, environmental constraints, and technological limitations. These challenges underscore the need for more advanced and intelligent construction approaches to improve the efficiency, safety, and reliability of tunnel construction projects. To overcome these challenges, the industry is attempting to integrate intelligent construction approaches into tunnel engineering. For instance, the application of machine learning algorithms has been shown to significantly enhance the reliability and sustainability of tunnel construction projects. Additionally, the use of digital tools can automate manual tasks and facilitate active data management, leading to time and cost savings as well as increased accessibility [21]. Techniques such as machine learning, computer vision, stochastic processes, soft computing, and digital modeling have been pivotal in enhancing efficiency, safety, and decision-making throughout the construction lifecycle [22]. The application of current intelligent construction techniques is shown in Fig. 4.

Tunnel construction always faces complex underground conditions, where unexpected geological conditions, such as weak rock formations, high groundwater pressure, and fault zones, can lead to significant delays and increased construction costs. These unforeseen circumstances often necessitate design modifications and the implementation of additional support systems to ensure structural stability, calling for the first need for intelligent techniques in this field. On the other hand, as tunnel boring machines (TBM) take most of the underground excavation job, the operation of the TBM is another essential factor affecting the quality of the construction. Traditional methods often rely on manual labor and less advanced machinery, which can limit the precision and speed of excavation. This reliance increases the likelihood of human error and may not adequately address complex geological conditions [23]. With these concerns, the current application of intelligent construction mostly falls into problems related to soil excavation and TBM operation.

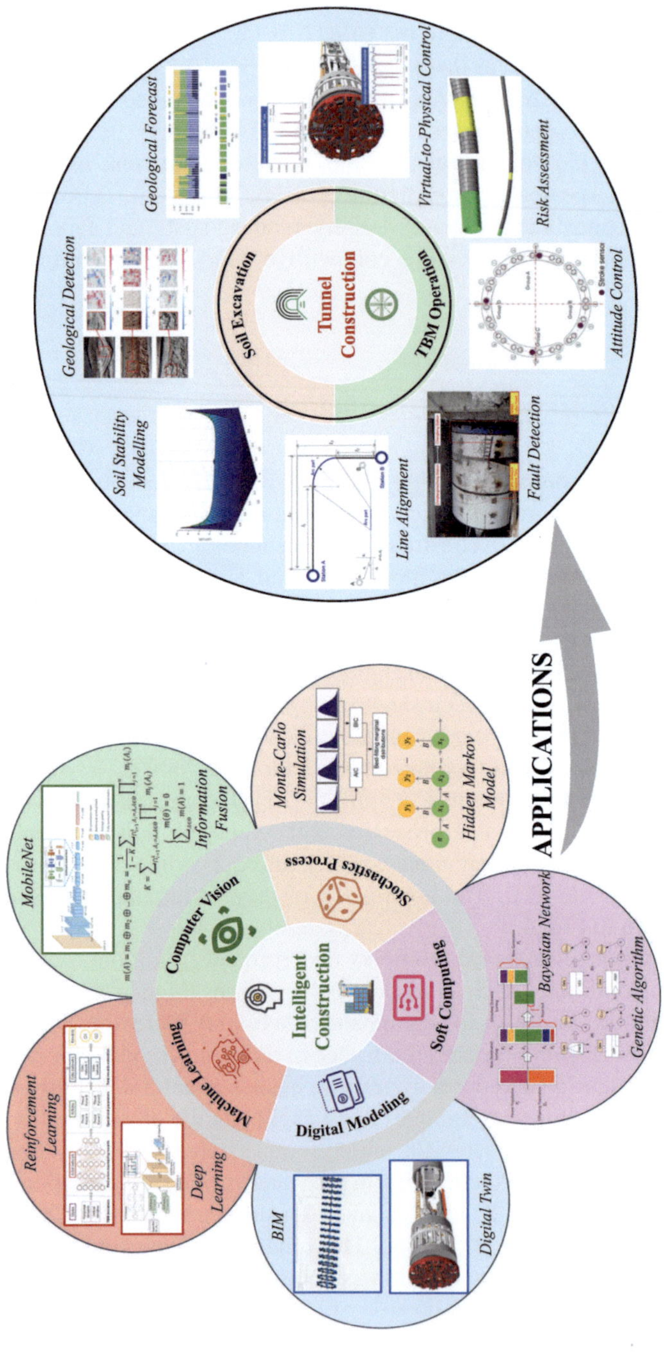

Fig. 4 Application of intelligent construction techniques in tunnel construction

Specifically, applications of machine learning include predicting TBM performance metrics like penetration and advance rates, as well as forecasting surface settlement to mitigate potential risks. The use of various machine learning techniques in TBM tunneling, emphasizes data processing, algorithm selection, and optimization methods to improve predictive accuracy [24]. Computer vision analyzes images and videos from construction sites, these systems can detect structural anomalies, monitor equipment usage, and ensure compliance with safety protocols. For instance, computer vision algorithms can automatically identify rock fractures in tunnel faces, aiding in geological condition assessment and informing excavation strategies [25]. Stochastic processes are utilized to model the inherent uncertainties in tunnel construction, such as variable geological conditions and equipment performance. Particularly, stochastic modeling can aid in simulation of the geological uncertainties, allowing for proactive mitigation measures [26]. Soft computing techniques, including neural networks, fuzzy logic, and genetic algorithms, have been applied to address complex problems in tunnel construction. The application of soft computing methods in tunneling, noting their effectiveness in handling the uncertainties and nonlinearities characteristic of underground construction. Lastly, digital modeling, particularly through BIM and digital twin technologies, has revolutionized tunnel construction by providing comprehensive virtual representations of physical assets. These models facilitate better planning, real-time monitoring, and predictive maintenance.

To better illustrate the application of intelligent construction approaches in tunneling works, a keyword network of related studies is plotted in Fig. 5. The size of the circle presents the weight of the keyword, and the color represents the average year that the study with the keyword was published. Similarly, keywords like "machine learning", "deep learning," and "artificial intelligence" are still prominent in tunnel construction. Notably, the keyword "machine learning" appears more than 400 times over 2000 studies, showing its significant generalization and applicability in tunnel construction. Additionally, it can be shown that the application of intelligent techniques follows the path from soft computing (neural networks and optimization) to machine learning, deep learning, and the application of digital twins appears most recently. Considering the time that these techniques were developed and applied in other fields, there is still great potential to develop more intelligent tunnel construction processes. This is still a young topic, and some more up-to-date techniques remain to be explored and show their value in tunnel construction.

5 Structure of the Book

This chapter briefly introduces the concept, development, and techniques of intelligent construction approaches and reviews their application in tunnel construction. The substantial benefits of intelligent approaches in tunnel construction are summarized.

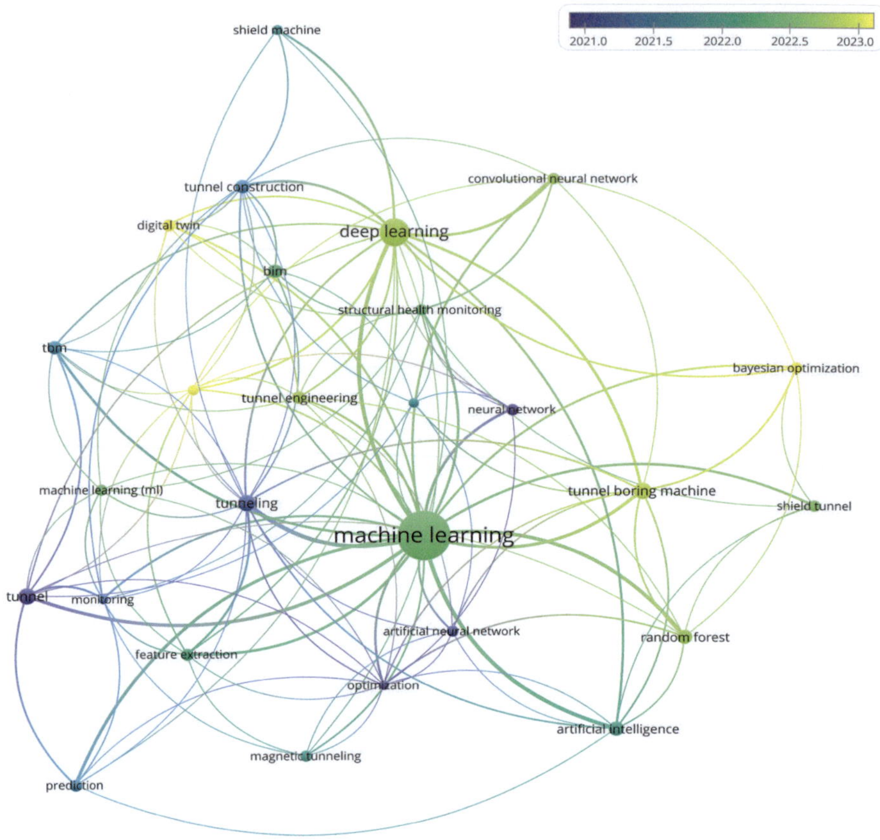

Fig. 5 Keyworks network of studies related to intelligent construction in tunneling works

Chapter "Advance Geological Forecast" describes the advanced geological forecast and its application in tunnel construction. Prevalent advanced geological forecasting methods are reviewed. A novel technique named the online hidden Markov model (OHMM) is particularly developed to predict geological risks in TBM-excavated tunnels. OHMM dynamically updates as new data becomes available, making it well-suited for real-time applications. The model parameters are estimated online using readily obtainable observations, enabling adaptive learning. An observation extension mechanism is incorporated to allow short observed data sequences to be extended to match the length of a complete sequence. A case study on a tunnel excavation project in Singapore demonstrates the effectiveness of this approach in assessing geological risks. By enabling early-stage risk prediction and long-term forecasting with minimal historical data, the OHMM approach optimally utilizes site investigation data, improving the efficiency and safety of tunneling operations.

Chapter "Geological Features Characterization" describes the geological features characterization and its application in tunnel construction. Several competitive geological features characterization methods are reviewed. Particularly, this chapter presents the DTW-Kmedoids hybrid algorithm, a time series clustering approach that combines Dynamic Time Warping (DTW) for similarity measurement and K-medoids for robust classification while maintaining low computational complexity. This method enables automatic perception of geological conditions ahead of the TBM without requiring prior knowledge of geological categories. To validate the effectiveness of this time series clustering method, a case study was conducted on a TBM tunneling project in Singapore. Practically, this method leverages large-scale TBM operational data to deliver accurate geological feedback, providing actionable insights for adjusting TBM operations.

Chapter "Geological Condition Detection" describes geological condition detection and its application in tunnel construction. Common geological condition detection methods are reviewed. Specifically, this chapter introduces a computer vision-based approach for identifying geological conditions encountered during TBM operations. The proposed method integrates multiple MobileNet base models with Dempster-Shafer Theory (DST) for information fusion, enabling reliable geological classification based on images of excavated muck transported on the TBM's conveyor belt. SHAP analysis is applied to enhance model interpretability, providing insights into the key image features that influence geological classification. A case study using data from Singapore's Circle Line 6 tunnel project was conducted for validation. Real excavation images were pre-processed, split, and used for model training and evaluation. The results demonstrate the method's superiority in improving geological risk assessment and real-time TBM operation adjustments.

Chapter "Tunnel Face Reliability Modelling" describes tunnel face reliability modeling and its application in tunnel construction. A copula-based methodology is developed for conducting robust probabilistic analyses of tunnel excavation face stability during construction. The method constructs and simulates the bivariate joint probability distribution function using a copula function and performs reliability analysis through the Monte Carlo (MC)-copula method. This approach provides an effective tool for accurately modeling dependency structures among variables and improving reliability assessments, especially in scenarios with incomplete probability information. The practicality and effectiveness of this approach are demonstrated through a case study on the Wuhan metro system, providing a more reliable and data-efficient estimation of excavation face stability in tunnels with limited observations.

Chapter "Tunnelling Equipment Selection" describes the tunneling equipment selection and its application in tunnel construction. A novel soft computing method, which is named cloud-TOPSIS, is designed to address uncertainty in the selection of the optimal tunnel excavation method. TOPSIS (Technique for Order Preference by Similarity to Ideal Solution) is a widely used multi-criteria decision-making (MCDM) method that analyzes numerical decision matrices. The cloud model is integrated into TOPSIS to capture the fuzziness and randomness in the assessment process, allowing for more realistic decision-making under uncertainty. A case study on a tunnel section of the Wuhan Metro Line 2 is conducted to validate the applicability and effectiveness of the proposed approach. This approach advances both theory and practice by offering a robust, uncertainty-aware decision-making framework that can be applied to tunnel excavation projects and other complex engineering problems.

Chapter "TBM Performance Prediction" describes the TBM performance prediction and its application in tunnel construction. Some well-known TBM performance prediction methods are reviewed. Specifically, this chapter proposes the C-GRU method for multi-step TBM attitude prediction, which combines a 1D convolutional neural network (CNN) and gated recurrent unit (GRU) network to preserve all essential features of tunneling process data and achieves high forecasting accuracy over multiple steps. The random balance design Fourier amplitude sensitivity test (RBD-FAST) is used to quantify the first-order sensitivity between input features and model predictions. To validate its effectiveness, the C-GRU model was applied to a tunnel project in Singapore, demonstrating its ability to provide accurate multi-step TBM attitude predictions. The C-GRU method demonstrates significantly higher accuracy and robustness when applied in engineering contexts.

Chapter "Advanced TBM Attitude Control" describes the advanced TBM attitude control and its application in tunnel construction. Several techniques used for TBM attitude control are reviewed. A deep reinforcement learning (DRL) approach that considers the spatial-temporal dynamics of the TBM is developed for position control in tunneling operations. The environment model is designed using the long-short-term memory (LSTM) network, to simulate the spatial-temporal dynamics of the TBM within its working environment, providing precise forecasting to support position control. By integrating spatial-temporal dynamics with TBM excavation into DRL, the method leads to more efficient and reliable position control for intelligent tunneling operations. The effectiveness of the proposed method is validated using real-world tunnel data from Singapore, demonstrating improvements in predictive accuracy that collectively advance intelligent system capabilities in complex real-world tunneling projects.

Chapter "Synchronous Excavation and Segment Assembly" describes the synchronous excavation and segment assembly (SESA) technique and its application in tunnel construction. Fundamental concepts and different techniques of SESA are introduced. A Bayesian optimization extreme gradient boosting (BO-XGBoost) model is particularly developed to enable real-time hydraulic pressure detection. An online Archimedean optimization algorithm (AOA) is additionally applied to enable real-time hydraulic pressure optimization. To validate the feasibility of the proposed framework, a SESA TBM experimental bench is constructed, and the methodology is tested through field experiments. The results demonstrate that the online optimization method effectively reduces hydraulic cylinder stress concentration, improving operational stability. This approach provides a theoretical foundation for advancing synchronous excavation and segment assembly technology, contributing to more efficient and reliable TBM operations.

Chapter "TBM Fault Detection" describes TBM fault detection and its application in tunnel construction. TBM failure mechanism and TBM fault detection framework are introduced. A systematic approach has been developed for assessing the risk of shield cutter head failures in tunnel construction. This method consists of four major steps: failure mechanism exploration, dynamic fault tree (DFT) construction, discrete-time Bayesian network (DTBN) formation, and risk analysis. The DTBN eliminates common challenges associated with Markov models, such as state-space explosion and error-prone conversion processes. To validate its applicability, a tunnel case study in China was conducted, showing its capability of performing both predictive analysis and diagnostic analysis. This approach provides decision support for both feedforward and feedback control using quantitative risk assessment, ensuring stable and efficient TBM operations over time.

Chapter "Tunnel Line Alignment" describes the tunnel line alignment and its application in tunnel construction. A genetic algorithm (GA) based approach for multi-objective optimization method is designed to satisfy conflicting objectives in tunnel line design. During the optimization process, a set of trade-off solutions is generated along the Pareto front, from which the solution closest to the ideal point is selected as a reference for decision-making. The method's applicability is validated through a real-world tunnel line alignment project in China, where the tunnel radius and depth are considered decision variables, and investment, headway, and comfort serve as optimization objectives. The result indicates that this approach can effectively perform multi-objective optimization, offering practical solutions across different scenarios and providing project owners with a more informed decision-making framework.

Chapter "Safety Risk Assessment" describes the safety risk assessment and its application in tunnel construction. This chapter proposes a hybrid approach that enhances risk management in complex underground projects by integrating the strengths of Building Information Modeling (BIM) in data processing and visualization with the capabilities of Dempster-Shafer (D-S) evidence theory in handling vagueness and uncertainty. By integrating with Dynamo, the approach enables direct risk visualization, allowing users to quickly perceive the current risk status and implement timely measures. A case study demonstrates the effectiveness of this approach in assessing and managing tunnel construction risks. The results show that this hybrid approach offers real-time risk perception, visualization, and dynamic data updates, enabling users to proactively mitigate potential risks. This capability significantly improves the safety, efficiency, and reliability of risk management in tunnel construction projects.

Chapter "Risk Mitigation Planning" describes risk mitigation planning and its application in tunnel construction. Various risk control techniques are reviewed. An intelligent decision support framework that leverages a deep neural network (DNN)-based method is developed for both risk prediction and optimization in shield tunnel excavation. The proposed framework follows three key stages: data preparation, DNN-based risk prediction, and gradient descent-based optimization (GDO) based risk control. By deeply analyzing the tunnel-soil-building interaction, this approach provides valuable insights to ensure the structural safety of both the tunnel and its surrounding buildings. The proposed approach has been validated in a case study involving the Wuhan metro tunnel project. This approach advances both theoretical and practical aspects of risk management in shield tunnel excavation by enhancing prediction accuracy, optimizing risk control strategies, and improving decision-making reliability.

Chapter "Digital Twin Platform Development" describes the digital twin platform development and its application in tunnel construction. This chapter introduces a digital twin model that enables real-time physical-cyber interaction for monitoring the operating status and improving the overall performance of TBM. By leveraging IoT sensors installed on the TBM, real-time operational data is transferred from the physical machine to the digital model. The collected data is analyzed using a deep learning approach that integrates Graph Convolutional Networks (GCN) and Long Short-Term Memory (LSTM) for performance estimation. To further optimize operations, an online multi-objective optimization (MOO) approach is proposed to determine optimal operating parameters, enhancing TBM performance under varying conditions. The applicability of the approach is validated through a case study using real data from a Singapore tunnel excavation project, demonstrating its potential for intelligent TBM performance optimization.

Chapter "Conclusions and Future Trends" summarizes the conclusions and points out future directions. The goal of intelligent construction to deliver automatic, smart, reliable, and sustainable tunnel construction is underlined. The prevalent techniques of intelligent construction are summarized. Promoting intelligent construction can benefit the industry from different perspectives, including standardization, digitization, intellectualization, collaboration, and customization. Several emerging trends of intelligent construction in tunnel construction are put forward, paving a more affordable and effective way to meet the ever-growing demand for digitalization, informatization, and green transition.

References

1. Bello SA, Oyedele LO, Akinade OO, Bilal M, Delgado JMD, Akanbi LA, Ajayi AO, Owolabi HA (2021) Cloud computing in construction industry: Use cases, benefits and challenges. Autom Constr 122:103441
2. Shi J, Pan Z, Jiang L, Zhai X (2023) An ontology-based methodology to establish city information model of digital twin city by merging BIM, GIS and IoT. Adv Eng Inform 57:102114
3. Criado-Perez C, Shinkle GA, Höllerer MA, Sharma A, Collins C, Gardner N, Hank Haeusler M, Pan S (2022) Digital transformation in the Australian AEC industry: prevailing issues and prospective leadership thinking. J Constr Eng Manag 148:05021012
4. Lo H-W (2023) A data-driven decision support system for sustainable supplier evaluation in the Industry 5.0 era: A case study for medical equipment manufacturing. Adv Eng Inform 56:101998
5. Zhang L, Li Y, Pan Y, Ding L (2024) Advanced informatic technologies for intelligent construction: a review. Eng Appl Artif Intell 137:109104
6. Arabshahi M, Wang D, Sun J, Rahnamayiezekavat P, Tang W, Wang Y, Wang X (2021) Review on sensing technology adoption in the construction industry. Sensors 21:8307

7. Ramírez-Moreno MA, Keshtkar S, Padilla-Reyes DA, Ramos-López E, García-Martínez M, Hernández-Luna MC, Mogro AE, Mahlknecht J, Huertas JI, Peimbert-García RE, Ramírez-Mendoza RA, Mangini AM, Roccotelli M, Pérez-Henríquez BL, Mukhopadhyay SC, Lozoya-Santos JD (2021) Sensors for sustainable smart cities: a review. Appl Sci Basel 11:29
8. Furmonas J, Liobe J, Barzdenas V (2022) Analytical review of event-based camera depth estimation methods and systems. Sensors 22:26
9. Mafukidze HD, Mishra AK, Pidanic J, Francois SWP (2022) Scattering centers to point clouds: a review of mmwave radars for non-radar-engineers. IEEE Access 10:110992–111021
10. Pang JY, Wang SB, Tang ZF, Qin YM, Tao XF, You XH, Zhu JK (2022) A new 5G radio evolution towards 5G-Advanced. Sci China Inf Sci 65:45
11. Wang Q, Zhu XQ, Ni YY, Gu L, Zhu HB (2020) Blockchain for the IoT and industrial IoT: a review. Internet Things 10:9
12. Eras L, Domínguez F, Martinez C (2022) Viability characterization of a proof-of-concept Bluetooth mesh smart building application. Int J Distrib Sens Netw 18:15
13. Schweiger G, Nilsson H, Schoeggl J, Birk W, Posch A (2020) Modeling and simulation of large-scale systems: a systematic comparison of modeling paradigms. Appl Math Comput 365:14
14. Topuz B, Alp N (2023) Machine learning in architecture. Autom Constr 154:15
15. Polson N, Sokolov V (2020) Deep learning: computational aspects, Wiley interdiscip. Rev Comput Stat 12:17
16. Wang HN, Liu N, Zhang YY, Feng DW, Huang F, Li DS, Zhang YM (2020) Deep reinforcement learning: a survey. Front Inform Technol Elect Eng 21:1726–1744
17. Chen WX, Ren HT, Zhou WJ (2023) Review of multi-objective optimization in long-term energy system models. Glob Energy Interconnect 6:645–660
18. Zanker M, Bures V (2022) Knowledge management as a domain, system dynamics as a methodology. Syst Basel 10:24
19. Canonico R, Sperlì G (2023) Industrial cyber-physical systems protection: a methodological review. Comput Secur 135:21
20. Tao F, Xiao B, Qi QL, Cheng JF, Ji P (2022) Digital twin modeling. J Manuf Syst 64:372–389
21. Li T, Li X, Rui Y, Ling J, Zhao S, Zhu H (2024) Digital twin for intelligent tunnel construction. Autom Constr 158:105210
22. Li J, Jing L, Zheng X, Li P, Yang C (2019) Application and outlook of information and intelligence technology for safe and efficient TBM construction. Tunn Undergr Space Technol 93:103097
23. Lin P, Zhang L, Tiong RLK, Song X (2025) Physics-guided deep reinforcement learning for improved operation of TBM considering kinetic inertia stability. Adv Eng Inform 63:102943
24. Shan F, He X, Xu H, Armaghani DJ, Sheng D (2023) Applications of machine learning in mechanised tunnel construction: a systematic review. Eng 4:1516–1535
25. Fu X, Tiong RLK, Zhang L (2025) EMNet: an ensemble deep learning approach for geological condition detection in tunnel excavation. Expert Syst Appl 261:125484
26. Lin P, Zhang L, Tiong RLK (2023) Multi-objective optimization ensembled generative design for large-diameter tunnel passing underneath existing tunnels. Appl Soft Comput 144:110448

Intelligent Geological Environment Perceptions

Advance Geological Forecast

1 Introduction

Geohazards, including collapse, water inrush, and landslide, have a significant impact on tunnel construction, resulting in project delays, cost overruns, and safety concerns. From 2002 to 2018, there were a total of 97 geohazard events during tunnel construction in mountainous regions of China, resulting in 393 fatalities, 467 injuries, and 51 missing people [1]. Prior knowledge of geological conditions is crucial in preventing such adverse events and minimizing associated losses, including the excavation methods and optimization of tunnel design in the pre-construction phase, prediction of operational parameters during construction, and allocation of emergency resources in response to geohazards [2].

Traditional techniques used to gather geological data can be classified into invasive methods and non-invasive methods. Invasive methods are typically employed during earth expeditions to extract core samples for geological analysis or to conduct in-situ measurements [3]. Non-invasive methods, which are based on the geophysical properties of the ground, offer high spatial resolution. However, none of the methods can provide a viable solution for accurately detecting volumes of the areas that would be covered by planned tunnel routes. In particular, the invasive method yields relatively accurate information but at a sparse sampling rate and with limited detection resolution. Non-invasive methods offer high spatial resolution but lack the same level of accuracy due to measurement errors and errors resulting from signal processing. Therefore, the tunneling industry currently needs a method to describe geological information and accurately assess corresponding risks to support tunnel construction effectively.

Recently, a multitude of machine learning (ML) methods have been employed to address the challenge of predicting geological conditions ahead of tunnel boring machines (TBMs). However, the efficiency of ML techniques is often limited by the adequacy of measurements for the system being modeled. Tunneling projects are massive undertakings that can extend over several or even tens of years, which is

L. Zhang et al., *Intelligent Construction in Tunnels*, Lecture Notes in Civil Engineering 679, https://doi.org/10.1007/978-981-96-8785-5_2

significantly longer than the time needed for model updating. This creates a dilemma where high-resolution geological conditions are desired before excavation, but data about these conditions arrive incrementally as the excavation progresses. It has been shown that additional data from the excavation of a 200-m tunnel section at the inlet and outlet locations can greatly enhance the prediction performance of rock mass rating (RMR) for each chainage of the tunnel [4]. However, the updating process was only performed once and relied on data that could be updated with chainage-specific information. Furthermore, accurately estimating the underlying geological risk remains challenging despite the ease of recording geological observations.

When it comes to modeling potential uncertainties, the hidden Markov model (HMM) stands out as a cutting-edge approach for analyzing sequential data. The HMM is a generative model that focuses on understanding the underlying mechanisms and dynamics of observation generation driven by hidden states [5]. The HMM method has shown promise in computer systems, but its potential in tunnel construction remains unexplored and worth investigating. Instead of focusing solely on improving computing efficiency, prioritizing the enhancement of multi-step forecasting ability can provide sufficient time for the TBM to respond to the operations.

Online learning, a subset of ML methods, offers adaptability to streaming data and real-time prediction by leveraging historical data. Unlike batch learning, online learning utilizes each sample as it is observed. This updating process not only improves efficiency but also enables scalability in large-scale applications [6]. The scenario of online learning aligns perfectly with the need to predict geological risks ahead of TBM tunnel excavation by continuously utilizing incoming data. To take advantage of these benefits, researchers have developed online learning methods for detecting geological anomalies based on TBM parameters [7]. However, the application still heavily depends on the availability of geological information, which significantly limits its usability.

To overcome limited data during the early construction stage, an observation extension mechanism is proposed to expand short observation sequences to the length of the full observation sequence, which characterizes the geological risk of the tunnel coverage region. In this chapter, a new advanced geological forecast approach is developed by utilizing streaming chainage data to infer hidden geological risk using an online hidden Markov model (OHMM). This approach advances the field of geological risk prediction by providing a high-resolution method (per chainage/ring) with the following contributions: (1) the systematic OHMM approach allows for frequent model updating based on new data for each chainage/ring, and (2) the observation extension mechanism incorporates feedback between the updated OHMM prediction and true states from sparse borehole samples, making it suitable for early construction applications.

2 Geological Condition Forecast Techniques

Tunnel excavation plays a crucial role in infrastructure development as it enables passage through various geological formations. However, the geological risks associated with tunnel construction present significant challenges, including delays, increased costs, and safety concerns. Traditionally, borehole logging has been prevalently used to record geological information. However, this method has limited spatial resolution in the horizontal plane and is restricted by the sampling distance.

To address these limitations, conventional geological models have been developed for geology modeling and risk assessment through mapping techniques. For instance, Xiong et al. [8] proposed a geological model that visualizes geological boundaries, such as strata and discontinuities, using data from regional field surveys. In addition to earlier 2D geological models, which focus on cross-sections, more recent 3D geological models connect multiple cross-sections to provide an additional dimension along the path connecting separate cross-sections [9]. These visualization models offer a clear understanding of the geological distribution, including information about strata and discontinuities. A detailed comparison of traditional geological prospecting techniques is presented in Table 1.

In recent years, the advancement of ML techniques has greatly motivated research in geological hazard analysis from presentation to prediction. Gaussian process regression, support vector machine (SVM), and decision tree (DT) techniques and their variations were employed to estimate rock quality for tunnel construction [4]. A variant of the K-means algorithm was also developed to identify geology clusters by analyzing various TBM data channels, such as cutter head speed, cutter head torque, thrust, and TBM advance rate [10]. As an example, Xu et al. [11] employed random forest (RF), SVM, and AdaBoost to predict geological conditions in a super-large SPBM tunnel. These methods have demonstrated high applicability and performance in accurately predicting geological conditions during tunneling.

Table 1 Comparison of traditional geological prospecting techniques

Methods	Invasive	Applications	Range	Resolution
Borehole logging	Yes	Vertical mapping	Up to 12 km	Sparse
Seismic methods	No	Oil exploration, reservoir characterization, subsurface stratigraphy	500–2000 m	High
Electromagnetic methods	No	Obtaining structural heterogeneities and lithological interfaces, evaluating geo-mechanical parameters	3–10 m	Very high
Electrical and induced polarization methods	No	Determining the geological discontinuities	Around 54 m	High, decreases with depth

With the growth of computational power, deep learning has drawn tremendous attention and has proven to have high promise in geological condition forecasting, especially in tunnel construction. For instance, Fu et al. [12] integrated a long short-term memory (LSTM) network with a graph convolutional network (GCN) to forecast the geological condition from TBM parameters. Wang et al. [13] developed an incremental generative adversarial network (iGAN) for geological reconstruction based on TBM operational data [14] employed convolutional neural networks (CNNs) to develop a muck analysis system based on image segmentation and classification. The advances in deep learning techniques show significant advantages in predicting geological conditions by handling complex, non-linear relationships and large datasets. It excels in incorporating spatial and temporal patterns, improving predictive accuracy, making them ideal for dynamic geological environments and large-scale simulations.

However, predicting geological conditions ahead of the TBM involves not only identifying the geological conditions but also determining the size and location of changes to support on-site decision-making [15]. Most geological prediction models rely on pre-excavation data and do not account for continuously arriving data, which is recorded with a spatial resolution of a chainage during excavation. Failure to consider this new data can lead to a failure to identify valuable information that helps constrain uncertainties. To sum up, a significant limitation in predicting geological risk in tunnel excavation is the availability and utilization of data. Traditionally, ML methods have failed to address uncertainties stemming from new data chainages. Additionally, while borehole data remains the most direct and accurate way to describe geological conditions, effectively utilizing this data remains an ongoing concern in this field.

3 Advanced Geological Forecast with Online HMM

The tunnel construction, being a megaproject, often takes months to years to complete. Therefore, it is practical to have the capability to inform geological risk ahead of the TBM with a short forecasting distance of several meters. To address these challenges, we deploy and update an HMM to inform the hidden geological risk of the TBM face based on available geological observations, including on-site observations and sampling analysis results. This developed method is referred to as OHMM. Moreover, an observation extension mechanism is proposed to overcome the issue of limited data at an early construction stage by utilizing sparse borehole samples. The framework of OHMM, which includes in-situ data accumulation, the OHMM learning algorithm, the OHMM feedback from sparse samples (i.e., observation extension mechanism), and performance evaluation, is illustrated in Fig. 1.

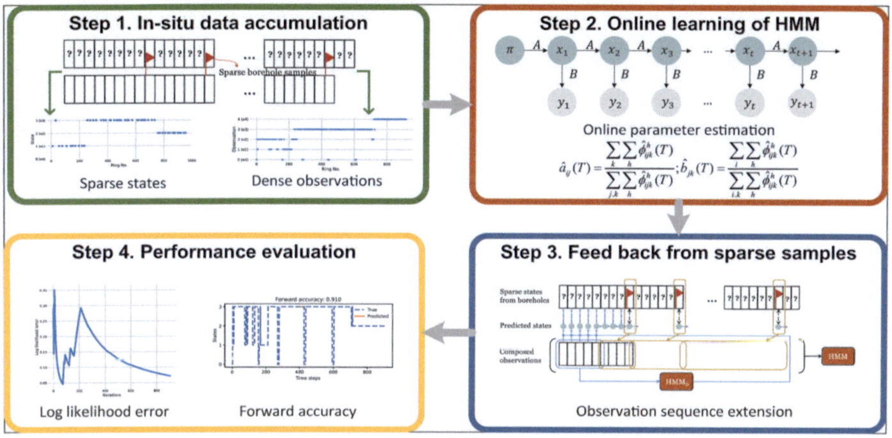

Fig. 1 A framework of the proposed OHMM in predicting geological risk

3.1 *In-Situ Data Accumulation*

Information regarding the geology ahead of the TBM is crucial in minimizing construction accidents and machine defects during the construction process [7]. During the pre-construction phase, samples from boreholes are carefully chosen and analyzed to provide preliminary geological information about the conditions in the area. However, these cylindrical drilling holes only represent a small section in the horizontal plane where the tunnel lies. This is illustrated by the small intersection part compared to the large horizontal plane in Fig. 2a. Once the construction process commences, the TBM excavates along the planned tunnel path, with the tunnel being built ring by ring. The arrangement of rings for a particular tunnel section is shown in Fig. 2b. The sparse drilling holes, obtained before excavation, are insufficient in characterizing the geological risks that guide the excavation process. Therefore, this chapter incorporates continuous in-situ geological observations, gathered during the excavation process, into the problem of inferring and predicting geological risks with fine resolution (i.e., one ring).

In practice, the construction process is diligently recorded in daily reports. Following good practice, these reports adhere to a consistent data input structure, which includes detailed information about the schedule, technical parameters, soil types, on-site photos, and more. Geological observations can be extracted accordingly. However, these reports typically do not include information regarding geological risks. Assessing geological risks, even with some guiding standards, is indeed a challenging task that necessitates expert knowledge and engineering experience. Throughout the construction process, the number of observations steadily increases while the geological risks remain unknown. The scenarios corresponding to the three stages (early, middle, and late stages) are depicted in Fig. 2c. Among all the rings, only a small number that intersect with the boreholes have their geological risks

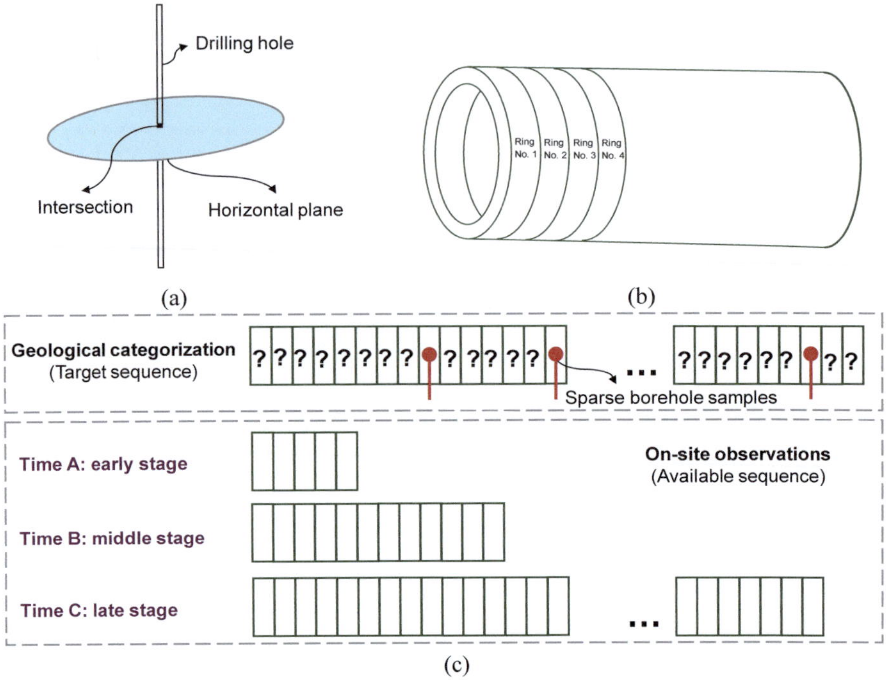

Fig. 2 Illustration of the problem: geological risk inference and prediction to inform excavation. **a** Schematic of limited geological information from drilling hole due to sparse samples; **b** schematic of tunnel rings; **c** gradually increased data volume of observations to infer desired yet unknown states from the early to the middle, and further to the late construction stages. Both sequences are recorded with one-ring (indicated by one rectangle) resolution, where empty rectangles indicate available data while question marks indicate unknown states; red flags indicate known states from borehole analysis, which has a multi-ring resolution and is sparser than the required

informed through borehole analysis, while the remainder remains unknown. In this chapter, the geological risks corresponding to the tunnel rings are explicitly labeled for validation purposes by employees closely involved in the project.

The proposed method aims to infer the unknown states of geological risks based on available observations and further extend the information from excavated rings to rings that are yet to be excavated. The model is one-dimensional, following the direction of the tunnel alignment, and operates at a fine spatial resolution. The necessary data, retrievable from construction daily reports adhering to a standard format, consist of (i) points pertaining to geological risks obtained from borehole analysis; and (ii) a sequence of geological observations whose length varies with the volume of excavation. The three assumptions of the model are listed below.

(i) The model does not include the depth of the tunnel.
(ii) Geological conditions for each ring in the cylindrical space are consistent, meaning that there is no variation.

(iii) Changes in geological conditions only occur between rings along the alignment direction, which is the direction of TBM advance if it is perfectly aligned.

3.2 *Online Learning of HMM*

A Markov chain, which serves as the foundation for HMM methods, is a memoryless stochastic process where the probability of a state depends only on the previous state in a first-order Markov chain. A higher-order Markov chain retains more memory about the states and can be either time-homogeneous or time-inhomogeneous. The Markov assumption for a kth order Markov model is modified and stated such that the probability of the current states, given all the previous states, is the same as the probability of the current state given the previous k states. In a time-homogeneous Markov chain, the probability of any state transition is independent of the time steps. In a time-inhomogeneous Markov chain, the transition probability is non-stationary and may vary with the steps. Transition matrices possess the property that a multi-step transition is equal to the product of step-by-step transition matrices. Mathematically, the k-step transition matrix follows Eq. (1), and its expanded form is shown in Eq. (2).

$$P_t^{(m)} = P_t P_{(t+1)} \ldots P_{(t+m-1)} \tag{1}$$

$$P_t^{(m)} = \begin{pmatrix} P(x_{t+m} = s_1 | x_t = s_1) & \cdots & P(x_{t+m} = s_N | x_t = s_1) \\ \vdots & \ddots & \vdots \\ P(x_{t+m} = s_1 | x_t = s_N) & \cdots & P(x_{t+m} = s_N | x_t = s_N) \end{pmatrix} \tag{2}$$

where $P_t^{(m)}$ is the m-step transition matrix from state x_t to state x_{t+m}, $P_t, P_{t+1}, \ldots, P_{t+m-1}$ are the transition matrices at steps $t, t+1, \ldots, t+m-1$, and the set $\{s_1, s_2, \ldots, s_N\}$ is the state space.

When states cannot be directly observed, the concept of hidden states is introduced alongside observable events in a parallel manner. This gives rise to a double-sequence model with inference capability known as HMM. Similar to a Markov chain, an HMM simulates a memoryless process that allows for multi-step transitions between hidden states and connections between hidden states and observations. The mathematical representation of an HMM is as follows: (i) state set $S = \{s_1, s_2, \ldots, s_N\}$; (ii) transition probability matrix $A = \left[a_{11} \ldots a_{ij} \ldots a_{NN}\right]$ with a_{ij} representing the probability of moving from state s_i to state s_j, and $\sum_{j=1}^{N} a_{ij} = 1$; (iii) observation set $O = \{o_1, o_2, \ldots, o_M\}$; (iv) emission probability matrix $B = \left[b_{11} \ldots b_{jk} \ldots b_{NM}\right]$ with b_{jk} being the probability of generating an observation o_k from a state s_j; and (v) initial probability distribution $\pi = [\pi_1 \pi_2 \ldots \pi_N]$. The HMM explains the generation process of observations, often denoted by the observation sequence $Y = y_1, y_2, \ldots, y_T$. Usually, S and O are specified according to the problem context; Y is the collected data; and A, B, π specify the model with inference capability.

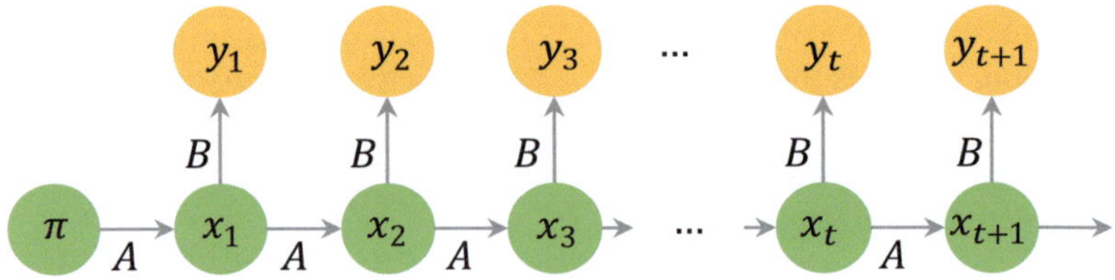

Fig. 3 Structure of a generic hidden Markov model

The OHMM is defined by the same parameters as the HMM, including state space S, observation space O, transition probability matrix A, emission probability matrix B, and initial probability distribution π. The difference between OHMM and HMM lies in the updating rules specified by recursion equations. A graphical illustration of the HMM structures is shown in Fig. 3. Besides the defining parameters, extra learning parameters are introduced for model updating of HMM as well as OHMM. The learning problem of OHMM is stated as below: Given the sequence of observations $Y = y_1, y_2, \ldots, y_T$, to infer the corresponding hidden states $X = x_1, x_2, \ldots, x_T$, the model parameters $\theta = (A, B, \pi)$ are to be learned. It should be mentioned that the parameters $\theta = (A, B, \pi)$ are sufficient to specify the observation generation process (i.e., observing Y). Details of updating rules are described below.

The HMM updates are based on the offline algorithm called the forward-backward algorithm. While OHMM updates its parameters by recursively updating the intermediate sufficient statistics. The transition probability of OHMM is estimated as the fraction of the probability of transiting from state s_i to state s_j and the probability of transiting from state s_i to any state. Similarly, the emission probability of OHMM is estimated as the fraction of the probability of observing o_k at state s_j and the probability of observing any observation at state s_j. Their mathematical forms are shown in Eq. (3).

$$\hat{a}_{ij}(T) = \frac{\sum_k \sum_h \hat{\phi}^h_{ijk}(T)}{\sum_{j.k} \sum_h \hat{\phi}^h_{ijk}(T)}; \hat{b}_{jk}(T) = \frac{\sum_i \sum_h \hat{\phi}^h_{ijk}(T)}{\sum_{i.k} \sum_h \hat{\phi}^h_{ijk}(T)} \tag{3}$$

where $\hat{a}_{ij}(T)$ is the estimated transition probability from state s_i to state s_j based on a sequence of length T, $\hat{b}_{jk}(T)$ is the estimated emission probability of observing o_k from state s_j based on a sequence of length T, and $\hat{\phi}^h_{ijk}(T)$ is the sufficient statistics representing the sum over the time index of the probability being at the state s_h and observing o_k at time index t with some previous state transition from state s_i to state s_j conditioned by observing the sequence of length T, i.e., $Y = y_1, y_2, \ldots, y_T$. The mathematical form of $\hat{\phi}^h_{ijk}(T)$ is expressed as:

$$\hat{\phi}^h_{ijk}(T) = \frac{1}{T} \sum_{t=1}^{T} \delta(y_t - o_k) \cdot P\big(x_{t-1} = s_i, x_t = s_j, x_T = s_h | Y\big) \tag{4}$$

where the pre-factor $\frac{1}{T}$ makes sure that its value does not diverge for an infinitely long sequence, T is the length of the sequence, $\delta(y_t - o_k)$ is the Kronecker delta function whose output is 1 for zero arguments and 0 otherwise, y_t is the observation at time index t, o_k is the value of y_t, and x_{t-1}, x_t, x_T are the states corresponding to the time index $(t-1)$, t, and T, respectively. The 4 indices of $\hat{\phi}^h_{ijk}(T)$ are the consecutive historical states s_i to s_j, the historical observation o_k from s_j, and the latest state s_h. Any irrelevant index is eliminated by summing over the full index range, that is, i, j, h from 1 to N, and k from 1 to M in the parameter estimation process.

The parameters are updated as the sufficient statistics are updated. The recurrence relation for the sufficient statistics corresponding to two consecutive time indices is given by Eq. (5), which represents the update based on a new sample. Model parameters are closely related to the introduced statistical variable.

$$\begin{aligned}\hat{\phi}^h_{ijk}(T) = &\sum_l \gamma_{lh}\big(y_T; \hat{\theta}(T-1)\big) \\ &\times \Big\{\hat{\phi}^l_{ijk}(T-1) + \eta(T)\big[\delta(y_T - o_k)\cdot g_{ij}(l,h) \\ &\cdot q_l(T-1) - \hat{\phi}^l_{ijk}(T-1)\big]\Big\}\end{aligned} \tag{5}$$

where $\hat{\phi}^h_{ijk}(T)$ and $\hat{\phi}^l_{ijk}(T-1)$ are the sufficient statistics corresponding to two consecutive time indices, $\gamma_{lh}\big(y_T; \hat{\theta}(T-1)\big)$ is a complicated probability term that considers the one-step transition of states, emission from the state, and the addition of a new observation. Its value can be calculated based on the latest model parameters. It is defined and computed as:

$$\begin{aligned}\gamma_{lh}\big(y_T; \hat{\theta}(T-1)\big) &\equiv \frac{P(y_T = o_k | x_T = s_h) P(x_T = s_h | x_{T-1} = s_l)}{P(y_T | y_1, y_2, \ldots, y_{T-1})} \\ &= \frac{\hat{a}_{lh}(T-1)\cdot \hat{b}_{hk}(T-1)}{\sum_{m,n} \hat{a}_{mn}(T-1)\cdot \hat{b}_{nk}(T-1)\cdot q_m(T-1)}\end{aligned} \tag{6}$$

with $q_m(T-1)$ being the probability of being state s_m given an observation sequence of length $T-1$, i.e., $y_1, y_2, \ldots, y_{T-1}$. It can be calculated recursively (will be given soon) and is defined as:

$$q_m(T-1) \equiv P(x_{T-1} = s_m | y_1, y_2, \ldots, y_{T-1}) \tag{7}$$

where $\eta(T)$ is a time-dependent discount factor, $\delta(y_T - o_k)$ is the Kronecker delta function whose output is 1 for zero arguments and 0 otherwise, y_T is the latest observation with time index T, o_k is the value of y_T, $g_{ij}(l, h)$ is a function defined as:

$$g_{ij}(l,h) \equiv \delta(i-l)\cdot\delta(j-h) \tag{8}$$

and $q_l(T)$ is the probability of being state s_l given an observation sequence of length T, i.e., $y_1, y_2, \ldots, y_T$. It is updated recursively by:

$$\begin{aligned} q_l(T-1) &\equiv P(x_{T-1} = s_l | y_1, y_2, \ldots, y_{T-1}) \\ &= \sum_m \gamma_{ml}\big(y_{T-1}; \hat{\theta}(T-2)\big) q_m(T-2) \end{aligned} \tag{9}$$

3.3 *Feedback from Sparse Samples*

During the early stages of construction, the observation sequence may not be long enough for the OHMM to converge. This can lead to overfitting problems due to the limited amount of observation data, thus adversely affecting performance in prediction. To address this issue and ensure the model is suitable for observation sequences of varying lengths that correspond to different construction stages, we propose a mechanism that explicitly extends short observation sequences to match the full sequence. This mechanism utilizes the most updated model parameter estimates and the sparse samples in advance. The convergence checking and observation sequence extension mechanisms are detailed below.

An OHMM with discrete observation values can be characterized by a transition matrix A and emission matrix B, which is usually denoted as $\theta = (A, B)$. To track the model updates due to the filled observation sequence, matrix distance (MD) is used to quantify the change of the two key matrices. The sum of the difference is divided by the number of distinct states in the state space (i.e., N) so that the resultant value is in the range [0, 1], where 0 means no change and 1 means that two matrices are completely different. The change in the transition matrix and emission matrix is defined in Eqs. (10) and (11), respectively. The normalized matrix distance is utilized to monitor the change in parameter estimates. A default criterion of 0.0005 is employed to determine if the OHMM has converged.

$$d(A(t+1), A(t)) = \frac{1}{N} \sum_{i=1}^{N} \sum_{j=1}^{N} \left| a_{ij}(t+1) - a_{ij}(t) \right| \tag{10}$$

where $d(A(t+1), A(t))$ is the change of the transition matrix from time step t to $t+1$.

$$d(B(t+1), B(t)) = \frac{1}{N} \sum_{i=1}^{N} \sum_{j=1}^{M} \left| b_{ij}(t+1) - b_{ij}(t) \right| \tag{11}$$

where $d(B(t+1), B(t))$ is the change of the emission matrix from time step t to $t+1$.

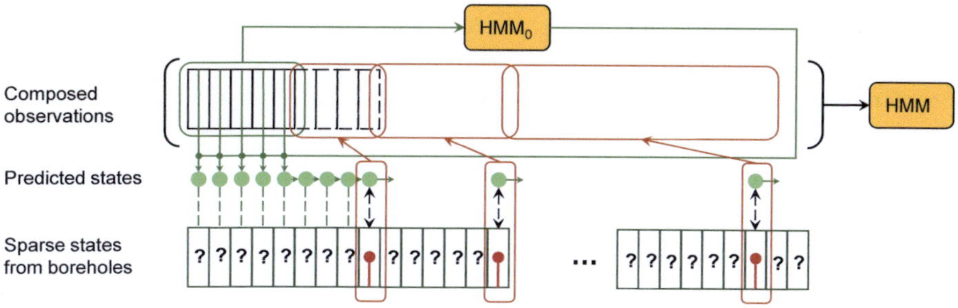

Fig. 4 A schematic illustrating the feedback from sparse samples. Available observation is used to train a preliminary HMM (denoted as HMM$_0$). For the rings with observations (indicated by rectangles with solid lines), the model HMM$_0$ is used to infer the states (indicated by non-connected circles) from the observations; for the rings without observations (indicated by rectangles with dashed lines), the most possible states (indicated by arrow-connected circles) are directly predicted based on the most recent state and the transition matrix of HMM$_0$. At the borehole position (indicated by red flags), the predicted state is compared to the known state, whose results are used to fill unavailable observations. Available observations (indicated by a blue rectangle) and filled observations (indicated by golden rectangles) together form the composed observations, which are then used to train the HMM model for prediction purposes

When the full observation sequence is not yet available, a preliminary HMM is trained using the limited short observation sequence that is available. This preliminary model is then used to infer the most probable states. At the locations where boreholes are situated and the states are known, the inferred states are compared to the known states. The comparison results are used as feedback to extend the short observation sequence to the length of the full observation sequence. The composed observation sequence consists of two parts: the available observations and the filled observations based on the feedback from the sparse states obtained from borehole samples and the inferred states from the preliminary HMM. This composed observation sequence is then used to train the HMM model for prediction purposes. The feedback process from the sparse borehole samples is illustrated in Fig. 4.

The principle of minimality, which is a primary focus in data repair and data cleaning, is employed in the observation sequence filling process. The filling of the observation sequence follows the principle of minimal change. In the context of HMM, the rule for data filling is as follows: if the next state (which may be several steps ahead) remains the same as the current state, the observation remains unchanged. However, if the next state is different from the current state, there is a single jump in the observation value at a random step from the current time index to the time corresponding to the next known state. The desired observation symbol is the one that is observed, and the next known state is most likely to occur at the corresponding time step. The mechanism for extending the observation sequence by leveraging HMM inference and known sparse states is shown in Algorithm 1.

Algorithm 1 Generation of composed observation sequence.

```
function extend_obs(model, Y_observed, pre_ring_no_idx, pre_ring_no_state)
        # model is an object of the initialized HMM model with defined functions
        seq_len=len(Y_observed) # seq_len in the range [1,T-1]
        # pre_ring_no_idx is the ring no. indices corresponding to sparse borehole samples
        # pre_ring_no_state is the states obtained from sparse borehole samples
        # find the index of pre_ring_no_idx where its element is large than seq_len for the first
time
        for j, idx in enumerate (pre_ring_no_idx) do
                if idx > seq_len
                        start=j
                        break
                end if
        model.fit(Y_observed) # model learning
        # get the ob so that the next state will be the same as the given state
        next_ob=model.fill_next_ob_from_state(pre_ring_no_state[start])
        Y_filled=repeat(next_ob, idx-seq_len+1) # get the filled obs
        Y=concatenate((Y_observed, Y_filled)) # get the composed seq
        for j=start, …, len(pre_ring_no_idx)-1 do
                if pre_ring_no_state[j+1]==pre_ring_no_state[j]
                        Y_filled_new=repeat(Y_filled[-1],
                        pre_ring_no_idx[j+1]-pre_ring_no_idx[j])
                else
                        model.fit(Y_observed)
                        next_ob=model.fill_next_ob_from_state(pre_ring_no_state[j+1])
                        # randomly pick the change position
                        jump_pos=randrange(pre_ring_no_idx[j], pre_ring_no_idx[j+1])
                        Y_filled_p1=repeat(Y_filled[-1], jump_pos- pre_ring_no_idx[j])
                        Y_filled_p2=repeat(next_ob, pre_ring_no_idx[j+1]-jump_pos)
                        Y_filled_new=concatenate(Y_filled_p1, Y_filled_p2)
                Y_filled=concatenate((Y_filled, Y_filled_new))
                Y=concatenate((Y_observed, Y_filled))
                end if
        end for
        return Y
```

3.4 Geological Forecast Performance Evaluation

To monitor the online learning process of OHMM, a log-likelihood error is defined to quantify the difference between the most probable states and the true states, as shown in Eq. (12). At each time step, using the updated model $\hat{\theta}(t)$ and uniform initial probabilities, the most possible hidden state sequence (up to the time step t) is discovered using the Viterbi algorithm. The obtained posterior probabilities are used to calculate the log-likelihood, which is compared to the log-likelihood corresponding to the true state sequence.

$$\lambda(t) = L(t) - L_{true} \tag{12}$$

where $\lambda(t)$ is the log-likelihood error, $L(t)$ is the average log-likelihood corresponding to the most probable states inferred from the model $\hat{\theta}(t)$, and L_{true} is the average log-likelihood corresponding to the true state sequence.

The developed OHMM can be deployed to infer the hidden geological risk based on available geological observations, as well as to predict the hidden geological risk for future time steps. These two tasks are referred to as inference and prediction, respectively. Inference ability is evaluated using backward accuracy, which involves inferring the hidden states based on the observation sequence. Prediction ability is assessed using forward accuracy, which extends the interpretation of hidden states to the next ΔT steps where observations are unavailable. Further details of these deployment scenarios are provided in Table 2. However, in practical terms, predicting future geological risks holds much greater significance. Early alerting of geological risks provides operators with sufficient time. Forward accuracy evaluates the states of future ΔT steps, as shown in Eq. (13).

$$\begin{aligned} Acc_{forward}(t, \Delta T) &= f(\hat{x}_{t+1}, \hat{x}_{t+2}, \ldots, \hat{x}_{t+\Delta T}; x_{t+1}, x_{t+2}, \ldots, x_{t+\Delta T}) \\ &= 1 - \frac{\sum_{i=t+1}^{t+\Delta T} |\hat{x}_i - x_i|}{\Delta T} \end{aligned} \tag{13}$$

where $f(\cdot)$ is the accuracy score for the predicted sequence and true sequence, ΔT is the steps ahead in forward prediction, $\hat{x}_{t+1}, \hat{x}_{t+2}, \ldots, \hat{x}_{t+\Delta T}$ denotes the predicted states of the future ΔT steps, and $x_{t+1}, x_{t+2}, \ldots, x_{t+\Delta T}$ denotes the true states of the future ΔT steps.

Table 2 The two deployment scenarios of the developed OHMM

Item	Scenario 1	Scenario 2
Name	Inference	Prediction
Description	Interpreting the hidden geological risk based on observation	Extending the interpreted geological risk from now to the near future
Input	$y_1, y_2, \ldots, y_t$	$y_1, y_2, \ldots, y_t$
Output	$\hat{x}_1, \hat{x}_2, \ldots, \hat{x}_t$	$\hat{x}_{t+1}, \hat{x}_{t+2}, \ldots, \hat{x}_{t+\Delta T}$
Performance metrics	Backward accuracy $Acc_{backward}(t)$	Forward accuracy $Acc_{forward}(t, \Delta T)$
Practical significance	Not that significant; To evaluate the practice of excavated sections, may contribute to engineering experience accumulation	Significant; To guide the excavation

4 Experimental Studies

4.1 Data and Resources

The tunnel case studied in this chapter is located in Singapore and comprises 915 rings. In geological investigations, earth materials at the site are typically described and classified based on their engineering uses. The widely recognized unified soil classification system (USCS) classifies soils according to their physical properties (e.g., grain size) and behavioral characteristics. The system consists of three major groups: (i) highly organic soils (e.g., peat), (ii) fine-grained soils (e.g., silts and clays), and (iii) coarse-grained soils (e.g., gravels and sands). In this chapter, soil types corresponding to sparse boreholes and per-ring records were obtained from the daily reports. The soil samples were classified as fluvial sand (FS), old alluvium (OA), fluvial clay (FC), marine clay (MC), and organic clay (OC), based on grain size and susceptibility to deformation under pressure.

The horizontal resolution of borehole samples (i.e., tens of rings) is coarser than that of tunnel rings (i.e., per ring), as illustrated in the spatial distribution of sparse drilling holes and tunnel rings. Figure 5 illustrates the side view of the region covering the tunnel. It can be observed that cylindrical drilling boreholes (indicated by long rectangular with filled patterns) are much sparser in space than tunnel rings (indicated by blue lines with arrow annotations). In other words, there are around 20 to 30 tunnel rings between every two consecutive boreholes.

Geo-materials with varying properties, including strength, density, and static and dynamic responses, demonstrate diverse stress paths and fracture behaviors that can pose challenges to tunnel construction. However, the relationship between soils and their geological risk is not well understood and is difficult to determine due to inter-class and intra-class variations. To facilitate interpretation, the geological risk is categorized into three levels: (i) Category 1 (low risk) consists of relatively hard soils, such as sands; (ii) Category 2 (medium risk) consists of fine-grained soils; and (iii) Category 3 (high risk) consists of soft soils. The observations are defined as shown in Table 3. To account for uncertainties in interpreting geological risk from soil observations, such as missing or multiple records, an additional category called "uncertain" is included. The hidden states representing geological risk are defined as shown in Table 4.

The training data in Fig. 6 display both sparse hidden states representing geological risk obtained from borehole analysis and dense observations obtained from on-site excavation. In the absence of excavation observations, geological risk state information is only available at the boreholes. As depicted in Fig. 6a, the state information is discrete in the unconstructed area, which is one of the initial challenges addressed in this paper using the sequence extension mechanism. Dense hidden states representing geological risk per ring are also collected for evaluation purposes; however, during the construction process, per-ring state information is not available. Consequently, the problem of predicting geological risk, which corresponds to the

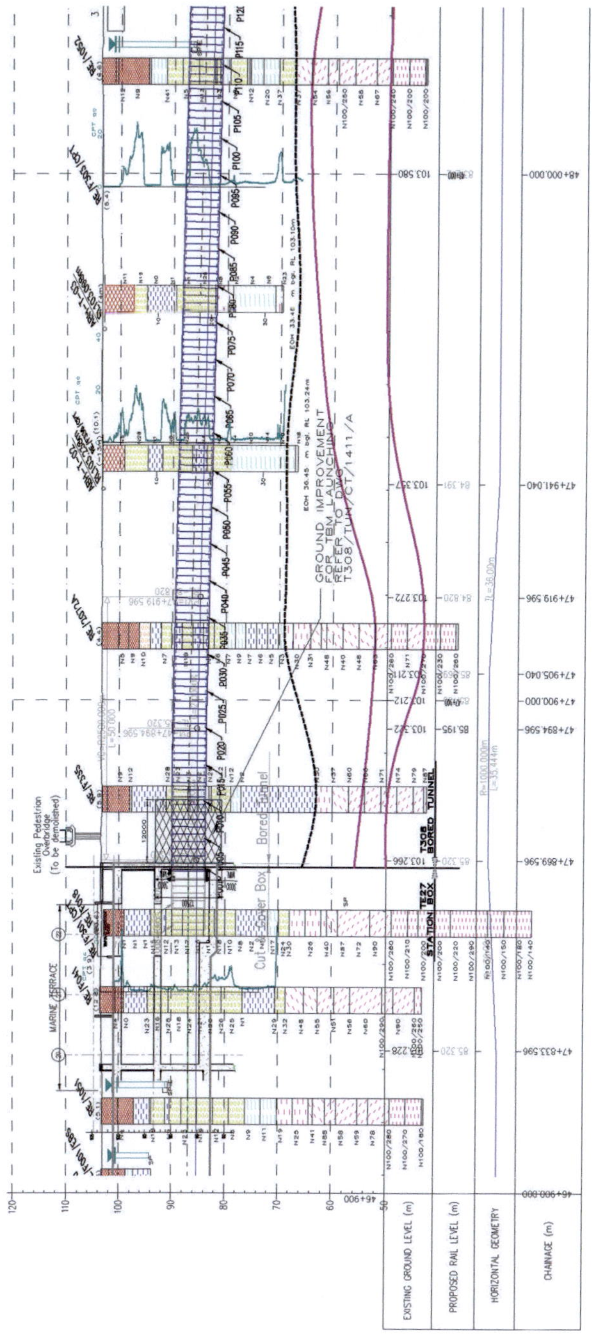

Fig. 5 Side view of the region covering the tunnel

Table 3 Observation events of soil conditions and their denotation symbols

Site record description (Observation events)	Symbols
A mix of FS and MC	o_1
FC	o_2
MC, a mix of OA and OC	o_3
OA	o_4
Missing or multiple records	o_5

Table 4 States of geological risk and their denotation symbols

Geological risk (states)	Symbols
Low	s_1
Medium	s_2
High	s_3
Uncertain	s_4

hidden states in the HMM framework, cannot be treated simply as a single time series prediction problem since the states of geological risk are typically not accessible.

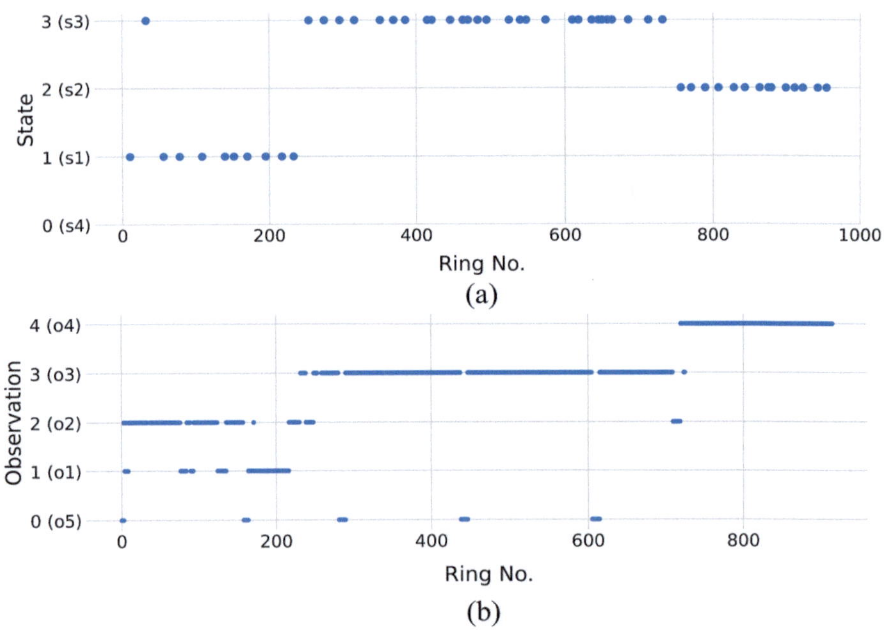

Fig. 6 Symbols and the encoded numeric values for **a** Table 2 obtained from borehole analysis; **b** dense observations per tunnel ring obtained from excavation

4.2 Model Development

Based on the defined observations (Table 3) and hidden states (Table 4), an OHMM is constructed to relate the on-site observations to the hidden states of geological risk. The state diagram and its emission patterns are shown in Fig. 7. Mathematically, the developed OHMM is specified by:

(i) State space of 4 states $S = \{s_1, s_2, s_3, s_4\}$;

(ii) A transition probability matrix $A = \begin{bmatrix} a_{11} & a_{12} & a_{13} & a_{14} \\ a_{21} & a_{22} & a_{23} & a_{24} \\ a_{31} & a_{32} & a_{33} & a_{34} \\ a_{41} & a_{42} & a_{43} & a_{44} \end{bmatrix}$;

(iii) Observation space of 5 observable events $O = \{o_1, o_2, o_3, o_4, o_5\}$;

(iv) Emission probability matrix $B = \begin{bmatrix} b_{11} & b_{12} & b_{13} & b_{14} & b_{15} \\ b_{21} & b_{22} & b_{23} & b_{24} & b_{25} \\ b_{31} & b_{32} & b_{33} & b_{34} & b_{35} \\ b_{41} & b_{42} & b_{43} & b_{44} & b_{45} \end{bmatrix}$;

(v) An initial probability distribution $\pi = \pi_1\pi_2\pi_3\pi_4$.

The learning of OHMM concerns updating the parameters $\theta = (A, B)$ till convergence. Three model parameters (i.e., A, B, π) and two sufficient statistics (i.e., $q_l(0)$ and $\hat{\phi}_{ijk}^h(0)$) are to be initialized for the training of the OHMM. Since the true values of the parameters are unknown, the initial probability distribution is set using a uniform distribution, the transition probability matrix is set with equal probabilities for any

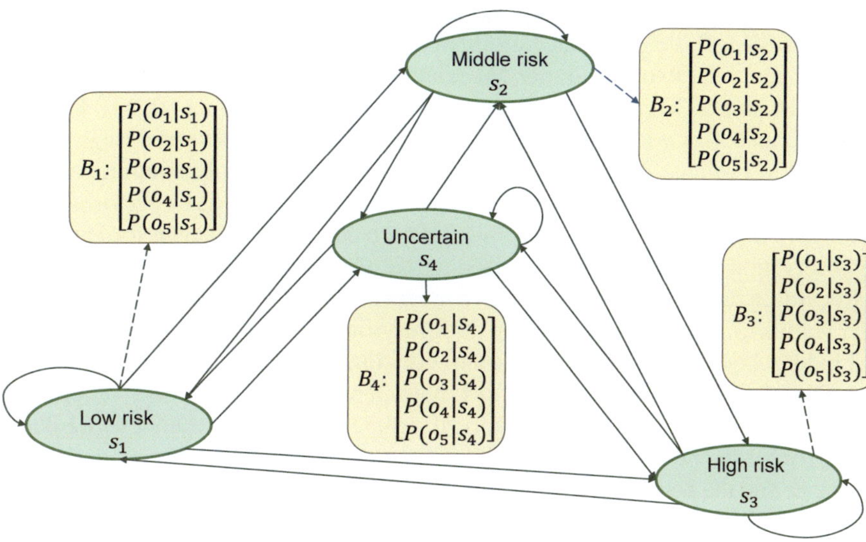

Fig. 7 State diagram of the OHMM with 5 observation events and 4 hidden states

state, and the emission probability matrix is set using a random matrix with a specific seed for consistency. The initial emission probabilities are represented as

$$B_0 = \begin{bmatrix} 0.136 & 0.093 & 0.053 & 0.299 & 0.420 \\ 0.066 & 0.272 & 0.233 & 0.019 & 0.410 \\ 0.175 & 0.335 & 0.347 & 0.070 & 0.072 \\ 0.264 & 0.106 & 0.048 & 0.259 & 0.322 \end{bmatrix}$$

Besides, the sufficient statistics $q(0)$ is initialized to be the same as π. The sufficient statistic $\phi(0)$ is initialized as zeros with the shape (N, N, N, len_obs), where N is the number of states in the state space and len_obs is the length of the observation sequence. The hyperparameter t_s is adjusted to ensure that sufficient statistics gather sufficient information about the system before being used to update the model parameters. By default, a value of 250 is used, which keeps the model static for the first 250 samples and then updates it for every subsequent sample after the 250th step.

4.3 Geological Forecast Results

To model the geological risk of the TBM face, the constructed OHMM is updated based on each observation representing the soil components per ring of the tunnel. To illustrate the OHMM learning process, the case of full observations is used as an example. Figure 8 presents the online estimates of the OHMM parameters. Additionally, Fig. 9 shows the probabilistic version of the inferred states obtained from the trained OHMM. To demonstrate the superior performance of the OHMM, a comparison is made with two traditional ML algorithms (namely neural network (NN) and SVM), an LSTM, and another generative model (the normal HMM). Furthermore, two instances with 300 observed rings (#300) and 600 observed rings (#600) are tested to evaluate the models' performance with limited observation data. All models are designed to predict 30 steps as a baseline. The OHMM's performance is thoroughly assessed in terms of its online nature and the mechanism for extending the observation sequence, in comparison to the selected competitors.

Regarding the training of competitive models, the NN is structured with an input layer featuring 3 neurons representing three historical states. Additionally, it consists of 3 hidden layers with 16 neurons each, utilizing ReLU activation, and an output layer with 4 neurons corresponding to 4 states, employing softmax activation. The NNs are trained by employing a categorical cross-entropy loss function, with a batch size of 16 and an epoch count of 100. The two parameters of SVM, the regularization parameter and the kernel coefficient, are determined and optimized by grid search in the range $\left[10^{-10}, 10^{5}\right]$ with 16 values spaced evenly on a log scale. The optimized

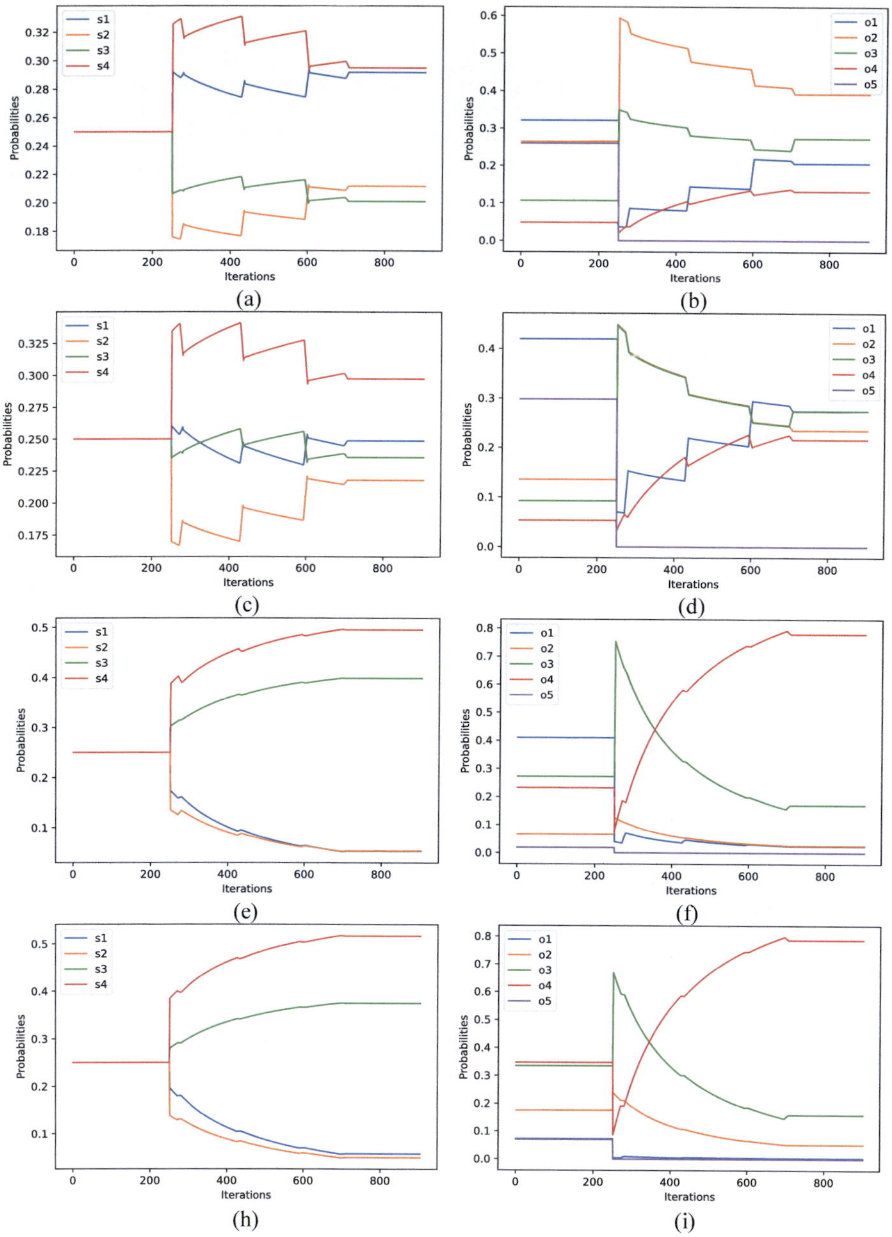

Fig. 8 Online estimates of the OHMM parameters: **a** transition probabilities from state s_1 to some state; **b** emission probabilities of observing some observation event at state s_1; **c** transition probabilities from state s_2 to some state; **d** emission probabilities of observing some observation event at state s_2; **e** transition probabilities from state s_3 to some state; **f** emission probabilities of observing some observation event at state s_3; **g** transition probabilities from state s_4 to some state; and **h** emission probabilities of observing some observation event at state s_4

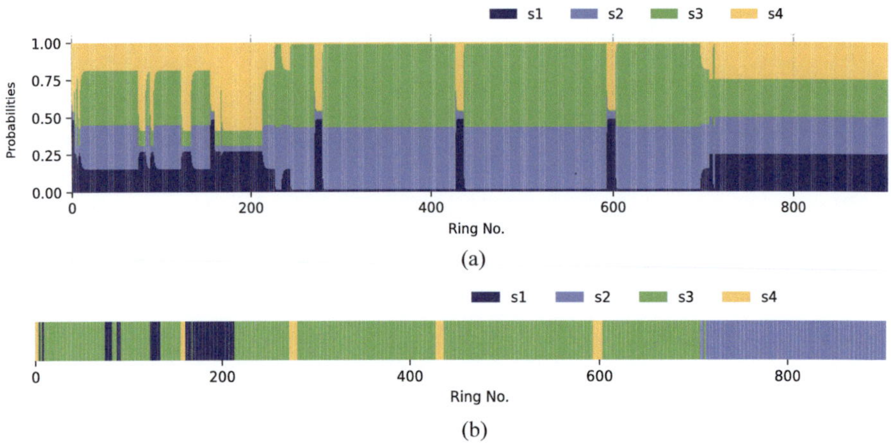

Fig. 9 Geological risk profile along the tunnel path: **a** probability distributions over the states, and **b** ground truth of the states

SVM for instance #300 has a regularization parameter $C = 100$ and a kernel coefficient $\gamma = 0.01$ for all 30 random trials. The optimized SVM for instance #600 has a set of parameters of $C = 1$ and $\gamma = 1$.

For the training of the LSTM model, the network consists of an embedding layer incorporating 10 historical steps, a spatial dropout layer with a dropout rate of 0.2, an LSTM layer including 16 neurons with respective dropout rates of 0.6 and 0.5 for regular dropout and recurrent dropout, and finally, a dense layer equipped with the softmax activation function, serving the purpose of classification. The LSTM model is trained by utilizing a categorical cross-entropy loss function, with a batch size of 32 and an epoch count of 100, employing the Adam optimizer. The average accuracy scores achieved throughout the 30 random trials, accompanied by significance tests, are presented in Table 5. Additionally, Fig. 9 provides a visual representation of the confusion matrices showcasing the predictions made by these algorithms, corresponding to the time instances 300 and 600, respectively. The categorical sensitivity and specificity of these prediction results can be observed in Table 6.

(1) The average likelihood of the learned OHMM approaches the average likelihood of the true model as the log-likelihood error tends to zero over time. In the context of OHMM, the mapping patterns between geological risk and soil components are represented by a space-varying emission probability matrix. Similarly, the transition of underlying states from one location to another is represented by a space-varying transition probability matrix. Converged OHMM can effectively represent the transition and emission behaviors of the system, which may vary from region to region in the case of geological risk prediction.

Table 5 Comparison with other methods

Time instance	Method	Input data		Performance
		Observation sequence (easy to be recorded) required	State sequence (hard to judge) required	Forward accuracy[a]
#300	NN	No	Yes	1.000
	SVM	No	Yes	1.000
	LSTM	No	Yes	0.833
	HMM	Yes	No	0.700
	OHMM	Yes	No	0.968
#600	NN	No	Yes	0.100
	SVM	No	Yes	0.100
	LSTM	No	Yes	0.892
	HMM	Yes	No	0.763
	OHMM	Yes	No	0.902

Note [a]Obtained from the average of 30 trials of training

Table 6 Categorical sensitivity and specificity for the prediction scenarios

Time instance	Method	Uncertain		Low risk		Medium risk		High risk	
		Sen	Spe	Sen	Spe	Sen	Spe	Sen	Spe
#300	NN	Nil	1	Nil	1	Nil	1	1	Nil
	SVM	Nil	1	Nil	1	Nil	1	1	Nil
	LSTM	Nil	1	Nil	1	Nil	1	1	Nil
	HMM	Nil	1	Nil	1	Nil	1	1	Nil
	OHMM	Nil	1	Nil	1	Nil	1	1	Nil
#600	NN	1	0	Nil	1	Nil	1	0	1
	SVM	1	0	Nil	1	Nil	1	0	1
	LSTM	0	1	Nil	0.87	Nil	1	0.87	Nil
	HMM	0	1	Nil	0.67	Nil	1	0.74	1
	OHMM	0	1	Nil	1	Nil	1	1	0

Note Sen—Sensitivity, Spe—Specificity

The initialization process plays a crucial role in the convergence speed, as it depends on the distance between the initial model and the true model. Under the chosen initialization schema, the variables are uniformly initialized, except for the emission probability matrix which is randomly drawn. The irregular trend observed during the first 250 iterations can be attributed to the control parameter t_s (Algorithm 1), which is set to 250. During this phase, the model remains static, and only sufficient statistics are updated. This can be observed from the iteration of parameters shown in Fig. 8. Early freezing of model parameters aids in stabilizing the training process, allowing sufficient statistics to gather enough information about the system before being used to update the model. From iteration 250 onwards, the log-likelihood error consistently decreases at a slower rate, indicating convergence.

(2) The trained OHMM is capable of accurately inferring the hidden states in certain cases, providing uncertain states when confronted with short sub-series that include new observation events. The raw probabilities associated with the defined four states vary with the ring number, as depicted in Fig. 9a, thus indicating a changing geological risk along the tunnel. By analyzing the OHMM, it is possible to retrieve the maximum probabilities for each tunnel ring, enabling a judgment of the state. Figure 9c illustrates a comparison between the OHMM's state determination and the ground truth of the states. Compared to the ground truth of the states (Fig. 9c), the state s_3 (i.e., high risk) can always be identified with high probability, which is of high significance as high risk is always under major concern in excavation processes. The states s_1 (i.e., low risk) and s_2 (i.e., medium risk) do not refer to the exact state but are semantically correct, especially in the last 100 rings when observing new events that do not last long and the probabilities of all states are relatively equal. Tracing back to the observations (Fig. 5b), the states s_1 and s_2 correspond to observations o_1 and o_4, respectively. By the definition (Table 3), categories o_1 and o_3 have some common soil components as o_3, where engineers distinguish them with certainty in the frame of categories while the model gives a judgment of "uncertain" with higher probabilities. For the rings with a missing record or multiple records, the model gives a judgment with a specific risk state rather than an uncertain state.

(3) The OHMM algorithm demonstrates superior performance compared to traditional ML techniques due to its online consideration of manner and the mining of hidden states. Traditional methods like NN and SVM lack stable performance in classification tasks and are highly affected by the dataset. Notably, there is a significant difference in performance between forward prediction accuracy at #300 and #600. In contrast, LSTM, HMM, and OHMM consistently exhibit stable forward prediction accuracy. Crucially, OHMM surpasses LSTM and HMM in forward prediction accuracy (0.968 for #300 and 0.902 for #600) because OHMM incorporates online consideration. Moreover, OHMM and HMM are generative models capable of inferring hidden states, while NN, SVM, and LSTM require historical hidden states to predict future hidden states. This characteristic makes OHMM more practical for real projects as it can be applied after initialization steps. Thanks to the observation sequence extension mechanism, the model reliably predicts geological risks based on site investigation data.

(4) The exceptional performance of OHMM stems from its consistently high sensitivity over risk classes for the upcoming state sequence. The confusion matrices in Fig. 10 illustrate the state classification results from the 1st random trial of the algorithms corresponding to the two instances. When predicting forward from this single trial, all five algorithms (NN, SVM, LSTM, HMM, and OHMM) accurately identify the geological risk as "high risk" for time instance

#300 (Fig. 10a–d). However, at time instance #600 (Fig. 10e–h), traditional methods like NN and SVM fail to identify "high risk," while LSTM, HMM, and OHMM can. Nevertheless, OHMM outperforms the others and exhibits the least misclassification. Table 6 displays the categorical sensitivity and specificity of these prediction results. In a multi-class classification problem, high sensitivity towards a specific risk level is preferred. OHMM maintains high sensitivity for the "high risk" in both instances and outperforms the other algorithms. This implies higher reliability when implemented in real projects.

(5) The OHMM demonstrates an accurate estimation of model parameters even when limited observation data is available, using the observation extension mechanism. Two instances mimic engineering scenarios at different stages of excavation and construction. In Fig. 11, changes in the model parameters and the final parameter are shown. The OHMM's learned parameters from the composed observation sequence converge at a similar speed to those learned from the full observation sequence. Additionally, the final estimated parameters based on the composed observation sequence display similar patterns to those based on the full observation sequence. The similarity of the matrix falls within a 99% confidence interval, as shown in Fig. 12. The OHMM model continuously updates itself with new incoming data, eliminating the need for a large amount of initialization data. This property makes the OHMM applicable in scenarios where only limited data is available, such as the early and middle stages of the construction process. As a result, the developed OHMM greatly extends the compatibility for on-site geological condition prediction throughout the construction process.

5 Merits Analysis

The previous section of this report discussed the overall performance of the OHMM. In this section, we will focus on studying the OHMM's ability to forecast multiple steps ahead and its performance at different stages of excavation. Initially, we select 6 equidistant time instances (ranging from 100 to 600 rings) to predict 30 rings. The results of this prediction are plotted in Fig. 13. It is worth noting that the key difference between these time instances lies in the volume of available data, with data volume increasing as the time index increases. At each time instance, the OHMM is trained using the available observations up to that point, with the length of the available observations corresponding to the number of excavated rings. Next, we use the model trained at each time instance to forecast different numbers of steps (20, 30, 50, 100, and 150). The forward accuracy, aggregated by the number of rings ahead, is presented in Fig. 14 and summarized in Table 7. The merits of the proposed OHMM approach are further discussed below.

(1) With the assistance of the observation sequence extension mechanism, the OHMM trained on the composed sequence demonstrates high accuracy in predicting scenarios after 200 rings. As the time index increases, the length of the observation sequence also increases. Specifically, OHMMs trained using

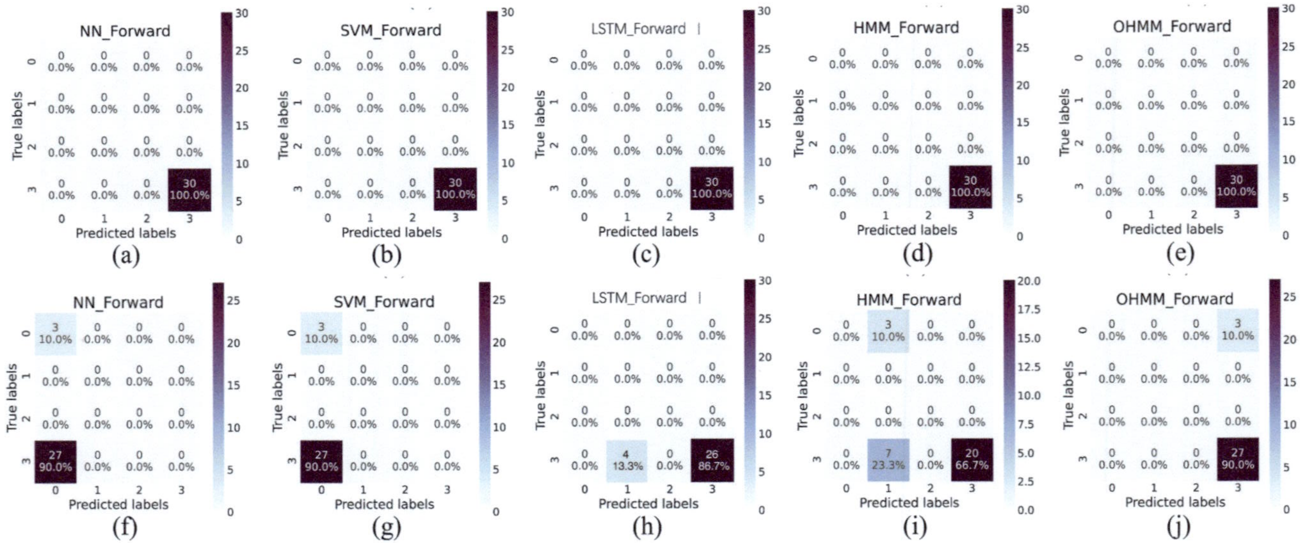

Fig. 10 Confusion matrixes for the forward predictions at time instance #300 from **a** NN; **b** SVM; **c** LSTM; **d** HMM; **e** OHMM, and time instance #600 from **f** NN; **g** SVM; **h** LSTM; **i** HMM; **j** OHMM. *Note* The numeric values 0, 1, 2, and 3 represent the states "uncertain", "low risk", "medium risk", and "high risk", respectively

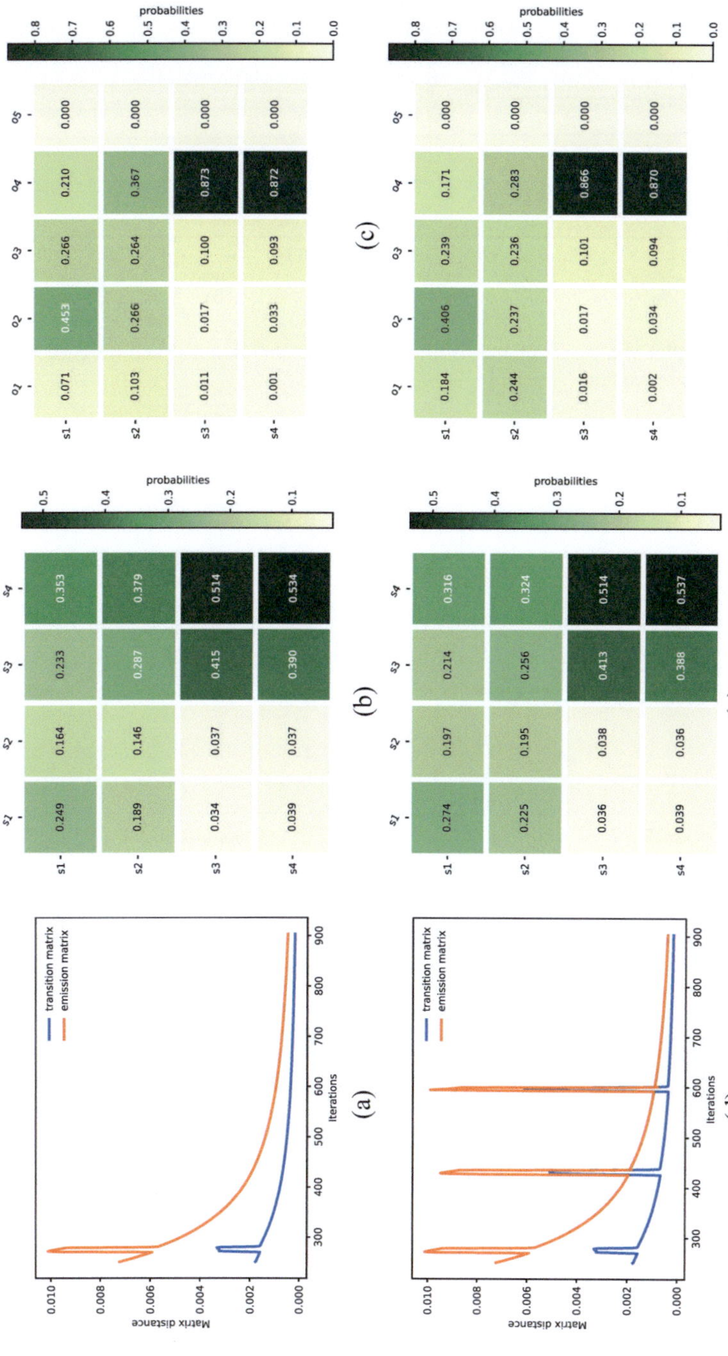

Fig. 11 OHMM parameter learning for several time instances $t = 300, 600$ corresponding to observation sequences with decreasing lengths. **a** Matrix distance between two consecutive parameter estimates for $t = 300$; **b** converged transition matrix for $t = 300$; **c** converged emission matrix for $t = 300$; **d** matrix distance between two consecutive parameter estimates for $t = 600$; **e** converged transition matrix for $t = 600$; **f** converged emission matrix for $t = 600$

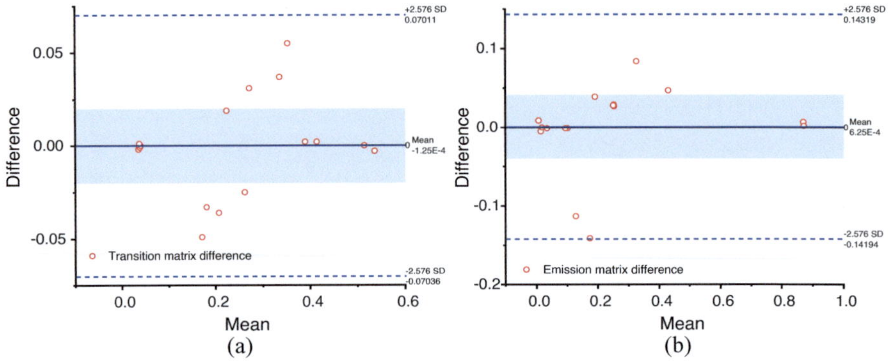

Fig. 12 Bland-Altman plot of **a** transition matrix difference and **b** emission matrix difference with $t = 300$ and $t = 600$ historical data

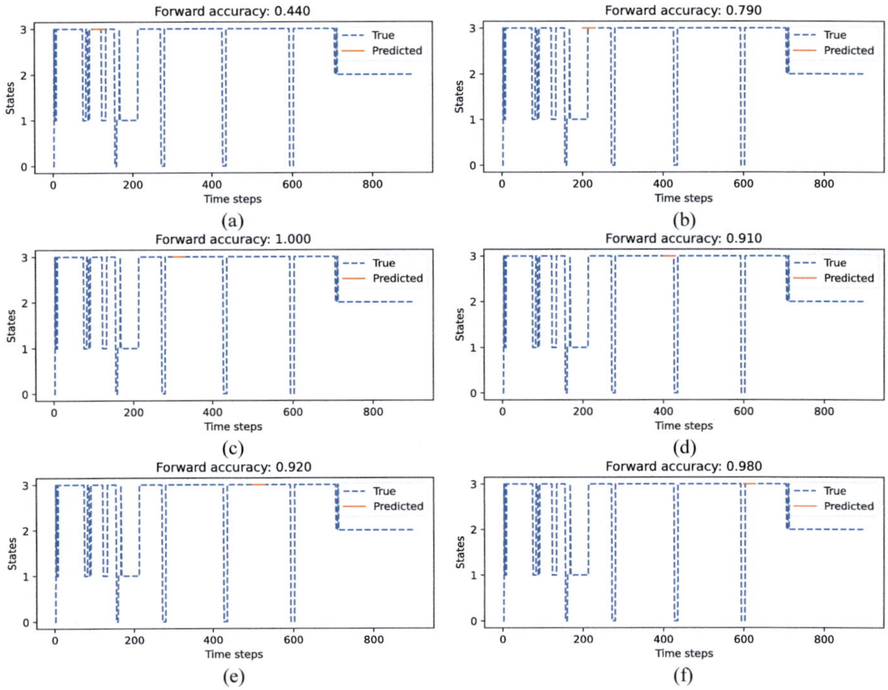

Fig. 13 Predicted hidden states for the next 30 samples from the OHMM at several time instances: **a** $t = 100$; **b** $t = 200$; **c** $t = 300$; **d** $t = 400$; **e** $t = 500$; **f** $t = 600$

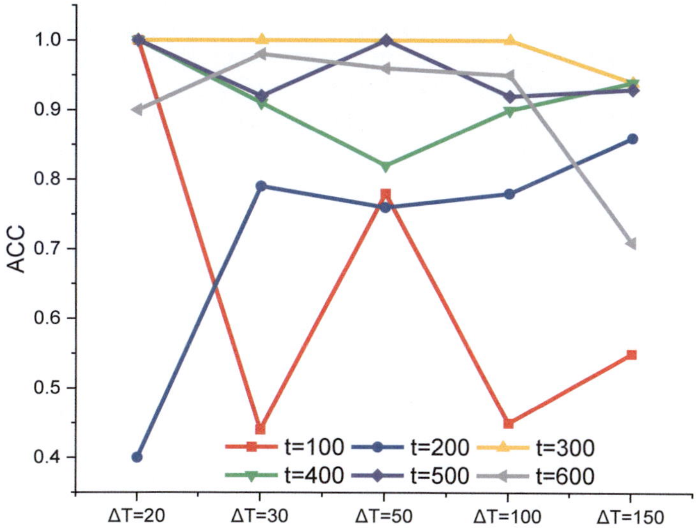

Fig. 14 Effects of the number of rings ahead (ΔT) on forward accuracy from the OHMM trained on full observations

Table 7 The accuracy of the OHMM predicting different numbers of rings ahead with different sizes of input data

Time instances	Number of rings ahead				
	$\Delta T = 20$	$\Delta T = 30$	$\Delta T = 50$	$\Delta T = 100$	$\Delta T = 150$
$t = 100$	1.00	0.44	0.78	0.45	0.55
$t = 200$	0.40	0.79	0.76	0.78	0.86
$t = 300$	1.00	1.00	1.00	1.00	0.94
$t = 400$	1.00	0.91	0.82	0.90	0.94
$t = 500$	1.00	0.92	1.00	0.92	0.93
$t = 600$	0.90	0.98	0.96	0.95	0.71

time instances $t = 300, 400, 500, 600$ can successfully predict the next 30 samples, achieving forward accuracy scores of 1, 0.867, 1, and 0.933, respectively. For the time instances $t = 100, 200$, the low forward accuracy scores are due to the frozen treatment of the OHMM. Additionally, it can be said that the models trained with $t = 100, 200$ have very high instability in forward predicting based on the results in Fig. 14. Therefore, even with the assistance of the observation sequence extension mechanism, a minimum amount of data for initialization is still necessary. This chapter has shown that this minimum amount is achieved at 200 rings. Therefore, if sufficient information is obtained during the early stages of construction, the OHMM can be effectively utilized to guide excavation in construction engineering.

(2) The OHMM demonstrates high accuracy in predicting geological risk up to a distance of 100 steps ahead while maintaining good performance in terms of stability and overfitting. The number of steps ahead can be considered as an indicator of the model's ability to predict future results. A larger number of steps indicates that the model can forecast over a greater distance, whereas a smaller number of steps allows for more precise patterns in each sample. The stability and overfitting of our OHMM model when forecasting different numbers of steps ahead using varying historical time steps are presented in Table 7. From the results, it can be observed that once over 200 historical data points are collected, the predictions of future patterns can achieve high accuracy even up to 150 steps ahead. The performance remains stable regardless of the changes in historical steps and the number of steps ahead. For a larger number of steps ahead, the forward accuracy preserves predictions for a larger range of samples and resists changes based on a few incoming samples. Selecting several steps ahead of 30 strikes a balance between the extent of foresight and the preservation of individual features in the forecasted samples.

6 Chapter Summary

The accurate estimation of geological risks is essential for preventing geohazards and ensuring efficient and safe construction processes. This chapter proposes a framework for applying an online version of HMM, known as OHMM, in the prediction of geological risk in TBM-excavated tunnels. The OHMM is tailored for the continuous nature of observational data, allowing it to adaptively update with each new piece of data. The parameters of the OHMM are estimated online based on relatively easily obtained observations. The trained OHMM can then be used to predict future hidden states based on the current state and their transition patterns. The prediction results can guide future excavation and construction in tunnel engineering. The performance of the model is evaluated based on forward prediction accuracy. This approach extends the short sequence of observed data to match the length of a complete sequence through an observation extension mechanism. The effectiveness of the OHMM, equipped with this observation extension mechanism, is demonstrated in a case study that models geological risks for a tunnel excavation project in Singapore.

The OHMM approach outperforms the traditional methods, including the hidden Markov model, long short-term memory network, neural network, and support vector machine, in predicting geological risks ahead of the tunnel boring machine. Notably, the OHMM can accurately forecast geological risks in areas yet to be constructed, using limited observational and site investigation data. This chapter advances geological risk prediction models by offering an online updating capability for tunnel excavation and construction projects. It enables early-stage risk prediction and provides long-term forecasts with minimal historical data requirements, maximizing the use of site investigation data.

References

1. Wang X, Lai J, Qiu J, Xu W, Wang L, Luo Y (2020) Geohazards, reflection and challenges in mountain tunnel construction of China: a data collection from 2002 to 2018. Geomat Nat Hazards Risk 11:766–785
2. Zhang L, Wang Y, Fu X, Song X, Lin P (2024) Geological risk prediction under uncertainty in tunnel excavation using online learning and hidden Markov model. Front Eng Manag, 1–20
3. Wang B, Sun D, Chen Q, Lin W, Li AW, Cao H (2020) Stress-state differences between sedimentary cover and basement of the Songliao Basin, NE China: in-situ stress measurements at 6–7 km depth of an ICDP scientific drilling borehole (SK-II). Tectonophysics 777:228337
4. Mahmoodzadeh A, Mohammadi M, Daraei A, Farid Hama Ali H, Ismail Abdullah A, Kameran Al-Salihi N (2021) Forecasting tunnel geology, construction time and costs using machine learning methods. Neural Comput Appl 33:321–348
5. Ravari A, Ghoreishi SF, Imani M (2024) Optimal inference of hidden Markov models through expert-acquired data. IEEE Trans Artif Intell 5:3985–4000
6. Hoi SCH, Sahoo D, Lu J, Zhao P (2021) Online learning: a comprehensive survey. Neurocomputing 459:249–289
7. Cao B-T, Saadallah A, Egorov A, Freitag S, Meschke G, Morik K (2021) Online geological anomaly detection using machine learning in mechanized tunneling. In: Barla M, Di Donna A, Sterpi D (eds) Challenges and innovations in geomechanics. Springer International Publishing, Cham, pp 323–330
8. Xiong Z, Guo J, Xia Y, Lu H, Wang M, Shi S (2018) A 3D multi-scale geology modeling method for tunnel engineering risk assessment. Tunn Undergr Space Technol 73:71–81
9. Peng S, Li Q (2021) Research on 3D geological modeling method of tunnel engineering computer based on geological cross-section. J Phys Conf Ser 1992:022112
10. Erharter GH, Marcher T (2020) MSAC: towards data driven system behavior classification for TBM tunneling. Tunn Undergr Space Technol 103:103466
11. Xu D, Wang Y, Huang J, Liu S, Xu S, Zhou K (2023) Prediction of geology condition for slurry pressure balanced shield tunnel with super-large diameter by machine learning algorithms. Tunn Undergr Space Technol 131:104852
12. Fu X, Wu M, Tiong RLK, Zhang L (2023) Data-driven real-time advanced geological prediction in tunnel construction using a hybrid deep learning approach. Autom Constr 146:104672
13. Wang H, Liu L, Song X, Zhang C, Tao DJAE-P (2022) AL-iGAN: an active learning framework for tunnel geological reconstruction based on TBM operational data. arXiv:2212.00965
14. Gong Q, Zhou X, Liu Y, Han B, Yin L (2021) Development of a real-time muck analysis system for assistant intelligence TBM tunnelling. Tunn Undergr Space Technol 107:103655
15. Sheil BB, Suryasentana SK, Mooney MA, Zhu H (2020) Machine learning to inform tunnelling operations: recent advances and future trends. Inst Civ Eng Smart Infrastruct Constr 173:74–95

Geological Features Characterization

1 Introduction

As urbanization accelerates, the demand for metro systems is increasing to provide rapid and eco-friendly transportation. The tunnel boring machine (TBM), a modern automated tool with a circular cross-section, is now preferred for urban tunnel excavation. TBMs offer advantages over traditional drilling and blasting, including high automation, continuous operation, minimal environmental disturbance, enhanced safety, and reduced costs. However, TBM performance can be affected by adverse geological conditions, which are often unknown before tunneling, leading to construction risks [1]. While studies have focused on understanding geological conditions along the tunnel axis to improve face stability and reduce failure probabilities, accurately predicting these conditions during continuous tunneling remains a challenge [2]. Our research aims to enhance the reliable perception of surrounding geological conditions at an early stage, providing crucial knowledge to improve the efficiency and intelligence of TBM tunneling.

Detecting geological information in deep tunnel engineering is challenging due to the complexity and ambiguity of the subsurface environment. Unknown geological conditions can lead to construction delays, cost overruns, and equipment jams, resulting in significant casualties and property losses [3]. While traditional methods of site investigation, such as borehole drilling and ground penetration radar (GPR), can uncover the geological structure and properties along the tunnel alignment, they could be somewhat slow, laborious, and costly [4]. Besides, emerging technologies, such as computed tomography and laser 3D scanning, have high initial costs and are not tightly related to the TBM operational performance, which tends to lack generalizability [5]. However, the surge in TBM driving data has created an opportunity for developing intelligent models using artificial intelligence (AI). These models can efficiently predict and classify geological conditions, enabling informed decisions about TBM operations. This research aims to develop data-driven techniques for

L. Zhang et al., *Intelligent Construction in Tunnels*, Lecture Notes in Civil Engineering 679, https://doi.org/10.1007/978-981-96-8785-5_3

accurately detecting geological conditions ahead of the tunnel face, replacing time-consuming manual assessments and enhancing the overall efficiency and safety of the tunneling process.

Various machine learning methods have demonstrated effective mapping capabilities for the interactions between tunnel boring machines (TBM) and the ground [6]. Many studies have focused on using supervised learning techniques to predict geological conditions based on TBM-related data, aiming to categorize rock masses with minimal errors [7]. However, a few issues remain with these existing studies. Firstly, most studies require appropriately pre-defined labels about the rock class as input, which is a prerequisite for learning and making correct decisions. The current method for obtaining the geological information relies on the soil investigation of boreholes, but the number of boreholes is too limited to provide an accurate interpretation of the target area [8]. Therefore, the preparation of the labeled data tagged with the ground truth is not an easy task. Manually assigning useful labels to raw data for developing a learning base normally takes a long time. Secondly, the rock mass classification scheme could be overly subjective in some cases. Field engineers with varying levels of expertise and knowledge may achieve different classification results for the same rock mass [9]. Because of this practical restriction, it is hard to consistently assure the labeling quality, which might lead to misinterpretation of the ground condition and unsatisfactory prediction outcomes. In other words, an apparent gap to be identified in supervised learning is the strong demand for high-quality labels associated with geological types, which may be difficult to obtain or perhaps unavailable in real engineering.

To minimize the need for labeled data and expedite computation, clustering algorithms can be employed to group unlabeled data into understandable structures based on shared similarities. This approach effectively uncovers hidden patterns in data without human intervention. Therefore, clustering algorithms are considered more practical for quickly producing a geology profile during the actual excavation process, even when geological type information is unavailable. Since lithology has a direct impact on the quality, progress, and safety of TBM excavation, trustworthy lithology identification is a major area of interest in TBM tunneling management. During tunneling, various sensors collect extensive TBM-related data linked to geological conditions. Machine learning algorithms can analyze this data to provide reliable estimates of the geological state at the tunnel face. Thus, the objective is to develop an effective time series clustering algorithm for continuous geological condition perception during TBM excavation, extracting temporal patterns from large volumes of monitoring data.

2 Time Series Clustering

Known as a special kind of clustering, time-series clustering is a specialized form of clustering that focuses on grouping sequences of data points indexed in time order. This technique is particularly useful for identifying patterns and trends over time, enabling the extraction of interpretable temporal patterns from complex data. Time-series clustering techniques aim to accommodate these characteristics by measuring similarity in a way that considers the temporal aspect, allowing for the effective analysis of sequences with varying lengths and time distortions. Among the clustering methodologies, distance-based, model-based, and deep learning-based approaches have garnered significant attention due to their efficacy in handling the unique challenges posed by time series data.

Distance-based clustering methods rely on predefined distance or similarity measures to evaluate the closeness between time series instances. Among these methods, the dynamic time wrapping (DTW) method is the most famous and prevalent. DTW is a robust method for measuring similarity between time series, effectively handling time distortions better than the Euclidean distance metric. It is preferred for capturing similarity in unsynchronized, varying-length time series, and seeks the optimal non-linear alignment between two sequences with minimal global cost [10]. Beyond DTW, other measures like Longest Common Subsequence (LCSS), Edit Distance on Real Sequences (EDR), and Time Warp Edit Distance (TWED) have been proposed. These measures offer robustness to noise and outliers by focusing on the alignment of similar subsequences within the time series [11]. As indicated by previous studies, distance-based methods are intuitive and straightforward to implement.

Model-based clustering approaches assume that the data is generated by a mixture of underlying probabilistic models, with each component representing a distinct cluster. These methods aim to infer the parameters of these models to best fit the data. Common methods include hidden Markov models (HMMs), autoregressive models, and Gaussian mixed models (GMMs) [12]. Among these algorithms, HMMs group series based on transition probabilities, while GMMs assume time series belong to different Gaussian distributions. Autoregressive models use AR, ARMA, and ARIMA to model the time series data and cluster based on the model parameters. Model-based methods offer a probabilistic framework for clustering, providing not only cluster assignments but also uncertainty estimates. However, they often require assumptions about the data distribution and can be computationally demanding, particularly for complex models [13].

The advent of deep learning has introduced powerful tools for learning complex representations of data, leading to the development of deep learning-based clustering methods for time series. These approaches aim to learn latent representations that capture the underlying structure of the data, facilitating more effective clustering. Two representative deep learning networks used in time series clustering are recurrent neural networks (RNNs) and convolutional neural networks (CNNs) [11]. For instance, as a prevalent type of RNN, the long short-term memory (LSTM) networks are well-suited for sequential data due to their ability to capture temporal dependencies [14]. Besides CNNs and RNNs, autoencoders have also been employed for time series clustering in recent years. As a type of neural network designed for unsupervised learning, autoencoders consist of an encoder that maps input data to a latent space and a decoder that reconstructs the original data from this representation, demonstrating improved accuracy in grouping similar time series [15].

Despite the importance of incorporating time series analysis into traditional clustering algorithms, relevant research about time series clustering in the field of tunnel construction remains few. Since a definition of time series data can be presented as an ordered collection of data points obtained over time [16], the in-situ TBM tunneling data composed of continuous sequences can also be regarded as time series data. Regarding this, the task of dividing the given TBM data in such an engineering background can be reasonably described as time series clustering. To fill this research gap, a high-performance time series clustering algorithm is developed to perceive the geological condition of each ring during the actual TBM excavation process.

3 Geological Features Characterization Framework

This framework, shown in Fig. 1, enables reliable perception of geological types ahead of TBM using three steps: TBM time-series data collection, DTW-Kmedoids clustering, and result analysis. Its ultimate goal is to automatically perceive the geological condition into four types based on data from the TBM monitoring system, which can provide real-time insight into construction progress. The clustering outcomes are expected to provide further guidance for controlling TBM operations and reducing construction risks while improving efficiency.

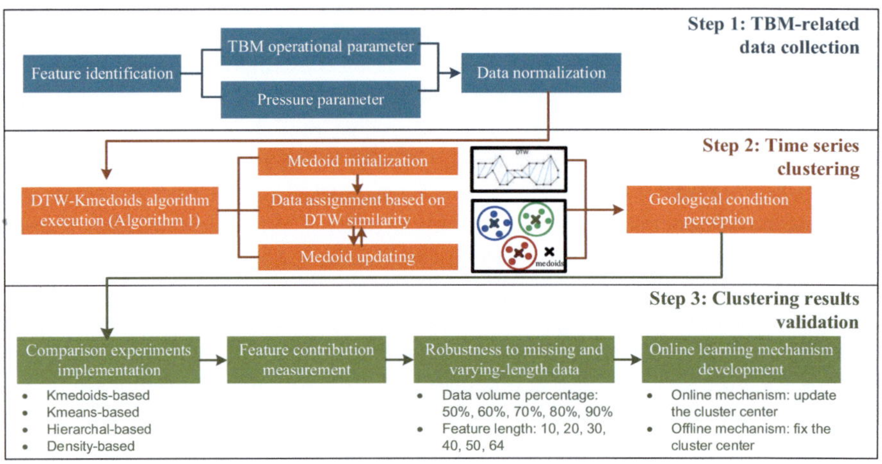

Fig. 1 Framework of the proposed DTW-Kmedoids-enabled clustering

3.1 Data Collection and Processing

TBM's logged data stores operational parameters in separate CSV files for each tunnel ring, including tunneling details, errors, and null values. This data covers mining date and time, TBM geometry, key operational parameters, soil pressure, grouting conditions, and excavated soil information. Sensors automatically collect data during excavation, generating large data sequences reflecting TBM operation and conditions.

The records contain errors, zero values, and noise. To extract useful information for clustering analysis, data preprocessing is required. Key steps include data cleansing, rescaling, noise removal, and reframing. Let the input data $x_n = \{x_1, x_2, x_3, \ldots, x_n\}$, where $x_1, x_2, x_3, \ldots, x_n$ denotes the TBM operational parameters. Equations (1) and (2) are defined to perform data cleansing.

$$F = f(x_1)f(x_2)f(x_3)\ldots f(x_n) \tag{1}$$

$$f(x_i) = \begin{cases} 1 & x_i \neq 0 \\ 0 & x_i = 0 \end{cases} \tag{2}$$

where $f(x_i)$ is the determining function for these parameters and F is the multiply result for all the determining functions. If any attribute in the input data x_i is either 0 or null, the determining function will result in $f(x_i) = 0$ with their multiply $F = 0$. The input data will be recognized as invalid and removed accordingly if $F = 0$.

To enhance model performance, outliers from machine or monitoring errors must be removed. A simple and effective method is calculating the standard score for each data point. The data will be identified as outliers and removed accordingly if the standard score of any attributes x_{ki} exceed m. The detailed procedure is expressed in Eqs. (3–5).

$$x_{ki} = \frac{x_i - \tilde{x}_i}{S} \tag{3}$$

$$f(x_{ki}) = \begin{cases} 1 & x_{ki} \leq m \\ 0 & x_{ki} > m \end{cases} \tag{4}$$

$$F_k = f(x_{k1})f(x_{k2})f(x_{k3})\ldots f(x_{kn}) \tag{5}$$

where x_{ki} is the standard score for each data point, $\tilde{x}_i$ is the mean value for the attribute, S is the standard deviation for the attribute, $f(x_{ki})$ is the determining function for the data point, m is the datum for outlier removal, and F_k is the multiply for the input data x_n.

With the outliers removed, data standardization is necessary since the TBM operational parameters have different magnitudes which may affect the performance of the LSTM model. In this research, all attributes are rescaled to the range of 0 to 1 by using Eq. (6).

$$x_j = \frac{x_i - x_{\min}}{x_{\max} - x_{\min}} \tag{6}$$

where x_j is the scaled data, and x_i is the original data from the training sample. $x_{\max}$ and $x_{\min}$ are the maximum and minimum values from the training sample, respectively.

3.2 DTW-Kmedoids Clustering

Given two temporal sequences $X = [x_1, x_2, \ldots, x_m], Y = \left[y_1, y_2, \ldots, y_n\right]$, a pairwise distance matrix between all points in two time series can be firstly established as Eq. (7), followed by identifying the optimal path that minimizes the cumulative distance Eq. (8).

$$\begin{bmatrix} d(x_1, y_1) & d(x_1, y_2) & \cdots & d(x_1, y_n) \\ d(x_2, y_1) & d(x_2, y_2) & \cdots & d(x_2, y_n) \\ \vdots & \vdots & \ddots & \vdots \\ d(x_m, y_1) & d(x_m, y_2) & \cdots & d(x_m, y_n) \end{bmatrix} \tag{7}$$

where $d(x_m, y_n)$ is the Euclidean distance between the mth element in X and the nth element in Y.

$$DTW(X, Y) = \min_{\pi} \sqrt{\sum_{(i,j)\in\pi} d\left(x_i, y_j\right)^2} \tag{8}$$

where $\pi = [\pi_1, \ldots, \pi_K]$ donates an aligned path satisfying the following constraints:

(1) Boundary constraints: $\pi_1 = (1, 1)$, $\pi_k = (m, n)$.
(2) Monotonicity and continuity constraint: For $\pi_k = (i_k, j_k)$ and $\pi_{k+1} = (i_{k+1}, j_{k+1})$, their time series indexes should be written as $i_k \leq i_{k+1} \leq i_k + 1$ and $j_k \leq j_{k+1} \leq j_k + 1$.

For a more detailed explanation, the first constraint means that the warping path needs to start from the lower-left corner (1,1) and end at the upper-right corner (*m*,*n*) of the matrix. The second constraint indicates the allowable steps for a current position, which are its reachable neighborhoods at the upper, upper right, and right positions, respectively. The aligned path is not allowed to go back or jump in the time index. Later, the neighborhood owing to the shortest distance to the current position will be appended to the warping path. Finally, the optimal path is determined using dynamic programming, which can be expressed as a recursive function in the form of Eq. (9).

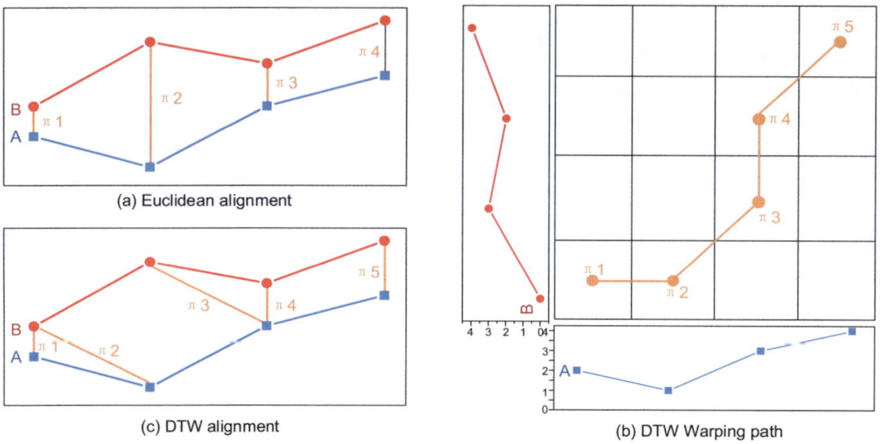

Fig. 2 Example of Euclidean distance and DTW

$$D(X_i, Y_j) = \pi\left(x_i, y_j\right) + \min\begin{cases} d\left(x_i, y_{j+1}\right) \\ d\left(x_{i+1}, y_j\right) \\ d\left(x_{i+!}, y_{j+1}\right) \end{cases} \tag{9}$$

To facilitate comprehension, an example of DTW computation is given below. Given two time series $A = [1, 2, 3, 4]$ and $B = [0, 3, 2, 4]$, their Euclidean alignment is shown in Fig. 2a with a total distance of 5. Then, the distance matrix in Fig. 2b can be obtained as a basis for determining the warping path. As a result, the DTW alignment in Fig. 2c reaches the DTW distance of 4, which is smaller than the Euclidean distance.

Another thing to notice is that DTW has a variant named soft-DTW, where a differentiable learning loss is formulated to calculate the soft minimum of all alignment costs [17]. The definition of soft-DTW is provided in Eqs. (10) and (11). In particular, $\min^{\gamma}$ stands for the soft-minimum function, which takes the place of the *min* operator in DTW. If $\gamma = 0$, the formulation of soft-DTW will turn out to be the classical DTW.

$$DTW_{\gamma}(X, Y) = \min_{A \in \mathcal{A}}^{\gamma}\langle A, \Delta(X, Y)\rangle \tag{10}$$

$$\min_{A \in \mathcal{A}}^{\gamma} f(A) = \begin{cases} \min_{A \in \mathcal{A}} f(A) \\ -\gamma \log \sum_{A \in \mathcal{A}} \exp\left(-\frac{f(A)}{\gamma}\right) \end{cases} \tag{11}$$

where $\gamma \geq 0$ represents a temperature parameter, and $A \in \mathcal{A}$ is the constraint.

As for the clustering task, a hybrid time series clustering algorithm integrating DTW and Kmedoids is developed to extract valuable patterns from multivariate sequential data. The clustering performance is greatly affected by the choice of distance measure and clustering method. DTW accurately measures distances between observations, accommodating varying lengths, while Kmedoids (or Partitioning Around Medoids) divide time series into distinct clusters with interpretable centroids. Unlike Kmeans, which calculates cluster centers as averages, Kmedoids selects actual data points as centers, enhancing interpretability. This makes Kmedoids particularly effective for clustering time series data.

For a given dataset with a set of sequences ordered by time, the DTW-Kmedoids algorithm works directly on the time series, attempting to perform the unsupervised partitioning of the dataset into several clusters with respect to their DTW-based similarity measurement. The detailed execution flow of DTW-Kmedoids is presented in the pseudo-code of Algorithm 1. The important steps are as follows:

(1) To specify the number of k cluster medoids instead of cluster centroids.
(2) To randomly select k cluster medoids from all data points for initialization.
(3) To calculate the distance metric DTW for assigning data points to their closest medoids.
(4) To design a loop for constantly updating the k medoids under an optimal strategy of minimizing the distance between each cluster medoid and its corresponding data points.
(5) To repeat the iteration until the cost function formulated by the sum of the DTW distance of all the time series pairs within the cluster reaches its minimum.

In consequence, k new medoids along with the new data points belonging to it can be recognized, resulting in k non-overlapping partitions. Sequences with similar variation tendencies are believed to be easily categorized into the same cluster in an unsupervised manner based on Kmedoids. After that, to make Kmedoid more suitable for time series data, the DTW or soft-DTW distance is taken as the distance metric to replace the Euclidean distance, thereby enabling similarity calculations between objects. In short, the core process of DTW-Kmedoid involves updating the medoids by maximizing the intra-group similarity and minimizing the inter-group similarity across the prepared time series data that is calculated by DTW distance. It is expected to group time series data that share similar properties and separate dissimilar time series data apart.

Noticeably, the hybrid DTW-Kmedoids algorithm is created to inherit the advantages from both DTW and Kmedoids. The superiorities of DTW-Kmedoids can be summarized as: (1) It is simpler and more computationally efficient than commonly used hierarchical clustering techniques. (2) It is good at processing high-dimensional data and is less sensitive to outliers. (3) It is outstanding in dealing with temporal drift and varying-length sequences. (4) It is possible to produce more accurate results even under some degrees of uncertainty.

Algorithm 1 DTW-Kmedoids.

Step 1: Initialization

1.1 Randomly select k points as the medoids out of all data points

1.2 Assign every remaining data point to its nearest medoid using the DTW-based similarity measurement $DTW(m_i, p_i)$ (where m_i represents the medoids, p_i indicates the data points)

Step 2: Medoid updating

2.1 While the cost function $c = min \sum_{i=1}^{k} \sum_{m_i} DTW(m_i, p_i)$ decreases:

2.2 Swap the medoid m_i and data point p_i

2.3 Assign every data point to the nearest new medoid m_i

2.4 Recalculate the cost function of the total DTW distances from all data points to their medoids. If the sum is larger than the previous one, stop the while loop

---

3.3 *Clustering Performance Validation*

Since clustering validation is an essential task in cluster analysis, several clustering evaluation metrics have been developed to quantify the clustering performance. This research focuses on the alignment of clustering results with pre-investigated realities. Thus, we employ two widely used external validation measures: the Rand index (RI) and Fowlkes-Mallows index (FMI), which assess the quality of discovered clusters by comparing outputs to the ground truth class.

To be more specific, RI measures the similarity between two assignments under the comparison of unordered pairs of samples. As given in Eq. (12), RI counts pairs assigned to the same clusters from the predicted result and ground truth. FMI in Eq. (13) is defined in the form of the geometric mean of the pairwise precision and recall, which can be regarded as a way to measure the confusion matrix. The higher value of these two metrics means a better clustering performance.

$$RI = \frac{a+b}{\binom{n}{2}} \tag{12}$$

where $\binom{n}{2} = \frac{n(n-1)}{2}$ stands for the number of unordered pairs in a set of n samples, a is the number of times a pair of samples belonging to the same clusters, and b is the number of times a pair of samples in different clusters.

$$FMI = \sqrt{\frac{TP}{TP+FP} \cdot \frac{TP}{TP+FN}} \tag{13}$$

where *TP* is the number of true positives, *FP* is the number of false positives, and *FN* is the number of false negatives.

4 Experimental Study

4.1 Data and Resources

The proposed time series clustering approach is validated in a tunnel excavation project for a fully underground 4-car MRT system in Singapore, located between Telok Kurau Road and Marine Crescent (see Fig. 3). The tunnel runs along the Singapore coastline through soft soil, posing challenges for design and excavation, making geological conditions vital for construction safety. To prepare for excavation, soil investigation boreholes were drilled every 20–30 segment rings over 30–50 m to collect samples and assess geological conditions. Four dominant geology types were identified: marine clay (Type I), a mix of fluvial sand and marine clay (Type II), fill (Type III), and fluvial sand (Type IV). Table 1 summarizes their engineering properties, providing essential geotechnical parameters for evaluation and planning, along with illustrative photos.

The tunnel section in this study was constructed using a TBM with a cutter head diameter of 6.67 m and a main machine length of 9.81 m. The TBM employs a circle-reinforced structure consisting of several segment rings with an inner diameter of 5.8 m and a width of 1.4 m. The engineering route comprises 953 segment rings, with each ring being approximately 1.5 m long. For this case study, the time series from the first 377 tunnel rings are taken to validate the clustering performance. The geology type ahead of the tunnel face in each segment ring can be perceived in the

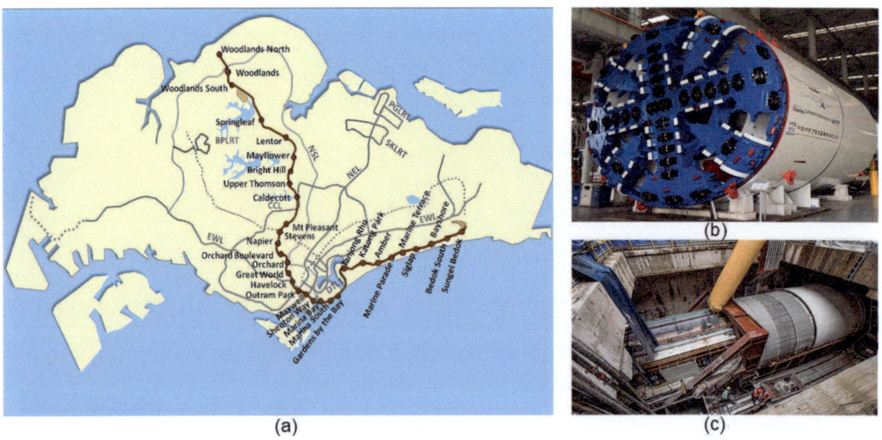

Fig. 3 The project profile about: **a** layout plan; **b** TBM; **c** construction site

Table 1 Summary of the four dominant kinds of geological conditions

Geological condition	Label	Description	Recommended bulk unit weight	Recommended Young's modulus	Recommended permeability	Photos
Marine clay	I	It is the type of soft soil in coastal area to cause settlement and instability	15.5 kN/m^3 (above + 85 mRL) 16.5 kN/m^3 (below + 85 mRL)	400 c_u (undrained)	1×10^{-8} m/s	
A mix of fluvial sand and marine clay	II	This is a mixed type of soil condition with fluvial sand and marine clay	–	–	–	

(continued)

Table 1 (continued)

Geological condition	Label	Description	Recommended bulk unit weight	Recommended Young's modulus	Recommended permeability	Photos
Fill	III	Fill materials can be varied, such as sand, wood, brick, and others	19 kN/m^3	10 Mpa (drained) 12 Mpa (undrained)	1×10^{-6} m/s	
Fluvial sand	IV	Its soft degree is similar to medium dense yellow grey silty	19 kN/m^3	In the range of [15 MPa, 25 Mpa] (drained)	1×10^{-6} m/s	

practical application. It is believed that the proposed time-series clustering approach can be reasonably extended to the entire tunneling process, where the excavation length is around 1.5 km and the excavation depth is between 19.3 and 29.3 m. More clearly, Fig. 4 illustrates the geological profile along the tunnel alignment of this project, and Fig. 5 displays the geological conditions in the study area from the No. 1 ring to the No. 377 ring. It is evident that the number of segment rings in the types I, II, and IV takes up 95.5% of the total rings. Only 17 rings experience the fill soil stratum. That is to say, most parts of this tunnel pass through soft clay and sand.

A TBM monitoring system has been established with various smart sensors to collect data on seven key parameters: thrust force (x1), cutter head torque (x2), cutter head rotational speed (x3), screw rotational speed (x4), penetration rate (x5), average chamber pressure (x6), and average soil pressure (x7). These parameters reflect the dynamic interaction between the TBM and surrounding geology during excavation, helping to reveal changes in geological conditions [18]. The selection of these variables is justified by their usefulness in tracking the evolution of TBM excavation over time. Variables x1–x5 represent TBM operating parameters that adapt to the subterranean environment, while x6 and x7 are more easily obtainable and provide insights into soil behavior and machine-environment interaction. The statistical descriptions of these variables are summarized in Table 2, highlighting their significance in tracking TBM excavation evolution.

During excavation, streaming data is acquired at 20 mm intervals, primarily regulated by the thrust cylinder system, which adjusts speed through deviation correction. Built-in stroke sensors indicate the TBM's position. Early and final stage data often contain outliers, complicating the reflection of actual conditions. To ensure data quality, the first and last four sets are removed, resulting in approximately 61 sets of recorded time series data per segment ring, totaling 22,997 samples for analysis. The time series dataset for the monitored tunnel section is structured as $7 \times 61 \times 377$. Each segment ring is about 1.5 m long, allowing predictions of geological conditions on a ring-by-ring basis. Figures 6 and 7 illustrate time series data for rings No. 4 and No. 212, while the remaining 375 rings exhibit similar data.

4.2 Intelligent Data Clustering

To perceive potential geological conditions, the unsupervised DTW-Kmedoid algorithm outlined in Algorithm 1 is applied to the dynamically increasing TBM-related dataset. The hybrid DTW-Kmedoids identifies intrinsic patterns based solely on similarity, without prior training or known labels. By clustering the unlabeled dataset, unsupervised learning can effectively discover hidden patterns in data without manual intervention, enabling real-time operation without needing extensive samples. This method is particularly valuable for extracting insights from large volumes of collected TBM data, as objects within the same cluster tend to share similar characteristics.

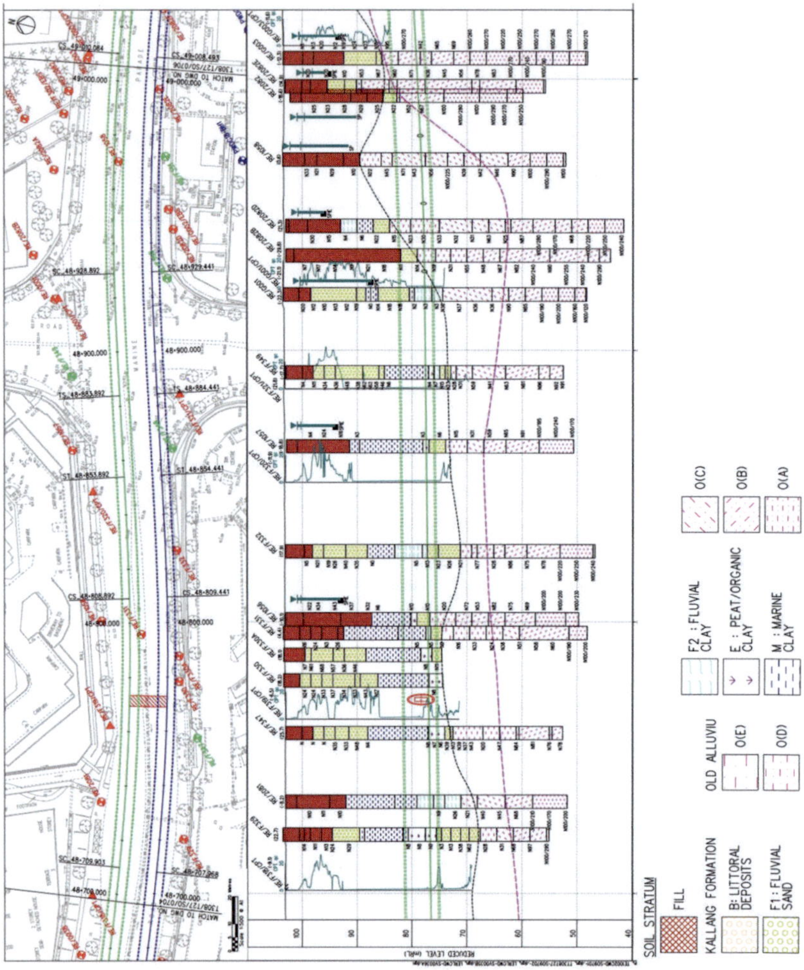

Fig. 4 Geology profile of the tunnel alignment

Fig. 5 A view of geological conditions from the segment ring No. 1 to No. 377

Table 2 Statistical description of determined inputs and outputs

Feature type	Variable	Name	Unit	Mean	Std	25% percentile	75% percentile
TBM operational parameter	x1	Thrust force	kN	20,706.14	4353.40	19,021.5	22,935.5
	x2	CHD torque	kN • m	1569.26	956.89	761	2459
	x3	CHD rotational speed	rpm	1.41	0.29	1.17	1.70
	x4	Screw rotational speed	rpm	3.17	0.35	2.91	3.42
	x5	Penetration rate	mm/min	18.73	10.20	10	26
Pressure parameter	x6	Average chamber pressure	bar	2.71	0.55	2.31	3.06
	x7	Average soil pressure	bar	2.57	1.73	1.2	3.8

To verify the effectiveness of the proposed time series clustering algorithm, comparison experiments are conducted with three algorithms: Kmedoids, DTW-Kmedoids, and softDTW-Kmedoids. While all three group time series data have similar shapes, Kmedoids relies on Euclidean distance, whereas the others use DTW as the core metric. The algorithms are implemented in Python 3.8, with parameter

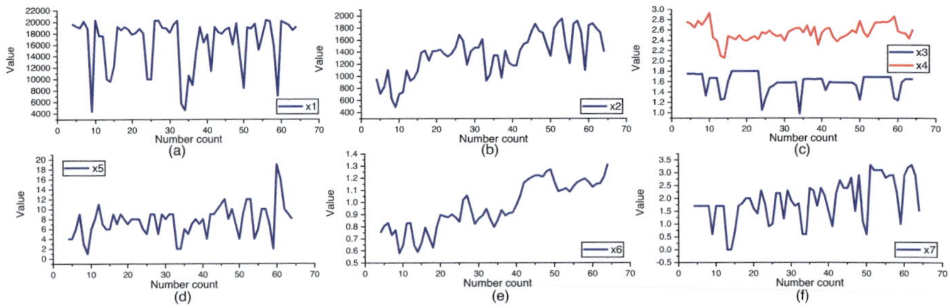

Fig. 6 The collected time series data in ring No. 4: **a** trust force x1; **b** CHD torque x2; **c** CHD rotational speed x3 and screw rotational speed x4; **d** penetration rate x5; **e** average chamber pressure x6; **f** average soil pressure x7

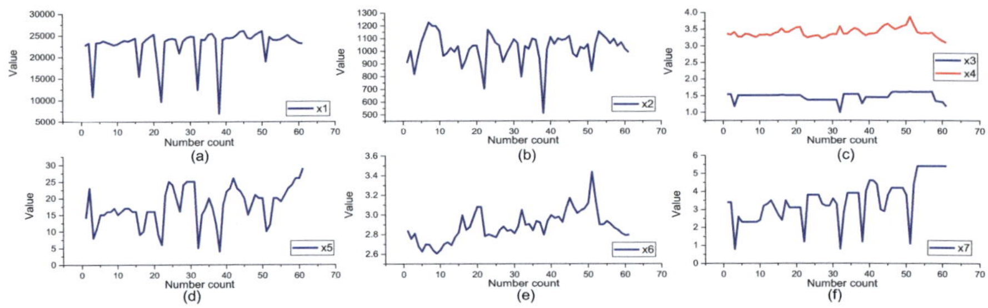

Fig. 7 The collected time series data in ring No. 212: **a** trust force x1; **b** CHD torque x2; **c** CHD rotational speed x3 and screw rotational speed x4; **d** penetration rate x5; **e** average chamber pressure x6; **f** average soil pressure x7

settings detailed in Table 3. The maximum number of iterations is varied (100, 200, 300, 400, and 500), but due to the limited data size, clustering performance does not significantly differ across iterations.

In consequence, Figs. 8, 9, and 10 demonstrate the qualitative outcomes of the three clustering algorithms using time series data of feature 1 as an example. Each subfigure displays a collection of time series with the highest degree of similarity that can be ascribed to a cluster. Apart from the cluster 3, feature 1 exhibits distinct characteristics across the other clusters. A 4×4 confusion matrix in Fig. 11 provides a visual representation of clustering outcomes compared to true labels, where the diagonal elements indicate correctly clustered points. The totals along the diagonal for DTW-Kmedoids and softDTW-Kmedoids are 259 and 257, respectively, nearly double that of Kmedoids. Both DTW-Kmedoids and softDTW-Kmedoids effectively distinguish geology type I, but they struggle with geology type III due to limited data. The main distinction between the two is that DTW-Kmedoids are more likely to misclassify type II as type III, while softDTW-Kmedoids often misclassify type II as type I. Kmedoids, on the other hand, fail to accurately represent the geological conditions. Overall, DTW-Kmedoids and softDTW-Kmedoids provide more reliable

Table 3 Parameter setting for the developed DTW-Kmedoids algorithm

Parameter	Description	Value	Reason for the value setting
Number of clusters	The number of clusters to form as well as the number of medoids to generate	4	There are four different types of geological conditions in this tunneling project
Maximum number of iterations	The maximum number of iterations when fitting	300	The default value is 300 in the scikit-learn package (Python)
Random state	For the random number generator	100	/
Data normalization method	To rescale features into a given range	Minimum-maximum scaler	It helps to preserve the shape of the original distribution

clustering results, highlighting their effectiveness in capturing geological profiles, primarily due to the suitability of the DTW distance metric for measuring time series similarity.

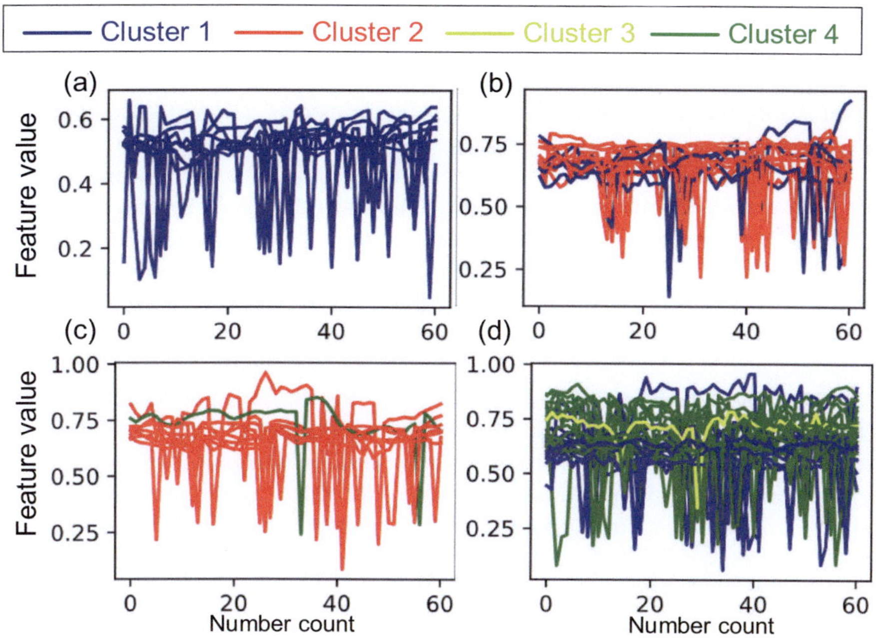

Fig. 8 Time series of feature 1 after Kmedoids algorithm in: **a** cluster 1; **b** cluster 2; **c** cluster 3; **d** cluster 4

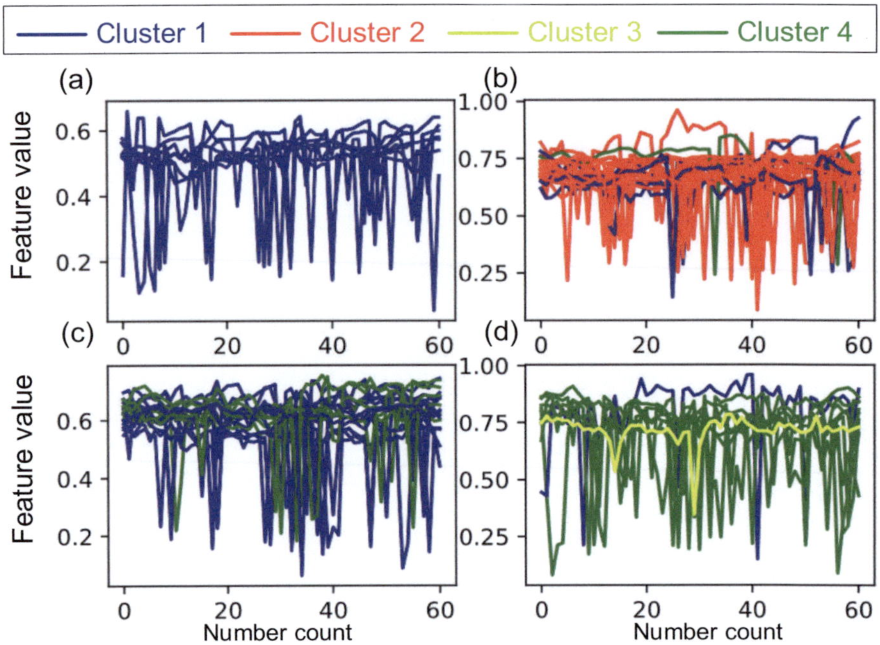

Fig. 9 Time series of feature 1 after DTW-Kmedoids algorithm in: **a** cluster 1; **b** cluster 2; **c** cluster 3; **d** cluster 4

Noticeably, one parameter called the number of clusters k is bound to greatly affect the clustering performance. To further compare operations in different configurations of presented systems, the number of clusters k has been set as $\{2,3,4,5\}$, while keeping all other parameters constant as in Table 3. The cost function of Kmedoids aims to minimize inter-cluster distance and maximize intra-cluster distance, allowing the optimal k to be determined by achieving the largest RI and FMI scores for k clusters. As shown in Fig. 12, the quantitative findings are evaluated by two assessment metrics from Eqs. (12) and (13), which can provide additional evidence to determine the optimal number of clusters. When the dataset is partitioned into four independent portions by the DTW-Kmedoids technique, the greatest values of RI and FMI are 0.737 and 0.587, respectively. Hence, the optimal number of clusters in this case study is 4, which is basically consistent with the actual engineering situation. Besides, the DTW-Kmedoids algorithm outperforms the other two algorithms at the cluster number of 2, 3, and 4, indicating the suitability of the proposed DTW-Kmedoids for this case. As for DTW-Kmedoids, the second greatest clustering performance is observed for the cluster number of 3 with the value of RI and FMI 0.031 and 0.001 lower than the maximum, suggesting that the prepared time series dataset owns either 3 or 4 clusters.

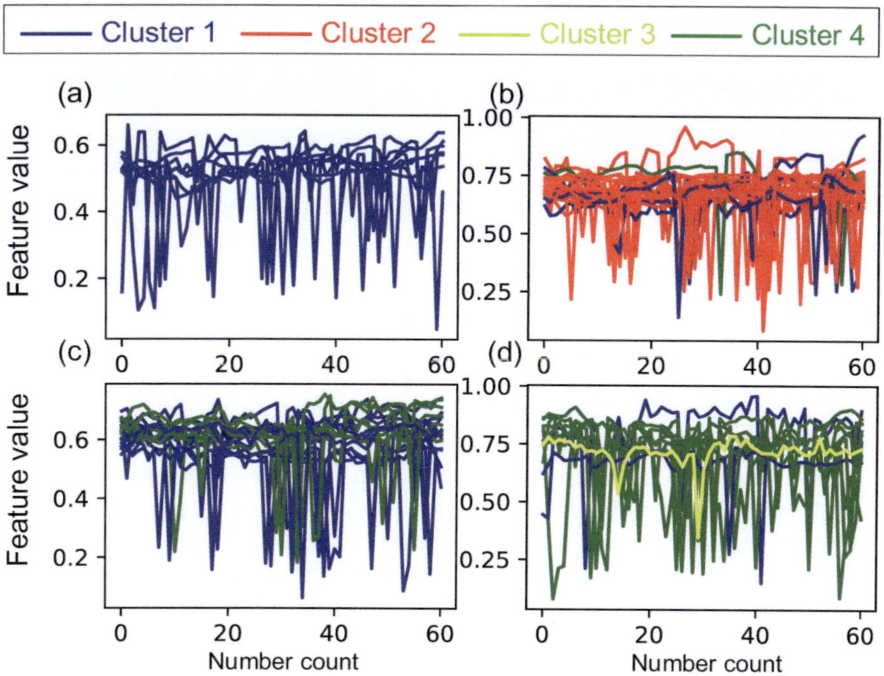

Fig. 10 Time series of feature 1 after softDTW-Kmedoids algorithm in: **a** cluster 1; **b** cluster 2; **c** cluster 3; **d** cluster 4

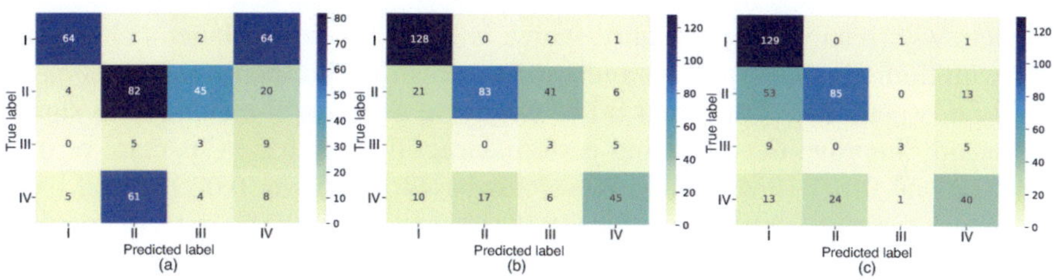

Fig. 11 Confusion matrix about the clustering results: **a** Kmedoids; **b** DTW-Kmedoids; **c** softDTW-Kmedoids

4.3 Clustering Result Analysis

The DTW-Kmedoids algorithm analyzes the time series dataset to provide insights into geological conditions during tunneling. Through comparison experiments, feature influence measurement, and robustness tests, the superiority of the proposed DTW-Kmedoids in this case study can be demonstrated in a rigorously evidence-based manner. The clustering results have been evaluated using RI and FMI as defined in Eqs. (12) and (13), respectively, and the findings can be further analyzed as follows.

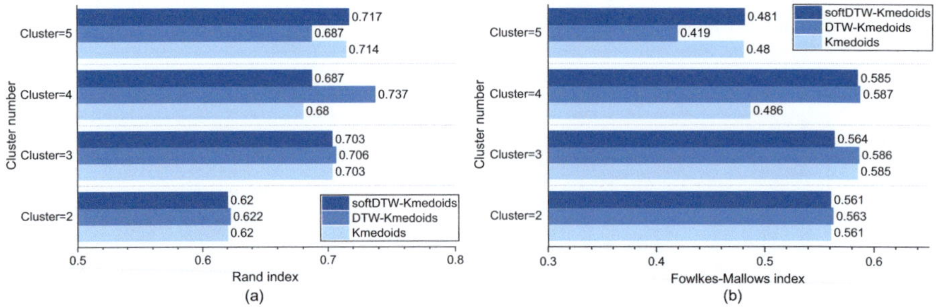

Fig. 12 Clustering performance in different cluster numbers measured by two evaluation metrics: **a** Rand index (RI); **b** Fowlkes-Mallows index (FMI)

(1) Kmedoids-based algorithms can deliver greater clustering performance than three state-of-the-art methods in this case, including the Kmeans-based algorithms, hierarchical-based algorithms, and density-based algorithms. Both Kmedoids and Kmeans are partitioning methods that use a predefined number of clusters, while the hierarchical approach assumes a nested structure and the density-based method connects high-density data areas. Table 4 outlines the comparative results under the cluster number of 4. DTW-Kmedoids achieve the best clustering results as underlined in bold, achieving an RI and FMI of 0.061 and 0.108 higher than basic Kmeans, respectively. The partitioning methods are generally good at discovering spherical-shaped clusters, making them more suitable for small- and medium-size datasets in this case. In contrast, hierarchical-based methods and density-based methods tend to generate clusters with relatively low quality, which is likely due to the latter's challenges with high-dimensional data and varying densities. Furthermore, the comparison experiments prove that DTW can serve as a useful technique to significantly improve the clustering performance, allowing for an increase of 0.05 (RI) and 0.101 (FMI) in the Kmedoids-based algorithm, 0.02 (RI and FMI) in the Kmeans-based algorithm, 0.016 (RI) and 0.039 (FMI) in the hierarchal-based algorithm, 0.011 (RI) and 0.033 (FMI) in the density-based algorithm. The finding confirms that the combined use of DTW and Kmedoids is better suited to the time series data collected from this tunneling project. It can further validate the rationality and practicality behind Algorithm 1 in successfully measuring the similarity between two temporal sequences. The four clusters derived from DTW-Kmedoids are more consistent with the actual geological condition.

(2) TBM operational parameters enhance the DTW-Kmedoids algorithm's ability to provide quick feedback on geological conditions during shield tunneling. To better understand the clustering mechanism, we quantified the influence of features by removing one at a time in several clustering tasks. As shown in Fig. 13a, excluding feature x1 reduced the RI and FMI by 6.50% and 16.86%, respectively. The penetration rate proved to be the most critical feature, causing reductions of 7.57 and 21.92% in RI and FMI. The influence degree of x1, x2,

Table 4 Evaluation of clustering performance among different categories of clustering algorithms

Category	Method	Evaluation metric	
		Rand index	Fowlkes-Mallows index
Kmedoids-based	Kmedoids	0.687	0.486
	DTW-Kmedoids	**0.737**	**0.587**
	softDTW-Kmedoids	0.680	0.564
Kmeans-based	Kmeans	0.676	0.479
	DTW-Kmeans	0.696	0.499
	softDTW-Kmeans	0.692	0.490
Hierarchal-based	Agglomerative	0.662	0.441
	DTW-Agglomerative	0.678	0.460
	softDTW-Agglomerative	0.675	0.458
Density-based	DBSCAN	0.672	0.452
	DTW-DBSCAN	0.683	0.485
	softDTW-DBSCAN	0.677	0.459

x3, and x4 is very similar and closely follows x5. On the contrary, since the improvement of overall clustering performance from features associated with the pressure parameters is considerably limited, it is reasonable to conclude that these two features are of less importance in geological perception. Moreover, Fig. 13b compares the clustering performance in three scenarios with different types of parameters. Compared with all parameters, the use of TBM operational parameters reduces the RI and FMI by 4.21 and 10.39%, while pressure parameters drop the RI and FMI to the minimal value of 0.677 and 0.443. This suggests TBM operational parameters provide more geological information, making them preferable for reliable and timely assessments when multivariate data collection is challenging.

(3) The proposed DTW-Kmedoids and softDTW-Kmedoids methods effectively handle time series data with missing values, demonstrating robustness in clustering geological conditions under high uncertainty. To assess their performance with incomplete inputs, we conducted comparison experiments at missing value rates of 50, 60, 70, 80, and 90%, repeating the data deletion process 100 times. Clustering results under different percentages of incomplete inputs are provided in Fig. 14 and Table 5. As the percentage of missing data increases, the mean and median RI and FMI decrease, while standard deviations slightly increase. These findings indicate that missing data can weaken clustering performance but the impact on DTW-Kmedoids and softDTW-Kmedoids remains limited. Remarkably, even with 50% missing data, both algorithms maintain reasonable geological condition perception, achieving acceptable RI and FMI values. The hybrid approach proves robust in extracting meaningful patterns from incomplete datasets, adaptable to more complex engineering scenarios with higher

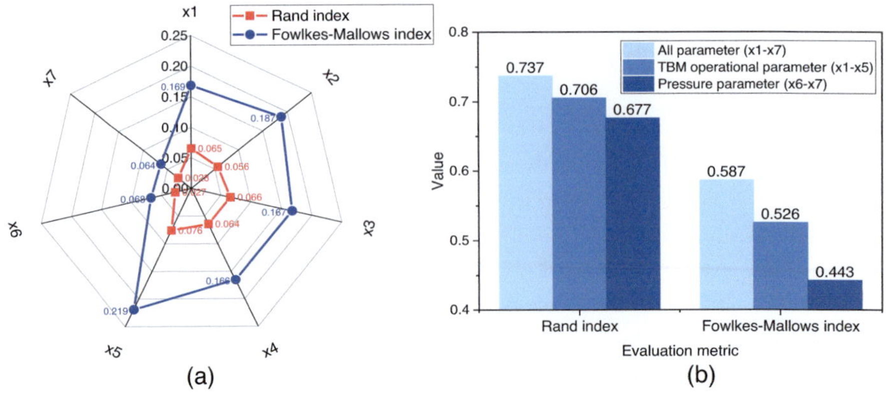

Fig. 13 Feature influence measurement under DTW-Kmedoids: **a** reduction of evaluation metric value when a feature is moved from the dataset; **b** clustering performance under three scenarios

uncertainty. Notably, under 10% missing data, DTW-Kmedoids achieves clustering performance similar to complete datasets, with an RI of 0.718 and FMI of 0.551, consistently outperforming softDTW-Kmedoids by at least 0.006 in RI and 0.008 in FMI.

(4) Using the previous 40 data points from each ring in the DTW-Kmedoids algorithm yields satisfactory clustering results for identifying forward geological conditions. The prepared time series dataset has a fixed feature length of 64. To explore how data quantity affects clustering performance, comparative experiments are conducted with previous data numbers set to 10, 20, 30, 40, 50, and 64. As shown in Fig. 15, increasing data length raises the RI from 0.685 to 0.737 and the FMI from 0.468 to 0.587. Notably, significant improvements occur when the previous data number reaches 30 or 40, with values for 40 being just 0.003 and 0.004 lower than the full observation sequence. These results demonstrate that

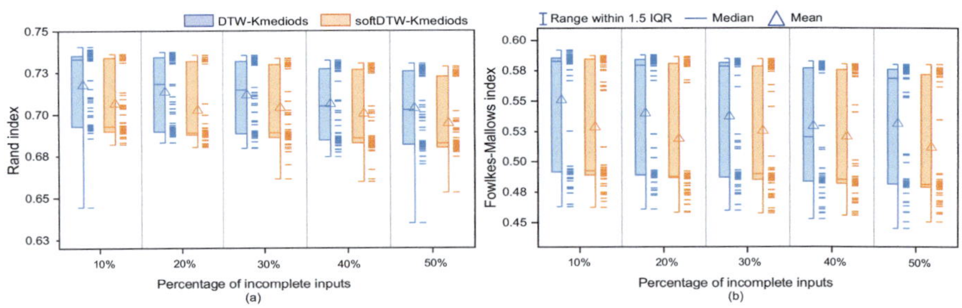

Fig. 14 Boxplot of clustering performance under different data volume percentages evaluated by: **a** Rand index (RI); **b** Fowlkes-Mallows index (FMI). *Note* IQR is the abbreviation of the interquartile range

Table 5 Comparison of clustering performance through 100 experiments for each setting of incomplete input percentage

Evaluation metric	Incomplete input percentage (%)	DTW-Kmedoids		softDTW-Kmedoids	
		Mean (std)	IQR	Mean (std)	IQR
Rand index	10	0.718 (0.021)	[0.693, 0.735]	0.706 (0.021)	[0.690, 0.734]
	20	0.713 (0.021)	[0.689, 0.734]	0.702 (0.021)	[0.688, 0.732]
	30	0.712 (0.021)	[0.688, 0.732]	0.704 (0.022)	[0.686, 0.730]
	40	0.706 (0.022)	[0.684, 0.727]	0.700 (0.022)	[0.683, 0.727]
	50	0.704 (0.023)	[0.682, 0.726]	0.695 (0.020)	[0.680. 0.723]
Fowlkes-Mallows index	10	0.551 (0.046)	[0.492, 0.585]	0.529 (0.047)	[0.489, 0.584]
	20	0.540 (0.047)	[0.489, 0.584]	0.519 (0.047)	[0.487, 0.581]
	30	0.537 (0.047)	[0.487, 0.581]	0.525 (0.048)	[0.485, 0.578]
	40	0.529 (0.049)	[0.484, 0.577]	0.521 (0.048)	[0.482, 0.576]
	50	0.531 (0.048)	[0.481, 0.576]	0.512 (0.046)	[0.479, 0.572]

Note IQR is the abbreviation of the interquartile range

DTW-Kmedoids can effectively use only the previous 40 data points to achieve reliable clustering, confirming its robustness in handling incomplete inputs.

5 Merits Analysis

As sequential monitoring data arrives over time during tunneling, obtaining the entire dataset before project completion is challenging. Using DTW-Kmedoids with complete observation data can lead to delays in decision-making due to its inherent hysteresis. As a promising solution, a subfield of machine learning called online learning can be introduced to enhance the accuracy in the sequence of predictions or decisions through learning continuous data instances one by one at each time [19]. Notably, this emerging new concept differs from classical offline machine learning methods that primarily learn from a complete dataset at once. Online learning holds significant potential for optimizing decision-making processes in tunneling projects.

Fig. 15 Clustering performance of DTW-Kmedoids using different numbers of previous data in each ring

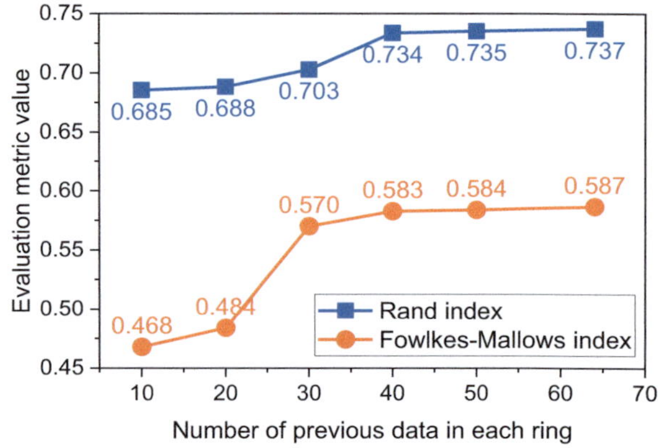

To enable dynamic clustering in ongoing construction projects, the proposed DTW-Kmedoids algorithm employs an online mechanism to handle sequential TBM monitoring data. Unlike offline learning, online learning iteratively updates clustering results as new time series data from the next TBM ring is received. This approach enhances the algorithm's effectiveness in detecting geological conditions by leveraging the accumulation of on-site observations for a better understanding of the stratum. Given that geological information significantly impacts excavation efficiency and safety, the real-time results from this online clustering are crucial for analyzing the complex construction environment. Additionally, it allows TBM operators to make timely adjustments to machine operations, aligning with engineering requirements. Further discussions on the online clustering process of DTW-Kmedoids follow.

(1) A series of comparative experiments were designed to evaluate offline and online clustering mechanisms using observations from the previous 100 TBM rings, with continuous excavation to gather new monitoring data. The key distinction between the two approaches is the selection of clustering centers. In offline clustering, subsequent tasks rely on fixed centers determined by the initial 100 rings, resulting in a static clustering process. In contrast, the online mechanism allows the clustering center to adapt based on the characteristics of incoming streaming data. The findings highlight the online clustering mechanism's potential for automatically detecting changing geological conditions during tunnel construction in an adaptive and reliable manner.
(2) The online mechanism for time series clustering enhances the reliability of assessing forward conditions. As shown in Fig. 16, the offline and online mechanisms accurately group 244 and 248 geological conditions, achieving accuracies of 64.72% and 65.78%, respectively. Both approaches provide valuable insights without label requirements, supporting the validity of the online clustering. However, both mechanisms tend to misclassify type II conditions as type III and type IV conditions as type II. The online mechanism outperforms the offline one, increasing RI and FMI values by 0.009 and 0.039, resulting in final

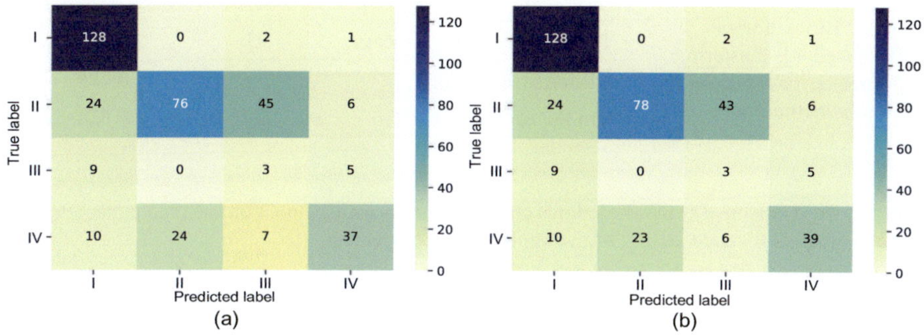

Fig. 16 Confusion matrix about the clustering results from two clustering mechanisms: **a** offline mechanism: fix the cluster center after 100 rings; **b** online mechanism: update the cluster center in each ring

values of 0.703 and 0.524. Thus, the online mechanism demonstrates practical benefits in improving clustering performance for real-time tunneling analysis.

(3) From the fitting line along with the 95% confidence band of the results in all rings No. 101–377, the clustering quality exhibits an obvious upward tendency as the volumes of available data grow. The results as shown in Fig. 17 imply that the DTW-Kmedoids algorithm can make continuous adjustments based on the newly acquired time series data, which can robustly identify the continuity property of the geology type during the TBM excavation process. This online mechanism offers significant advantages: it does not require complete temporal sequences and can dynamically update cluster centers based on streaming data, enabling real-time inference of geological conditions. Additionally, its ease of implementation, high computational efficiency, and scalability make it well-suited for big data analysis, ultimately improving tunneling control in underground construction (Table 6).

(4) The online clustering mechanism using the DTW-Kmedoids algorithm demonstrates strong robustness to incomplete time-series inputs. In additional experiments, incomplete input percentages of 10, 20, 30, 40, and 50% were tested, with clustering performance evaluated in Table 7. Results show that when the percentage of incomplete inputs is below 30%, the RI and FMI remain high,

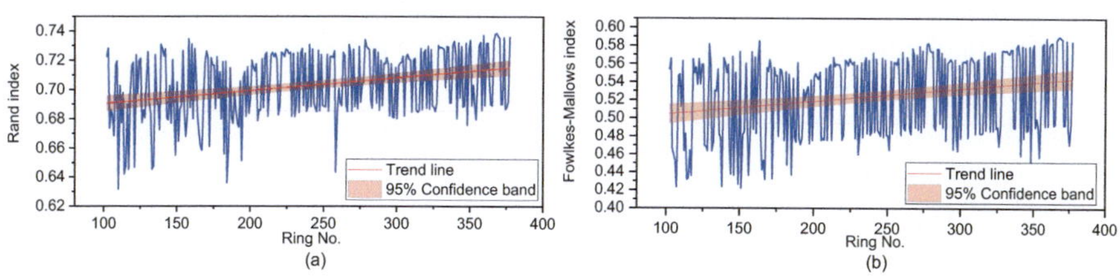

Fig. 17 Performance of online clustering mechanism based on DTW-Kmedoids in each ring evaluated by: **a** RI; **b** FMI

Table 6 Evaluation of DTW-Kmedoids clustering performance from two online clustering methods

Types of clustering process	RI	FMI
Offline mechanism	0.694	0.485
Online mechanism	0.703	0.524

Table 7 Comparison of DTW-Kmedoids clustering performance under online mechanism through 100 experiments for each setting of incomplete input percentage

Incomplete input percentage (%)	RI		FMI	
	Mean (std)	IQR	Mean (std)	IQR
10	0.702 (0.028)	[0.681, 0.722]	0.522 (0.053)	[0.487, 0.553]
20	0.700 (0.028)	[0.675, 0.713]	0.521 (0.053)	[0.483, 0.546]
30	0.700 (0.028)	[0.676, 0.713]	0.521 (0.053)	[0.484, 0.561]
40	0.696 (0.032)	[0.675, 0.711]	0.490 (0.056)	[0.456, 0.525]
50	0.692 (0.033)	[0.670, 0.710]	0.481 (0.056)	[0.442, 0.526]

Note IQR is the abbreviation of the interquartile range

above 0.700 and 0.521, respectively, closely matching the values of 0.703 and 0.524 for complete data. Even with 50% of the data, the algorithm achieves promising results, with RI and FMI only 0.010 and 0.041 lower than those for full data. This indicates that the online version of DTW-Kmedoids effectively detects geological conditions dynamically and reliably, even with incomplete inputs.

6 Chapter Summary

This chapter introduces the DTW-Kmedoids hybrid algorithm, which effectively clusters time series data with low computational complexity, using DTW for similarity measurement and Kmedoids for robust classification. This algorithm automatically perceives the geological conditions ahead of the TBM without prior knowledge of categories. Key contributions include the development of a novel algorithm that robustly handles complex, multi-dimensional, and incomplete time series data, enabling valuable pattern extraction without data labeling. Additionally, an online clustering mechanism updates clustering centers in real time, enhancing the ability to identify changing geological types as tunneling progresses. Practically, this approach leverages extensive TBM-related data to provide accurate geological feedback, facilitating actionable guidance for TBM operation adjustments, which is crucial for ensuring safety and efficiency in tunnel construction.

To validate the proposed time series clustering method, a case study was conducted in a TBM tunneling project in Singapore. Key findings include: (1) DTW-Kmedoids outperformed Kmedoids and softDTW-Kmedoids, achieving RI and FMI values of 0.737 and 0.587, respectively, with results closely aligning with actual geological conditions, aiding TBM operations. (2) The algorithm effectively handles incomplete time series data; even with half of the data, it reached RI and FMI values of 0.704 and 0.531. Using the previous 40 data points was sufficient for promising clustering results. (3) The online clustering mechanism mitigates lag in geological identification, allowing DTW-Kmedoids to continuously adapt to incoming TBM-related data and perceive the geological state ring by ring during tunneling.

The practical value behind the hybrid time series clustering is to automatically and quickly perceive lithology categories along the designed tunnel trail with no requirement of data labeling, possibly allowing for further inference of geological risk and guidance in safe excavation. This online clustering mechanism is well designed to adaptively update the cluster center, which can accomplish the dynamic recognition of the forward geological circumstances ring by ring.

References

1. Pan Y, Wu M, Zhang L, Chen J (2023) Time series clustering-enabled geological condition perception in tunnel boring machine excavation. Autom Constr 153:104954
2. Liu W, Chen EJ, Yao E, Wang Y, Chen Y (2021) Reliability analysis of face stability for tunnel excavation in a dependent system. Reliab Eng Syst Saf 206:107306
3. Liu Q, Wang X, Huang X, Yin X (2020) Prediction model of rock mass class using classification and regression tree integrated AdaBoost algorithm based on TBM driving data. Tunn Undergr Space Technol 106:103595
4. Li S, Liu B, Xu X, Nie L, Liu Z, Song J, Sun H, Chen L, Fan K (2017) An overview of ahead geological prospecting in tunneling. Tunn Undergr Space Technol 63:69–94
5. Zhang C, Liang M, Song X, Liu L, Wang H, Li W, Shi M (2022) Generative adversarial network for geological prediction based on TBM operational data. Mech Syst Signal Process 162:108035
6. Yang J, Yagiz S, Liu Y-J, Laouafa F (2021) Comprehensive evaluation of machine learning algorithms applied to TBM performance prediction. Undergr Space 7(1):37–49
7. Zhu M, Gutierrez M, Zhu H, Ju JW, Sarna S (2021) Performance evaluation indicator (PEI): a new paradigm to evaluate the competence of machine learning classifiers in predicting rockmass conditions. Adv Eng Inform 47:101232
8. Fu X, Wu M, Tiong RLK, Zhang L (2023) Data-driven real-time advanced geological prediction in tunnel construction using a hybrid deep learning approach. Autom Constr 146:104672
9. Sebbeh-Newton S, Ayawah PE, Azure JW, Kaba AG, Ahmad F, Zainol Z, Zabidi H (2021) Towards TBM automation: on-the-fly characterization and classification of ground conditions ahead of a TBM using data-driven approach. Appl Sci 11:1060
10. Pan Y, Zhang L, Unwin J, Skibniewski MJ (2021) Discovering spatial-temporal patterns via complex networks in investigating COVID-19 pandemic in the United States. Sustain Cities Soc 77:103508
11. Holder C, Middlehurst M, Bagnall A (2024) A review and evaluation of elastic distance functions for time series clustering. Knowl Inf Syst 66:765–809
12. Aghabozorgi S, Seyed Shirkhorshidi A, Ying Wah T (2015) Time-series clustering—a decade review. Inf Syst 53:16–38
13. Inoue T, Kubota K, Ikami T, Egami Y, Nagai H, Kashikawa T, Kimura K, Matsuda Y (2024) Clustering method for time-series images using quantum-inspired digital annealer technology. Commun Eng 3:10
14. Manduchi L, Hüser M, Vogt J, Rätsch G, Fortuin V (2019) DPSOM: deep probabilistic clustering with self-organizing maps. arXiv:1910.01590

15. Tavakoli N, Siami-Namini S, Adl Khanghah M, Mirza Soltani F, Siami Namin A (2020) Clustering time series data through autoencoder-based deep learning models. arXiv:2004.07296
16. Alqahtani A, Ali M, Xie X, Jones MW (2021) Deep time-series clustering: a review. Electronics 10:3001
17. Cuturi M, Blondel M (2017) Soft-DTW: a differentiable loss function for time-series. In: International conference on machine learning, PMLR, pp 894–903
18. Zhang Q, Liu Z, Tan J (2019) Prediction of geological conditions for a tunnel boring machine using big operational data. Autom Constr 100:73–83
19. Hoi SC, Sahoo D, Lu J, Zhao P (2021) Online learning: a comprehensive survey. Neurocomputing 459:249–289

Geological Condition Detection

1 Introduction

Construction automation in tunneling is one of the advanced topics that draws considerable attention from both industry and academia [1]. However, there are limited attempts to implement the autonomous tunnel boring machine (TBM) due to many constraints. One of the major challenges for tunnel construction is the heterogeneous geological conditions which are always associated with risks and hazards to construction safety [2]. Currently, the TBM parameters need to be manually adjusted by the human operators according to the encountered geological conditions and other ambient environment factors. In addition, it has been proven in many studies and accidents that a better understanding of the geological information ahead of tunneling greatly improves construction safety during tunneling [3]. Therefore, accurate detection of the geological condition is critical for the successful completion of the tunnel project.

The topic of geological condition detection has attracted considerable research interest, and the majority of current studies mainly focus on providing predictions for the geological conditions in front of TBM. These studies can generally be classified into 3 groups according to their adopted methodologies, namely the destructive methods, non-destructive methods, and intelligent methods [4]. The destructive methods detect the geological condition through soil investigation borehole prior to the tunnel excavation [5]. However, the number of boreholes is normally limited due to the cost issue, which estimation based on insufficient information may lead to low accuracy. The non-destructive methods mainly apply seismic, electrical, and magnetic methods to identify the potential nonhomogeneous soil conditions ahead of the TBM. These approaches can collect reliable information but have relatively low efficiency and high cost [6].

The intelligent methods attempt to forecast the advanced geological information by analyzing the TBM parameters through developed machine-learning or deep-learning models. They have been employed by many studies to analyze the interaction

L. Zhang et al., *Intelligent Construction in Tunnels*, Lecture Notes in Civil Engineering 679, https://doi.org/10.1007/978-981-96-8785-5_4

mechanism between TBM and geological conditions. Although some studies have shown positive results with high accuracy in the prediction, there are also criticisms that the mechanism of how deep-learning methods predict the geological condition is questionable [7]. The prediction normally assumes that the ground truth of existing geological conditions is made known, which requires constant human intervention to identify the excavated soil. In summary, the current approaches have their limitations and they are not suitable for automatic geological detection during tunnel excavation. That is to say, there is a need for further improvement or an alternative solution to better resolve the problem.

To counter the drawbacks of these methods, this study intends to propose an alternative approach by fully imitating how the human operator works. During tunnel excavation, there is a closed-circuit television (CCTV) installed to monitor the real-time condition of excavated mucks at the belt conveyor. These mucks generally represent the current geological condition, which is the main source of information for the TBM operator to adjust the key TBM parameters accordingly. Therefore, such an idea motivates us to propose a computer vision-based method that can automatically identify the encountered geological conditions through the images of the excavated mucks. Compared to the previous intelligent methods, the proposed approach enables automatic geological condition identification without the need for human intervention, which can serve as a fundamental step for autonomous TBM operation. That is to say, once the automatic geological condition identification is achieved, a subsequent intelligent model could be established which allows the TBM to self-adapt the key operation parameters according to the existing geological conditions. By integrating a suite of solutions for automatic geological conditions, TBM performance estimation and operating parameters optimization, autonomous TBM operation can then be realized under complex geological conditions [8]. In this chapter, a novel computer vision method is proposed for geological condition identification with high accuracy.

Convolutional neural networks (CNN) have been widely applied to perform the task of image recognition and have received positive results in many domains of science [9]. In civil and construction engineering, current research efforts mainly focus on crack detection, structural health monitoring, unsafe act identification, etc. During tunnel excavation, the excavated soil will generally have a different appearance in terms of surface texture, color, etc., which has great potential to be identified by computer vision algorithms. However, there is limited effort to investigate the possibility of using computer vision for intelligent tunnel construction, especially for geological identification. Among the many well-developed and pre-trained CNN architectures, MobileNet is a lightweight CNN structure that has proven its high efficiency in many image-processing tasks [10]. However, during tunnel excavation, the obtained soil image is always associated with unrelated noise and poor picture quality due to the complex working environment, which poses difficulty in achieving high classification accuracy. Therefore, simply applying the MobileNet model may not be sufficient to accurately identify the encountered soil.

To address this problem, necessary data pre-processing and augmentation techniques are applied to minimize the negative effect of noise. In this study, the idea of random crops is adopted to remove the unnecessary parts and standardize the image scale. Furthermore, the cropped images are randomly selected and trained with the concept of ensemble learning to further improve the accuracy as many studies have proven it can enhance the performance of deep learning models [11]. Under this concept, multiple MobileNet models can be established as the base models and the generated results can be fused with better generalization capability. However, when different results are presented by the base models, a reliable approach shall be introduced to fuse the information from these base models to achieve the best prediction result. In this regard, Dempster–Shafer theory (DST) can be adopted with its advantages in dealing with evidence from multiple sources and combining them with a certain degree of brief [12]. That is to say, the base models could be viewed as different observers and the predicted results are the evidence with uncertainties. DST is then employed for the information fusion process by finding the degree of belief for each class based on evidence provided by the observers. Therefore, we propose a novel method that integrates ensembled MobileNet models with DST, named EMNet to identify the soil encountered during tunnel construction with high accuracy. In addition, the Shapley additive explanations (SHAP) method is adopted to provide further interpretation of the developed model and determine which are the key features that assist the geological identification.

The key research questions are: (1) How to pre-process image data of small size to make it sufficient for deep-learning model training, (2) How to establish an appropriate ensemble deep-learning model to correctly identify the geological conditions during tunnel excavation, and (3) How to interpret the predicted results from the base model and provide estimation with high accuracy. To address these questions, a novel ensemble MobileNet method is proposed with DST for information fusion to provide reliable judgment on the encountered geological conditions. The key contributions of this study are: (1) a novel ensemble deep learning method is proposed which performs image classification with high accuracy; (2) The proposed method enables automatic identification of the geological condition encountered during tunnel excavation which serves as a fundamental step towards autonomous TBM driving; (3) By employing SHAP method, the developed deep-learning model is further interpreted to identify the key contributing factors for classification.

2 Related Studies

Detection of geological conditions in front of the TBM's cutterhead is critical to optimize TBM operation and reduce potential safety hazards. This topic has attracted great research interest with a large number of different approaches being proposed. These developed approaches can generally be grouped as destructive methods, non-destructive methods, and intelligent methods, where the detail of each method is elaborated below.

The destructive method is the approach that has been widely used in the industry. For tunnel construction with TBM, the geological information can mainly be obtained through the soil investigation borehole prior to tunnel construction. Engineering properties of the in-situ geological condition can be further determined through the test in the laboratory or in-situ testing on the collected soil samples. Although the geological information at the borehole location can be confirmed, the point area covered by the boreholes is limited, where the geological condition at the remaining areas can only be inferred based on expert judgment. As a result, there is limited information with high accuracy on this approach. An improved version of the destructive method is to apply borehole logging techniques to extend the detecting area by the boreholes. Sensors including sonic logs, gamma logs, borehole radar, etc., are implemented to obtain geological information [4]. However, these methods are mainly utilized in the drill and blast tunnels instead of TBM excavation.

Over the past decades, numerous non-destructive methods were developed to detect the geological condition ahead of the TBM excavation face. The key idea of the non-destructive method is to determine the soil condition or potential adverse geological condition without the need for additional horizontal drilling and interrupting TBM operation. Seismic, acoustic, or electromagnetic methods are the common non-destructive approaches that have been widely discussed in many research papers. For instance, Wei et al. [13] utilized the data from ground penetration radar and seismic sensors to detect the possible anomalous event ahead of TBM by generating images of the geological condition in front. Li et al. [14] proposed an integrated non-destructive method that combines the result from multiple sensors including seismic reflection, electromagnetic, and ground penetration radar to detect the water-bearing geological structures ahead of TBM. Although many studies have proven the high accuracy and strong capacity of their proposed non-destructive methods, these methods are not been widely applied in tunnel construction due to the relatively high cost of implementing the sensors and devices.

The rapid development of machine learning and deep learning techniques as well as the huge monitoring data produced by the modern TBM provides substantial opportunities for the exploration of intelligence methods. Over the past few years, many different intelligence-based methods have been developed. For instance, Shi et al. [15] predict the geological condition with a deep neural network model. Zhang et al. [16] combined the K-means++ method and support vector classifier to classify and predict the geological condition based on TBM operational data. Fu et al. [17] utilized the density-based spatial clustering of application with noise (DBSCAN) method to classify the TBM parameters and proved its strong relation with the geological conditions. Although many papers reported that the intelligence method can achieve very high accuracy, some limitations hinder the application in actual tunnel projects. Firstly, there is criticism that the prediction is actually achieved by taking the average result from the previous time states, where the mechanism of predicting the geological condition with past TBM parameters is not clear [7]. Secondly, the excavated mucks have to be constantly identified by humans and the geological information must be input manually for the prediction. Such practice hinders the objective of intelligent tunneling since there shall be minimum human intervention for autonomous TBM.

Recently, there have been some attempts to use computer vision-based methods for the identification of rock masses. For example, Gong et al. [18] adopted U-net to perform rock segmentation on the excavated mucks. Yang et al. [19] investigated the AlexNet for real-time rock mass classification during tunnel excavation. However, there are some limitations to the current computer vision-based methods. Most of the current studies focus on rock TBM and there are limited works to detect the soil condition during tunnel excavation. In addition, the current studies mainly adopt well-established CNN architectures, while there is a need to develop new methods for geological condition identification with better accuracy.

In summary, many approaches have been proposed to detect the geological condition during tunnel excavation, where destructive, non-destructive, and intelligent methods are the three categories. Although many studies have proven its efficiency, existing methods received some criticism with certain limitations. Therefore, this paper intends to investigate an alternative approach by proposing a novel computer vision-based approach to identify the encountered geological conditions during tunnel excavation.

3 Computer Vision Framework for Geological Condition Detection

Detection of the geological condition is critical for the success of tunnel construction, and we propose a novel framework to identify the geological condition directly based on the images obtained from the belt conveyor. The proposed approach has three main steps, namely data acquisition and preprocessing, development of the ensemble deep-learning model with information fusion DST, and model evaluation and interpretation by SHAP. The flowchart of the proposed ensemble deep learning approach is shown in Fig. 1.

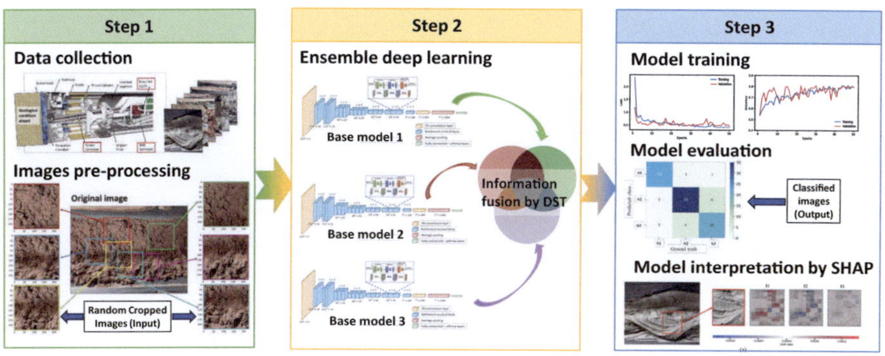

Fig. 1 Flowchart of the developed ensemble deep learning approach

3.1 Geological Images Acquisition and Processing

During the tunnel excavation, the removed soils are transported through the screw conveyor followed by the belt conveyor. Then a wagon is positioned at the rear part of the TBM to collect the mucks from the belt conveyor and transport it to the ground level. For better site monitoring and timely control of the tunnel excavation, modern TBMs normally install a CCTV to record the status of soil excavation on the conveyor belt. The video can be shown to the operator and engineers in real-time for decision-making and necessary adjustment of the TBM operational parameters. To simulate this process, photos taken from the belt conveyors are obtained, which can represent the actual condition of the excavated mucks. These photos can be analyzed and obtain useful information that reflects the geological condition of the present tunnel excavation. Considering that image-based technologies are always prone to environmental and human factors, the photos shall be taken in a stable position under a controlled environment. Under the context of shield tunneling, there must be sufficient lighting provision with no blockage of the CCTV and the camera shall be installed in a fixed position, focusing on the same area when capturing soil images. An illustration of the process and some collected photos are shown in Fig. 2.

However, the photos also include noises and unwanted objects which may affect the image classification process. The main source of noise includes part of the belt conveyor that was not covered by the soil, other objects at the edge of the photo, etc. In addition, the conventional MobileNet algorithm requests the images to be converted into numeric matrices with a standard size of 224*224 pixels and color decomposes into 3 channels, namely red, green, and blue (RGB) [20]. Since the size is much smaller than the original photos, information may be lost during the compression process and may result in compromised accuracy. Therefore, data pre-processing and augmentation are necessary to remove the noises and assist in improving the performance of the proposed model.

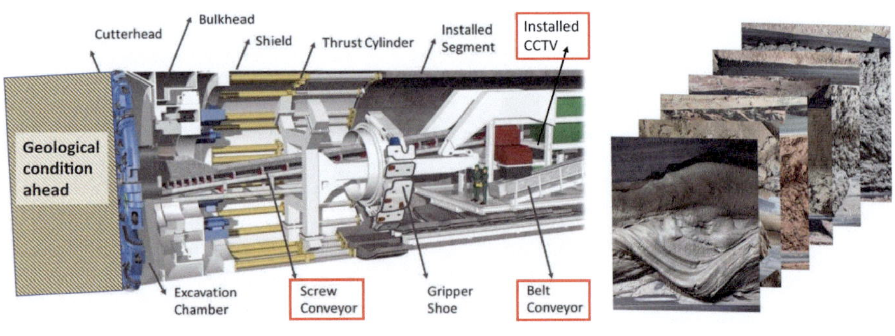

Fig. 2 Illustration of soil images collection process during TBM excavation for **a** sketch of the key TBM components and **b** samples of the collected images

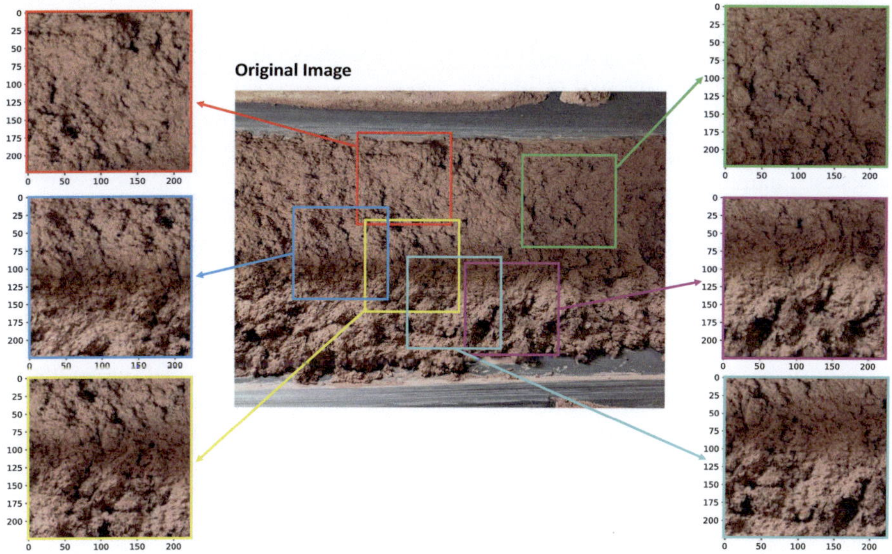

Fig. 3 Sample of the original image and the images for model training after data preprocessing

In this study, the random crop method is employed as the key data augmentation process. Since most of the useful information accumulates near the center portion of the images, the edge of 200 pixels is removed to minimize the disturbance of the noise. Next, random crops are performed by randomly extracting the squared shape images of a standard size of 400*400. This is to also avoid the impact of different image scales that are captured by various devices. With such a pre-processing method, subsequent computer-vision analysis can be performed even when the captured images have different sizes. For each image, random crops will be performed 10 times to ensure sufficient samples are extracted from the images and the cropped parts can cover most of the original image. The cropped images are then resized into the requested 244*244*3 format, with the numbers in the matrices rescaled into [0,1]. All obtained geological images are labeled based on the engineering judgment of the experts and the cropped images are labeled the same as their original photos. An illustration of random crops and images used for data training and testing is shown in Fig. 3.

3.2 Ensemble Model Development

The pre-processed images can be divided by following the conventional 80-20 split as the training and testing data for model training and evaluation. However, the training and testing images shall be randomly split to avoid bias and improve generalization. In addition, the cropped images from the same original photo must be kept in the same group. MobileNet V2 is adopted as the base model for the proposed ensemble deep learning method. Different from conventional convolutional networks, MobileNet V2 is recognized for three major initiatives, namely depthwise separable convolution, linear bottleneck, and inverted residues.

Different from the conventional convolution that requires the kernels to operate with all channels, the deep-wise separable convolution performs the calculation into two steps, namely depthwise (DW) convolution, and pointwise (PW) convolution. Consider the image with an input size of $w_i \times h_i \times c_i$ and request an output size of $w_j \times h_j \times c_j$, conventional convolution operation requires c_j number of filters of size $w_k \times h_k \times c_i$. In comparison, the DW convolution request c_i number of 2D kernels with the size of $w_k \times h_k$, and the kernels will perform convolution with their respective channels. The output information after DW convolution will be in the size of $w_j \times h_j$ and split into c_i channels. To ensure the information exchange from different channels, a special filter with a size of $1 \times 1 \times c_i$ and number of c_j is proposed to perform the PW convolution, which is to ensure the output feature size can be $w_i \times h_i \times c_j$ as needed. The main purpose of deep-wise separable convolution is to reduce the computational cost, where the cost for conventional convolution and DW convolution is shown in Eqs. (1) and (2), respectively.

$$Cost_1 = w_i \cdot h_i \cdot c_i \cdot w_k \cdot h_k \cdot c_j \tag{1}$$

$$Cost_2 = w_i \cdot h_i \cdot c_i\left(w_k \cdot h_k + c_j\right) \tag{2}$$

Given a neural network that consists of multiple convolution layers with activation functions, in which the entire information flow forms a manifold of interest. The dimension for the manifold of interest can be reduced through activation functions, e.g. Rectified Linear Unit (*ReLU*) function. However, previous studies have proven that there will be a higher loss in information for the input of low dimensions compared with the input of higher dimensions. To minimize the information loss during the transformation, the concept of linear bottleneck and inverted residuals is introduced. Consider the input tensor with a dimension of $h \times w \times d$, where d is a relatively small number. The input data will first pass through an expansion layer with PW convolution with an expansion factor of t followed by batch normalization and activation by *ReLU6*. The purpose of the expansion layer is to reduce the information loss when passing the activation function. Next, the tensor with a size of $h \times w \times dt$ will be transformed by DW convolution followed by batch normalization and activation by *ReLU6*. Lastly, the tensor will pass through a projection layer with PW convolution to reduce the size to $h \times w \times d'$ and batch normalization. However, the tensor will pass through a linear activation instead of a non-linear one to prevent information loss. Ultimately, the output tensor has a size of $h/s \times w/s \times d'$, where s is the stride factor to reduce the size of the tensor. If the output tensor has the same size as the inputs ($s = 1$ and $d' = d$), a residual connection is provided to add the information from the input to the output tensors. A detailed illustration of the bottleneck residual block is shown in Fig. 4.

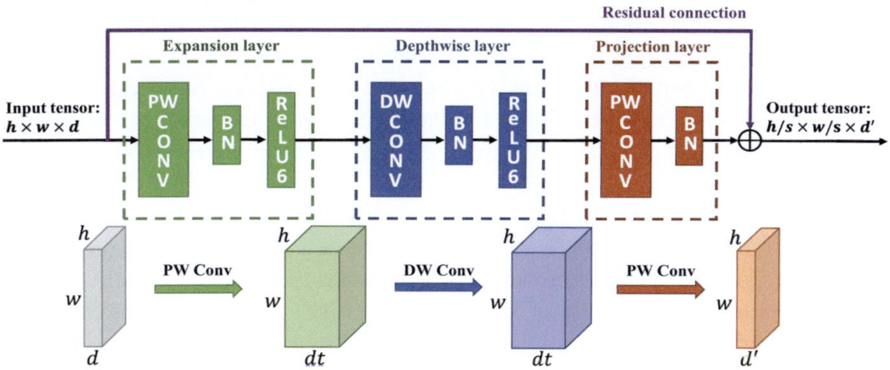

Fig. 4 Illustration of the bottleneck residual block structure and the size of tensors

Overall, the architecture of the MobileNet starts with a 2D convolution layer followed by 17 bottleneck residual blocks and ends up with the average pooling and fully connected layers as the output. The size of the 2D convolution layer is 224^2, which is the same as the size of the input images. The features of the images are extracted by the bottleneck residual blocks with the size gradually reduced to 7^2 by the blocks when the stride $s = 2$. An illustration of the MobileNet architecture is shown in Fig. 5.

The result from the single MobileNet model may suffer from compromised accuracy and high uncertainty. To further improve the accuracy and robustness of geological detection, an ensemble deep learning approach is proposed which incorporates the evaluation from multiple models with improved accuracy through information fusion. In particular, the original image is randomly cropped $10n$ times with the symbol n marking the number of the base MobileNet models involved. Each model is trained with 1 set of training data and evaluated by the respective testing data.

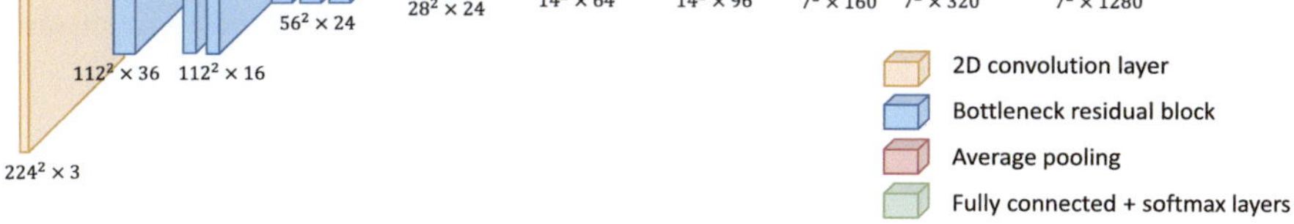

Fig. 5 Architecture of MobileNet

However, the original photos for training and testing for each model are the same, except the cropped images are at random positions.

Information fusion is necessary to conclude the observations from different base models. DS-theory is utilized for its capability to quantify the degree of belief from multiple sources with uncertainties. Let the frame of discernment $\Theta = \{A_1, A_2, A_3 \ldots, A_N\}$, where A_1 to A_N are mutually exclusive hypotheses, and N is the total number of hypotheses. In our case, N is the total number of geological conditions to be classified and A_1 to A_N represents each class. The basic probability assignment (BPA) is a mass function m that indicates the measure of belief for the hypotheses and mapping to [0,1]. In addition, the mass function of any empty set shall be 0, and the sum of the probability from all hypotheses shall equal 1. This property can be expressed mathematically by Eq. (3).

$$\begin{cases} m(\theta) = 0 \\ \sum\limits_{A \in \Theta} m(A) = 1 \end{cases} \tag{3}$$

where θ is the empty set and $m(A)$ is the BPA of A which represents the measure of belief for the set A based on all related evidence.

Consider the observation results from the base models as $m_1, m_2, m_3, \ldots m_n$, where n is the total number of models. The mass function for set A by model n can be regarded as $m_n(A)$. In order to fuse the information from the base models, Dempster's rule of combination can be applied to obtain the combined measure of belief for set A. The combination rule can be expressed by Eqs. (4) and (5).

$$\begin{aligned} m(A) &= m_1 \oplus m_2 \oplus \ldots \oplus m_n \\ &= \frac{1}{1-K} \sum_{\bigcap_{i=1}^{k} A_i = A, A \in \Theta} \prod_{j=1}^{n} m_j(A_i) \end{aligned} \tag{4}$$

$$K = \sum_{\bigcap_{i=1}^{k} A_i = A, A \in \Theta} \prod_{j=1}^{n} m_j(A_i) \tag{5}$$

where $m_1 \oplus m_2 \oplus \ldots \oplus m_n$ is the combination of the evidence, K is named as the conflict coefficient, A_i is the ith hypothesis and m_j represent the jth piece of evidence. The combined belief for each geological class can be used to determine the final classification in which the class with the highest possibility is chosen.

3.3 Model Evaluation and Interpretation

Model evaluation is critical to examine the performance of the proposed method. It can be conducted by evaluation metrics that are calculated based on the predicted results and the true label of the testing data. To evaluate the proposed model, this paper adopts the common evaluation metrics for classification tasks including accuracy, precision, recall, and F1-score. The mathematical expressions of the evaluation metrics are shown in Eqs. (6–9), respectively.

$$Accuracy = \frac{TP + TN}{N} \tag{6}$$

$$Precision = TP/(TP + FP) \tag{7}$$

$$Recall = TP/(TP + FN) \tag{8}$$

$$F1\text{-}score = 2 \times \frac{(Precision \times Recall)}{(Precision + Recall)} \tag{9}$$

where N is the sample size of testing data, FP denotes false positive and FN denotes false negative.

In addition to model evaluation, the model shall be further analyzed to understand which are the key features in the images that influence the classification. SHAP analysis is then adopted since it could provide reliable model interpretation through the Shapley values. Inspired by the game theory, the principal idea of the SHAP method is to establish a simpler explanation model which is the approximation of the original complex model. The mathematical expression of the simpler model for the original complex model f can be presented by Eq. (10).

$$g(z') = \phi_0 + \sum_{i=1}^{M} \phi_i z_i{'} \tag{10}$$

where $g(z')$ is the explanation model; $z' \approx x'$ which is the simplified input generated from the original inputs x and through the mapping function $x = h_x(x')$ under the condition that $g(z') \approx f(h_x(x'))$; ϕ_0 is the contribution with zero inputs, and M is the total number of the input features after simplification, ϕ_i is the contribution of feature i.

4 Experimental Studies

4.1 Data and Resources

This study utilized one of Singapore's Circle Line (CCL) 6 extension projects, C885 as a case study. C885 tunnel project consists of constructing two tunnel bounds (inner bound and outer bound) that connect Cantonment Station and Prince Edward Road Station. Data collected from the inner bound is utilized as the case study for this research. The tunnel has a total length of 1.003 km and a depth varying from 9 to 21 m. The permanent lining of the tunnel structures consists of 718 circular-shaped tunnel rings that are formed by 7 pieces of pre-cast concrete segments with a width of 1.4 m and thickness of 0.05 m. An earth pressure balance TBM (EPB-TBM) with an outer diameter of 6.68 m and a total length of 9.81 m is utilized for the tunnel construction. For this research, images of the newly excavated soil from the belt conveyor are obtained, which can represent the current state of the geological profile. The layout plan of the CCL6 extension line and C885 tunnel project, the site photo of TBM's earth transport system, and the completed tunnels are shown in Fig. 6a–c, respectively.

Soil investigation is conducted prior to the commencement of tunnel construction. For this project, geological information is mainly obtained through destructive methods, where vertical boreholes are drilled along the tunnel axis to examine the soil. These boreholes are drilled at intervals of 10 to 20 rings (15 to 30 m) with drilling depths from around 30 to 50 m. According to soil investigation results, the Kallang formation and the Jurong formation are the two major types of formations that are detected in this area. The engineering properties of the soil samples are mainly obtained through in-situ and laboratory tests, including standard penetration test (SPT-N) and a series of triaxial tests. In addition, a series of rock tests are conducted to determine the strength and characteristics of the encountered rocks. A summary of the engineering properties for each type of soil that was used for tunnel design is shown in Table 1, where d stands for the depth of the soil sample and N is the number obtained from SPT-N.

During the tunnel excavation, the removed soils are transported through the screw conveyor followed by the belt conveyor. Then a wagon is positioned at the rear part of the TBM to collect the mucks from the belt conveyor and transport it to the ground level. For this study, images showing the newly excavated mucks in the conveyor belts are collected and utilized for further evaluation to detect the geological profile. The encountered geological conditions are generally classified into 3 major types, which are labeled as S1, S2, and S3. Type 1 is mainly Kallang formation, Type 2 represents the SV/SVI of Jurong formation, and Type 3 represents the SIII/SIV of Jurong formation. Although both S2 and S3 belong to the Jurong formation, S3 soil has a higher rock content compared with S2. An illustration of the geological profile and typical soil images that represent each type of geological condition are shown in Fig. 7a–d, respectively.

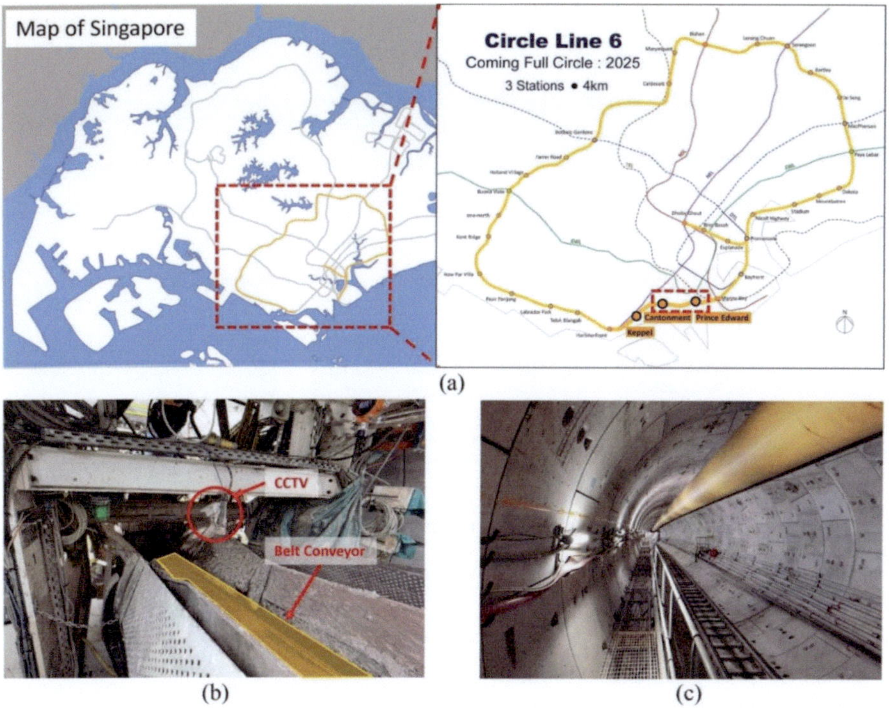

Fig. 6 Figures of **a** the layout plan of the CCL6 extension line and C885 tunnel project, **b** the site photo of TBM's earth transport system, and **c** the site photo of the completed tunnels

Table 1 Engineering properties of the encountered soil types

Engineering properties	Symbol	Unit	Type of soil (labels)		
			Marine clay (S1)	Jurong formation SV, SVI (S2)	Jurong formation SIII, SIV (S3)
Unit weight	γ	kN/m^3	16	20 ~ 21	21 ~ 24
Effective angle of friction	φ'	Degree	22	30 ~ 34	33 ~ 35
Effective cohesion	c'	kPa	0	5 ~ 10	15 ~ 50
Undrained shear strength	*Cu*	kPa	$15 + 1.3(d\text{-}5)$	5*N*	Above 350
Permeability	*Kd*	m/s	1×10^{-8}	1×10^{-7}	1×10^{-6}

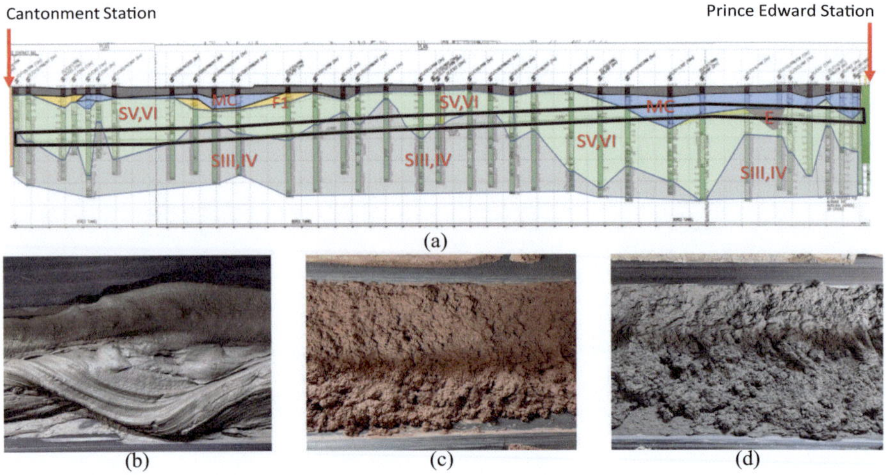

Fig. 7 Figures for **a** geological condition of the tunnel project in the case study, and sample photos of **b** S1 soil, **c** S2 soil, and **d** S3 soil

4.2 *Model Training and Validation Details*

The collected images are labeled and split for model training and testing purposes. In particular, a total of 449 images were collected with each image representing the soil condition encountered during the respective tunnel rings from around ring no. 70 to no. 530. fivefold cross-validation is employed to further examine the effectiveness of the proposed method. Following the proposed data pre-processing procedure, 10 random crops with a height and width of 400 pixels were performed for each image. In addition, these random corps were performed by removing the edge of 200 pixels on the original images, which avoids noise and unwanted objects. The randomly generated 4490 images were then resized to a size of 224^2 pixels and converted into the numerical array with the colors represented by the RGB channels. Finally, the numbers are rescaled into a range of [0,1] and these images are labeled according to the label of their original geological conditions.

The computational experiment was implemented by TensorFlow Keras under the hardware provided with a CPU of 12600KF, GPU of 3070, and RAM of 32 GB. A total of 6 MobileNets models are trained by the randomly cropped image set in order to confirm the optimum number of the base models. For illustration purposes, the training process including the result of the loss function and accuracy for the first MobileNets model is shown in Fig. 8a and b, respectively. As suggested by many studies, overfitting normally can be observed during the model training stage, where the accuracy for training data will keep increasing and that for validation data remains or even deteriorates along the epochs. Refer to Fig. 8 for the training process in our study, both loss function and accuracy for the training and validation data stabilized

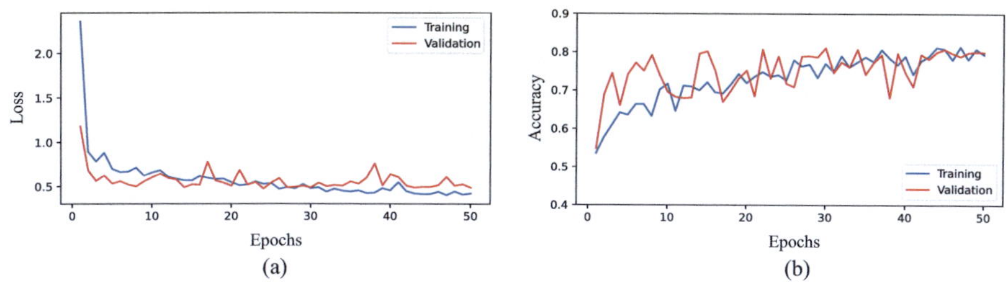

Fig. 8 Training process for **a** the result of the loss function for training and evaluation data and **b** the result of accuracy for the training and evaluation data

Table 2 Performance evaluation by EMNets with fivefold cross-validation

Fold	Accuracy	Precision	Recall	F1-score
Fold 1	0.933	0.938	0.929	0.933
Fold 2	0.809	0.808	0.804	0.806
Fold 3	0.775	0.830	0.807	0.795
Fold 4	0.798	0.827	0.807	0.804
Fold 5	0.860	0.862	0.882	0.861
Average	0.835	0.853	0.846	0.840

in the last 10 epochs. This result indicates that overfitting does not happen during the training stage.

The trained base models are utilized to predict the geological condition of the testing data. As the images from testing data are also randomly cropped with 10 samples from each image, a matrix under the size of 10×3 can be obtained with the column as the probability of each class. In our paper, the final mass function m_i for the ith base model is obtained by the unweighted average results from the 10 cropped samples. To perform cross-validation, 3 base models are trained and then fused through DST. Metrics including accuracy, precision, recall, and F1-score are adopted to evaluate the performances. The results for each fold and their average accuracy are presented in Table 2. Additionally, considering that the number of base models may also affect the performance of EMNet, further analysis is conducted to evaluate the impact using 6 base models and fold 5 as testing data. The results of different combinations as well as the average numbers are listed in Table 3.

4.3 Geological Condition Detection Results

The soil condition that is classified by the proposed method is compared with the ground truth to analyze the performance of the proposed EMNets. The selected metrics, including accuracy, precision, recall, and F1-score, are utilized for better evaluation. Furthermore, SHAP analysis is performed on some selected images for

Table 3 Performance evaluation by the proposed EMNets under a different number of base models

No. of models	Accuracy	Precision	Recall	F1-score
2 models	0.831	0.856	0.852	0.844
3 models	0.865	0.860	0.890	0.862
4 models	0.865	0.864	0.880	0.864
5 models	0.865	0.860	0.890	0.862
6 models	0.876	0.872	0.900	0.874
Average	0.860	0.862	0.882	0.861

a better understanding of the mechanism behind them. A detailed analysis of the results is presented below.

(1) The proposed method detects the encountered geological condition with high accuracy. Refer to Tables 2 and 3 and Fig. 9 for the prediction evaluation using cross-validation and under a different number of base models as well as the average results. The cross-validation reports a high average accuracy, precision, recall, and F1-score of 0.835, 0.853, 0.846, and 0.840, respectively. In terms of the impact of base models, the proposed method achieves a high value for all metrics regardless of the number of base models. In addition, EMNet generates the best performance when 6 base models are selected, with the metrics as high as 0.876, 0.872, 0.900, and 0.874 for accuracy, precision, recall, and F1-score, respectively. The performance slightly drops if only 2 base models are utilized, with the accuracy, precision, recall, and F1-score following 0.831, 0.856, 0.852, and 0.844, respectively. However, even though only 2 base models are selected, the classification accuracy of 0.831 indicates more than 80% of images can be correctly classified, which will be very helpful in promoting construction automation in tunneling.

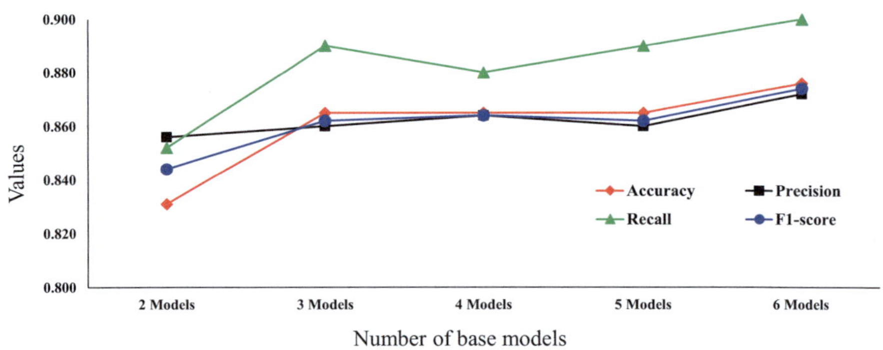

Fig. 9 Plot of the evaluation metrics under different numbers of base models

(2) Further analysis indicates that EMNet outperforms all the base models involved, which implies the importance of model ensemble with DST. The EMNet model constructed by 3 base models is chosen for further analysis. Refer to Table 4 and Fig. 10 for the accuracy, precision, recall, and F1 score performed by the 3 base models as well as EMNet, the results clearly show that EMNet provides the image classification with much higher accuracy. In particular, the classification by the three base models generates results that hover at around 0.8 with an average value of 0.791, 0.795, 0.852, and 0.782 for accuracy, precision, recall, and F1-score, respectively. By ensemble the 3 base models with DST, the performance significantly improves to above 0.85, with an accuracy of 0.865, precision of 0.860, recall of 0.890, and F1-score of 0.862, respectively, which indicates an improvement of 9.40%, 8.18%, 4.42% and 10.23% for the evaluation metrics, respectively. The confusion matrix for the predicted value and ground truth as well as the precision-recall (p-r) curve for the 3 base models and EMNets are presented in Fig. 11 for further analysis. The confusion matrix of the classification from the base models implies that the performance of the base models is relatively unstable. Specifically, base models 1 and 3 provide an accurate classification for S1 and S2 but are unable to identify S3 properly, while base model 2 accurately identifies S1 and S2 but generates larger errors for S2. In comparison, the proposed EMNets are able to take advantage of the

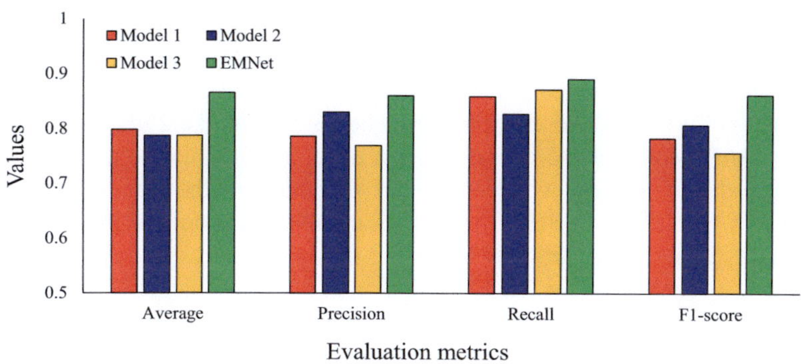

Fig. 10 Bar charts of the evaluation metrics for the base models and EMNet

Table 4 Performance evaluation for the base models and EMNet

Base models	Accuracy	Precision	Recall	F1-score
Model 1	0.798	0.786	0.859	0.783
Model 2	0.787	0.830	0.827	0.807
Model 3	0.787	0.769	0.871	0.756
Average	0.791	0.795	0.852	0.782
EMNet	0.865	0.860	0.890	0.862

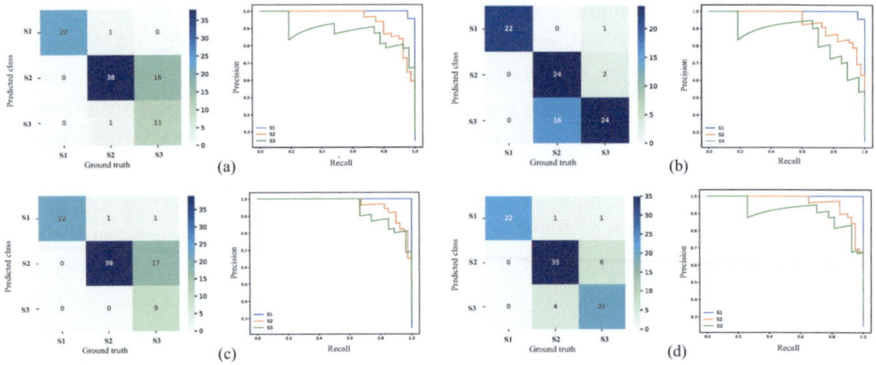

Fig. 11 Plots of the confusion matrix and p-r curve for **a** base model 1, **b** base model 2, **c** base model 3, and **d** EMNet

base models and provide relatively accurate estimations for both S2 and S3. The result demonstrates the importance of model ensemble with DST.

(3) SHAP analysis demonstrates that surface texture and color could be the key feature that assists the model for classification. For illustration purposes, three images, each from 1 type of soil, are selected to carry out SHAP analysis. Instead of classifying the images directly, images with a width and height of 400^2 pixels are cropped and resized into 244^2 for image classification. The original images, cropped images, and the SHAP analysis for S1 to S3 soil are presented in Fig. 12a–c, respectively. The colored markers under S1 to S3 are the SHAP values of the image under a different classification, where the red color indicates the area with positive SHAP values and the blue color represents the area with negative SHAP values. Generally, the proposed model is able to capture the necessary features and avoid unwanted ones. Referring to Fig. 12a, there are many positive SHAP values under S1, few negative SHAP values under S2, and only minimum SHAP values are shown under S3. The result generally shows that the feature that triggers the image to be classified as S1 is the area with a smooth surface. For Fig. 12b, there are no SHAP values under S1, many positive values under S2, and a large number of negative values under S3, which indicates that the algorithm is pretty sure that the image shall be classified under S2. It implies that the model captures the area with a brownish color and classifies it as S2 soil since there are positive SHAP values throughout the cropped image under S2. Similarly, the last photo is classified under S3 since it has the most positive SHAP values, whereas the area with dark grey credits positive SHAP values to S3 and negative values to S2.

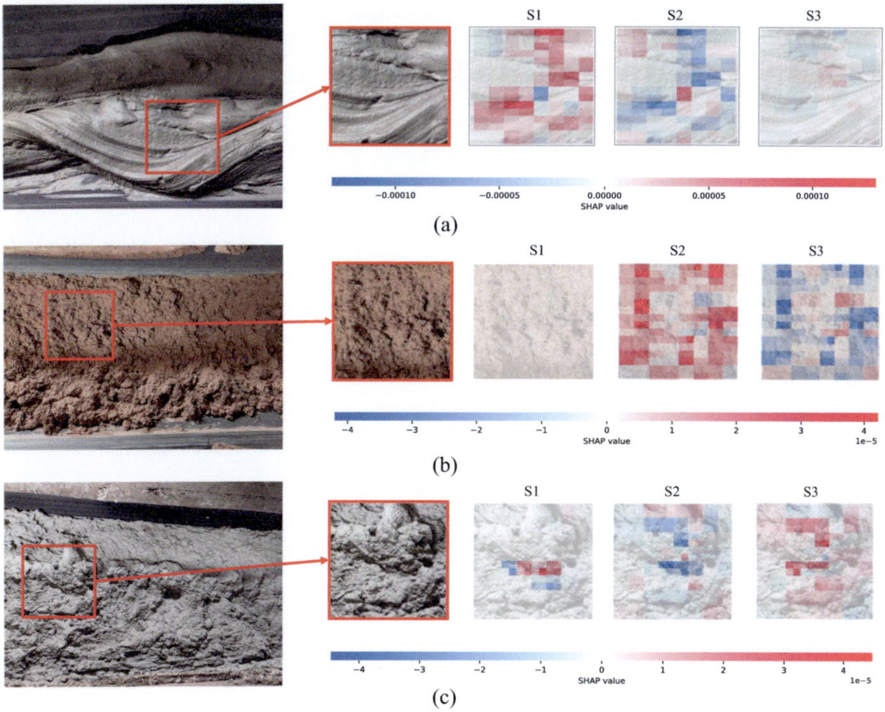

Fig. 12 Illustration of the SHAP analysis for the randomly selected images for **a** S1, **b** S2, and **c** S3

5 Merits Analysis

This paper proposed an ensemble deep learning approach for accurate identification of the geological conditions encountered during TBM excavation. To perform the image classification with the proposed EMNet, multiple MobileNets are established as the base models. Although the proposed model is well-performed, it is still worth comparing the performance with other state-of-the-art methods.

Among a large number of developed CNN architectures, residual neural networks (ResNet), visual geometry group (VGG), and GoogleNet are three of the most well-recognized architectures and have been successfully applied in many areas. To facilitate the comparison study, the experiment is conducted with the same training and testing images. Each original image is preprocessed with 10 random crops, resized into the height and width of 224^2 pixels. During the model training and evaluation, these cropped images are transferred into numerical matrices and the numbers are rescaled into [0,1]. The hyper-parameters are critical for the model's performance. For a fair comparison, the key hyper-parameters are set the same as the MobileNet model and all training processes are evaluated by accuracy. A detailed elaboration of hyper-parameters is displayed in Table 5, accordingly. To assess the models'

Table 5 Details of the hyper-parameters

Hyper-parameters	Selected function
Activation function	*Softmax*
Optimizer	*Adam*
Loss function	Sparse categorical cross-entropy
Evaluation metrics	Accuracy
Training epochs	50
Learning rate	0.001
Batch size	32

Table 6 Evaluation of the testing data

Models	Accuracy	Precision	Recall	F1-score
ResNet	0.636	0.603	0.429	0.501
VGG net	0.716	0.693	0.733	0.682
GoogleNet	0.682	0.641	0.697	0.587
MobileNet	0.798	0.786	0.859	0.783
Average	0.708	0.681	0.680	0.638
EMNet	**0.865**	**0.860**	**0.890**	**0.862**

performance, the same metrics including accuracy, precision, recall, and F1-score are utilized with the results presented in Table 6 and Fig. 13, respectively. Detailed classification results that are shown in the confusion matrix for each model are plotted in Fig. 14. These results are analyzed as follows.

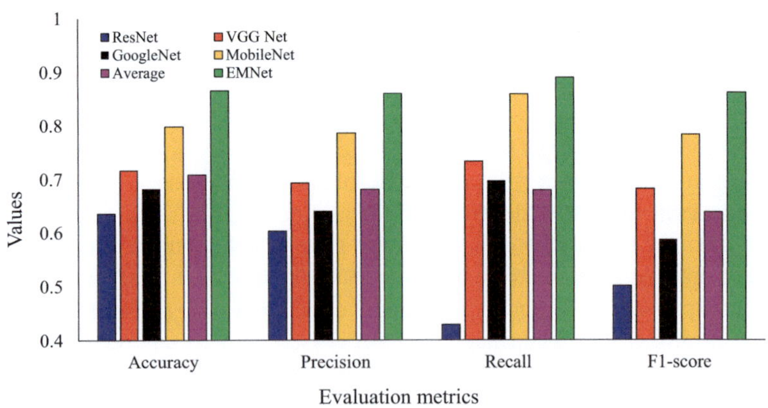

Fig. 13 Bar charts of evaluation metrics between EMNet and the state-of-the-art methods

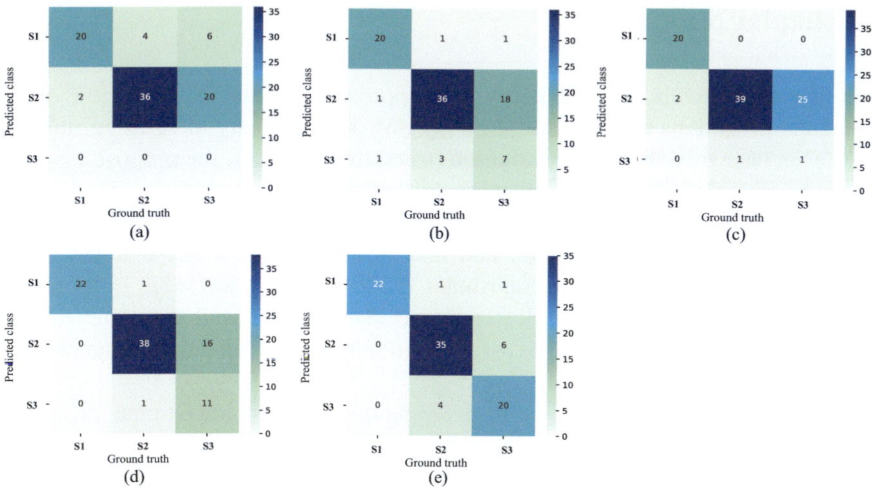

Fig. 14 Confusion matrix of the classification by **a** ResNet, **b** VGG, **c** GoogleNet, **d** MobileNet model 1, and **e** the proposed EMNet

The proposed EMNet outperforms the state-of-the-art methods. Refer to Table 6 for the evaluation of the different methods for accuracy, precision, recall, and F1-score. The results clearly indicate that the proposed EMNet achieves the highest accuracy among these methods. In particular, the accuracy, precision, recall, and F1-score by the state-of-the-art methods generally hovered at around 0.6 to 0.7 with average values of 0.708, 0.681, 0.680, and 0.638, respectively. In comparison, the classification by the proposed EMNets achieves an accuracy of 0.865, precision of 0.860, recall of 0.890, and F1-score of 0.862, respectively. It remarks an improvement of 22.18%, 26.33%, 30.98%, and 35.06% in the evaluation metrics, respectively. Such a result is also clearly reflected in Fig. 13, where the columns by EMNet are the highest among them all. A more detailed analysis regarding the actual classification by different algorithms is shown in the confusion matrix in Fig. 14a–e, respectively. The result indicates that most of the state-of-the-art methods are unable to distinguish the S2 and S3 soils. In particular, ResNet wrongly classifies all S3 soil into S1 and S2, VGG and GoogleNet misclassify more than 75% and 95% of S3 into S2, respectively, while MobileNet can only achieve around 40% accuracy in a worst-case scenario. In comparison, the proposed EMNet has much better performance, which can correctly identify the majority of S3 soil images.

6 Chapter Summary

This chapter presents a computer vision approach for the identification of the geological conditions encountered during TBM operation. By integrating multiple MobileNet base models with information fusion through DST, the proposed method is able to provide reliable identification of the geological condition based on the images of the excavated mucks on TBM's conveyor belt. SHAP analysis is further presented for model interpretation and understanding of the key features in the images. The developed method contributes to the state of knowledge in proposing a novel ensemble method for reliable image identification and the state of practice in proposing a useful tool that assists in identifying the encountered geological condition automatically.

A case study with data collected from Singapore's Circle Line 6 tunnel projects is conducted to verify the effectiveness of the proposed method. Real images collected during the tunnel excavation are pre-processed and split for the model's training and evaluation. The results indicate that the proposed method detects the encountered geological condition with an average accuracy of 0.860, precision of 0.862, recall of 0.882, and F1-score of 0.861. The proposed EMNet outperforms all the base models with an average improvement of 9.40%, 8.18%, 4.42%, and 10.23% for accuracy, precision, recall, and F1-score, respectively. The SHAP analysis results demonstrate that surface texture could be the key features that assist the model for the classification.

The proposed method outperforms the state-of-the-art CNN architectures including ResNet, VGG Net, and GoogleNet, with an average improvement of 22.18%, 26.33%, 30.98%, and 35.06% in terms of accuracy, precision, recall, and F1-score, respectively. Overall, the developed method contributes to the state of knowledge in proposing a novel ensemble method for reliable image identification and the state of practice in proposing a useful tool to identify the encountered geological condition automatically.

References

1. Fu X, Wu M, Tiong RLK, Zhang L (2024) Data-driven joint multi-objective prediction and optimization for advanced control during tunnel construction. Expert Syst Appl 238:122118
2. Li L, Liu Z, Shen J, Wang F, Qi W, Jeon S (2023) A LightGBM-based strategy to predict tunnel rockmass class from TBM construction data for building control. Adv Eng Inform 58:102130
3. Liu B, Guo Q, Liu Z, Wang C, Nie L, Xu X, Chen L (2019) Comprehensive ahead prospecting for hard rock TBM tunneling in complex limestone geology: a case study in Jilin, China. Tunnelling Undergr Space Technol 93:103045
4. Li S, Liu B, Xu X, Nie L, Liu Z, Song J, Sun H, Chen L, Fan K (2017) An overview of ahead geological prospecting in tunneling. Tunn Undergr Space Technol 63:69–94
5. Wei L, Khan M, Mehmood O, Dou Q, Bateman C, Magee DR, Cohn AG (2019) Web-based visualisation for look-ahead ground imaging in tunnel boring machines. Autom Constr 105:102830

6. Yin X, Liu Q, Huang X, Pan Y (2022) Perception model of surrounding rock geological conditions based on TBM operational big data and combined unsupervised-supervised learning. Tunn Undergr Space Technol 120:104285
7. Erharter GH, Marcher T (2021) On the pointlessness of machine learning based time delayed prediction of TBM operational data. Autom Constr 121:103443
8. Zhang L, Guo J, Fu X, Tiong RLK, Zhang P (2024) Digital twin enabled real-time advanced control of TBM operation using deep learning methods. Autom Constr 158:105240
9. Fu X, Tiong RLK, Zhang L (2025) EMNet: an ensemble deep learning approach for geological condition detection in tunnel excavation. Expert Syst Appl 261:125484
10. Sandler M, Howard A, Zhu M, Zhmoginov A, Chen L-C (2018) Mobilenetv2: inverted residuals and linear bottlenecks. In: Proceedings of the IEEE conference on computer vision and pattern recognition, pp 4510–4520
11. Ganaie MA, Hu M, Malik AK, Tanveer M, Suganthan PN (2022) Ensemble deep learning: a review. Eng Appl Artif Intell 115:105151
12. Guo K, Zhang L (2021) Multi-source information fusion for safety risk assessment in underground tunnels. Knowl-Based Syst 227:107210
13. Wei L, Magee DR, Cohn AG (2018) An anomalous event detection and tracking method for a tunnel look-ahead ground prediction system. Autom Constr 91:216–225
14. Li S, Song J, Zhang J, Wang C, Liu B, Liu F, Ma S, Nie L (2017) A new comprehensive geological prediction method based on constrained inversion and integrated interpretation for water-bearing tunnel structures. Eur J Environ Civ Eng 21:1441–1465
15. Shi M, Sun W, Zhang T, Liu Y, Wang S, Song X (2019) Geology prediction based on operation data of TBM: comparison between deep neural network and soft computing methods. In: 2019 1st international conference on industrial artificial intelligence (IAI). IEEE, pp 1–5
16. Zhang Q, Liu Z, Tan J (2019) Prediction of geological conditions for a tunnel boring machine using big operational data. Autom Constr 100:73–83
17. Fu X, Feng L, Zhang L (2022) Data-driven estimation of TBM performance in soft soils using density-based spatial clustering and random forest. Appl Soft Comput 120:108686
18. Gong Q, Zhou X, Liu Y, Han B, Yin L (2021) Development of a real-time muck analysis system for assistant intelligence TBM tunnelling. Tunn Undergr Space Technol 107:103655
19. Yang Z, He B, Liu Y, Wang D, Zhu G (2021) Classification of rock fragments produced by tunnel boring machine using convolutional neural networks. Autom Constr 125:103612
20. Nan Y, Ju J, Hua Q, Zhang H, Wang B (2022) A-MobileNet: an approach of facial expression recognition. Alex Eng J 61:4435–4444

Tunnel Face Reliability Modeling

1 Introduction

The shield tunneling method is widely favored in metro construction due to its high efficiency, safety, low cost, and minimal ground disturbance. During excavation, the shield machine applies support force to the excavation face, which can disturb the surrounding soil. Excessive support force may lead to instability at the excavation face, risking damage to nearby structures and impacting construction schedules [1, 2]. Due to the great complexities and uncertainties underlying the underground geological condition, developing a reliable model to accurately represent the behavior of the tunneling process and assess excavation risks remains a significant challenge.

Several methods have been employed to analyze excavation face reliability, primarily the limit equilibrium method [3] and numerical analysis [4]. Limit equilibrium methods have been used to determine critical collapse pressures and predict necessary support pressures under uncertainty. However, studies suggest these methods can overestimate failure probabilities and typically evaluate reliability based on single risk factors, neglecting interdependencies among various factors. To overcome these limitations, numerical analysis has been adopted, allowing for more comprehensive assessments. For example, finite element analysis has been used to estimate required support pressures based on load–displacement relationships, while finite difference methods have examined the impacts of soil strength and reinforcement on excavation stability. However, these numerical approaches can be time-consuming and may not account for correlations among multiple risk factors.

Real-world tunneling scenarios often involve interacting risk factors rather than isolated variables. Kim et al. [5] noted that the ground heave was likely to occur with a large supporting pressure in the tunneling excavation, indicating that there was indeed a dependency relationship between the support pressure and the ground settlement. Yang and Qin [6] concluded that insufficient supporting pressure would cause a large displacement of soil mass and the collapse of the excavation face, and both the support pressure and the ground settlement were the most significant

L. Zhang et al., *Intelligent Construction in Tunnels*, Lecture Notes in Civil Engineering 679, https://doi.org/10.1007/978-981-96-8785-5_5

factors for controlling the excavation face stability. Consequently, constructing a reasonable joint probability distribution for a more convincing reliability analysis is essential. Despite this need, consensus on how to represent the correlation between support pressure and ground settlement in tunneling remains elusive, largely due to geological uncertainties.

The copula function emerges as a flexible tool for modeling the joint distribution of multivariate data, effectively capturing the dependence structure among variables. Copulas allow for the separate modeling of marginal distributions and dependence structures, facilitating the capture of both linear and non-linear relationships. Although some studies have applied copula theory to reliability analysis in construction, no specific frameworks exist for modeling excavation face reliability while considering the interdependencies among system variables. Additionally, limited research has explored the impact of varying failure criteria on reliability assessments.

To address those limitations, this research intends to propose a copula-enabled approach to enhance the accuracy of excavation face reliability estimations by accounting for dependence structures, variations in failure criteria, and the quality of estimations based on data size. A significant challenge is the limited amount of data, often less than 30 observations, which can lead to biased statistics and unreliable results [7]. Sufficient data is critical for ensuring meaningful statistical analyses and rational reliability estimations. This chapter presents a copula-enabled approach for modeling and assessing the reliability of the dependent system with incomplete probability information. A bivariate framework consisting of the supporting pressure and the ground settlement is proposed to estimate the excavation face reliability in tunneling under limited observations. Monte Carlo simulation is utilized to model random uncertainties in measurements, enhancing the accuracy of reliability estimates.

2 The Copula Theory

The Copula is a powerful tool based on the Sklar's theorem [8], aiming to build a correlation among variables and to link the marginal distribution together to form multivariate distributions. In accordance with the Sklar's theorem, F $(x_1, x_2, \ldots, x_n)$ of n-dimensional random variables denotes the joint cumulative distribution function (CDF) with the marginal distribution functions $F_1(x_1), F_2(x_2), \ldots, F_n(x_n)$, as given in Eq. (1). When there are two random variables x_1, x_2, the bivariate joint CDF F (x_1, x_2) using copula can be expressed as Eq. (2).

$$F(\mathrm{x}_1, \mathrm{x}_2, \ldots, \mathrm{x}_\mathrm{n}) = C[\mathrm{F}_1(\mathrm{x}_1), \mathrm{F}_2(\mathrm{x}_2), \ldots, \mathrm{F}_\mathrm{n}(\mathrm{x}_\mathrm{n})] \tag{1}$$

$$F(x_1, x_2) = C(F_1(x_1), F_2(x_2); \theta) = C(z_1, z_2; \theta) \tag{2}$$

where C is a copula function to describe the correlation among x_i, and $F_i(x_i)$ is the corresponding marginal distribution function of x_i. When marginal distribution functions $F_i(x_i)$ and copula function C is known, the joint distribution $F(x_1, x_2, \ldots, x_n)$ could be constructed in order to fit the correlation structure to data. In the case of two variables x_1, x_2, $C(z_1, z_2; \theta)$ is the bivariate copula function, where z_1 and z_2 are the marginal distribution of x_1 and x_2, and θ is the copula parameter to illustrate the correlation between x_1 and x_2.

The construction of a multivariate joint distribution function involves two interdependent steps: (1) modeling the marginal distribution function from raw data, and (2) selecting a copula function to account for the dependence among random variables. These two steps do not need to be done in sequence. A key advantage of copulas is their ability to separate marginal distributions from the dependence structure, allowing for independent modeling of each, even when variables follow different types of marginal distributions. The accuracy of analysis results significantly depends on the chosen copula functions, which vary in their dependence characteristics, such as symmetry and tail dependence [9]. Contour plots as shown in Fig. 1 present the distinct difference of dependence structures underlying four different copulas (i.e., Gaussian, Clayton, Frank, and Gumble). These four copulas are used to construct the bivariate distribution of two random variables A and B with different marginal distributions, where A follows an $N(0,1)$ distribution and B follows a $t(3)$ distribution. These differences can lead to varying outcomes in decision analysis, highlighting the importance of appropriate copula selection.

3 Reliability Analysis Under Limited Information

In order to address dependency relations that are not well reflected in the traditional reliability analysis methods, a copula-based approach is proposed to support the estimation of the reliability in dependent systems. Figure 2 illustrates the flowchart of the developed approach. There are mainly three main steps incorporated, as elaborated below.

3.1 Identification of the Best Marginal Distribution Function

The marginal distribution function is used to describe the probability distribution of variables exactly. In order to construct the joint distribution by the copula function, the key point is to find the best-fitting marginal distribution among candidate distributions. Different variables in copula do not have to fit the same kind of marginal distribution function. This research focuses on four commonly used marginal distributions: namely, the log-normal distribution, truncated gumbel distribution, truncated normal distribution, and Weibull distribution. Table 1 lists a summary of candidate marginal distributions.

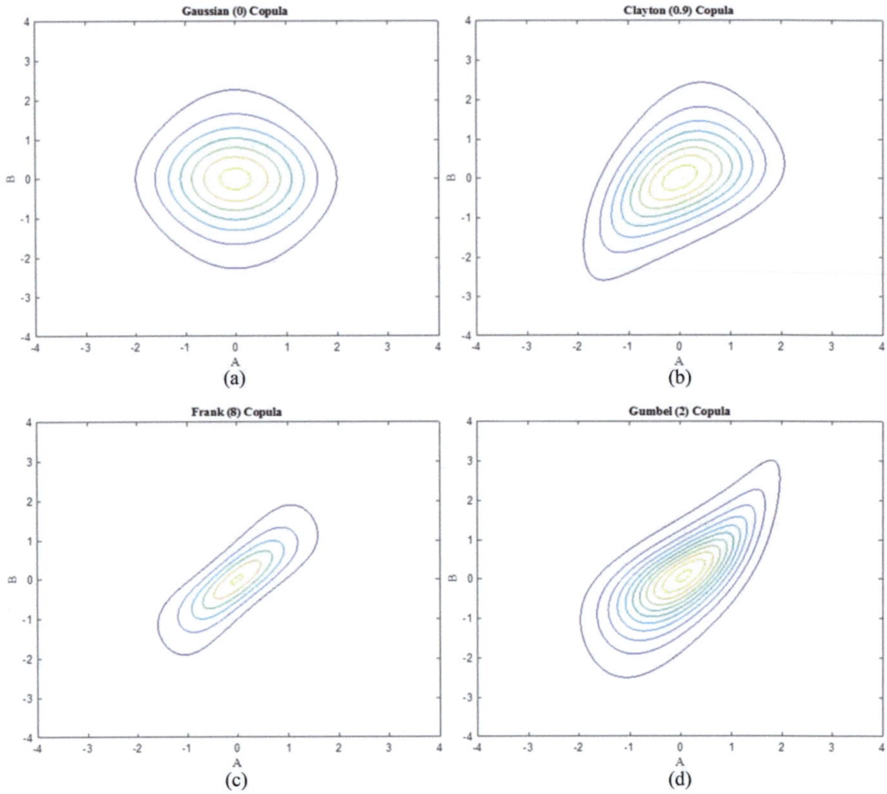

Fig. 1 Contour plots of bivariate pdf from four copulas: **a** Gaussian; **b** Clayton; **c** Frank; and **d** Gumble. *Note* A follows an $N(0,1)$ distribution and B follows a $t(3)$ distribution

Selection criteria, including Akaike information criterion (AIC) [10] and Bayesian Information Criterion (BIC) [11], are used to identify the best-fitting marginal distribution. Both criteria, based on the likelihood principle, determine the marginal distribution with the smallest AIC or BIC value, indicating the best fit for the data. Equations of AIC and BIC are given in Eqs. (3) and (4).

$$\mathrm{AIC} = -2\sum_{i=1}^{N} lnf(x; p, q) + 4 \tag{3}$$

$$\mathrm{BIC} = -2\sum_{i=1}^{N} lnf(x; p, q) + 2lnN \tag{4}$$

where p, q are distribution parameters associated with μ, σ respectively, $lnf(x; p, q)$ is the logarithm likelihood function.

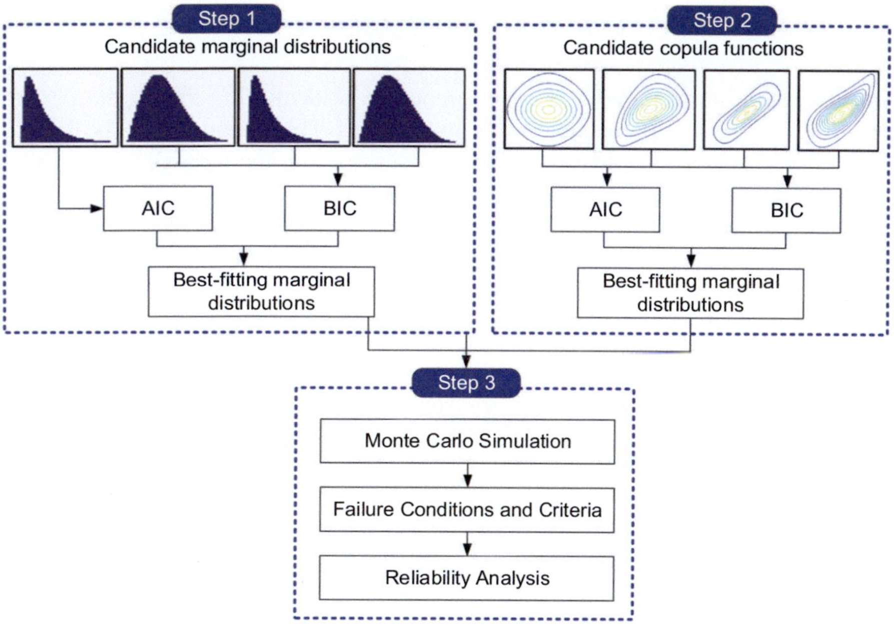

Fig. 2 Flowchart of the developed approach for reliability analysis

Table 1 A summary of candidate marginal distributions

Marginal distribution	Distribution function $f(x; p, q)$	Mean μ	Variance σ^2
Log-normal	$\frac{1}{qx\sqrt{2\pi}}\exp\left[-\frac{1}{2}\left(\frac{\ln x-p}{q}\right)^2\right]$	$\exp(p+0.5q^2)$	$[\exp(q^2)-1]\exp(2p+q^2)$
Truncated gumbel	$\frac{q\exp\{-q(x-p)-\exp[-q(x-p)]\}}{1-\exp[-\exp(pq)]}$	$p+\frac{0.5772}{q}$	$\pi^2/(6q^2)$
Truncated normal	$\frac{1}{q\sqrt{2\pi}}\exp\left[-\frac{1}{2}\left(\frac{x-p}{q}\right)^2\right]/[1-\Phi(-\frac{p}{q})]$	p	q^2
Weibull	$\frac{q}{p}\left(\frac{x}{p}\right)^{q-1}\exp\left[-\left(\frac{x}{p}\right)^q\right]$	$p\Gamma\left(1+\frac{1}{q}\right)$	$p^2\left[\Gamma\left(1+\frac{2}{q}\right)-\Gamma^2\left(1+\frac{1}{q}\right)\right]$

Note Φ is the cumulative distribution function of the standard normal distribution; Γ is the gamma function; *p*, *q* are distribution parameters

Overall, three major steps for determining the best-fitting marginal distribution include the following: (1) Calculation of mean μ and standard deviation σ of measured data; (2) Estimation of distribution parameters *p*, *q* based upon different types of candidate marginal distributions; and (3) Identification of the preferred marginal distribution for bivariate distribution according to the minimum value of AIC or BIC.

3.2 Identification of the Best-Fitting Copula Function

A critical step in constructing a copula function is identifying the copula parameter θ, which measures dependence among variables. This parameter can be defined using either the Pearson's linear correlation coefficient or the Kendall rank correlation coefficient. Pearson's coefficient is stable under perfect linear conditions, it is unreliable in non-linear situations. Therefore, this research employs the Kendall rank correlation coefficient τ_k to estimate the parameter θ, which reflects the dependence between random variables independently of their marginal distributions. According to the copula theory τ_k and the copula function are related as Eq. (7).

$$\tau_k = \binom{N}{2}^{-1} \sum_{i<j} sign\left[\left(x_{1i} - x_{1j}\right)\left(x_{2i} - x_{2j}\right)\right] \tag{5}$$

$$\text{sign} = \begin{cases} 1\left(x_{1i} - x_{1j}\right)\left(x_{2i} - x_{2j}\right) \geq 0 \\ -1\left(x_{1i} - x_{1j}\right)\left(x_{2i} - x_{2j}\right) < 0 \end{cases} \tag{6}$$

$$\tau_k = 4 \int_0^1 \int_0^1 C(z_1, z_2; \theta) dC(z_1, z_2; \theta) - 1 \tag{7}$$

where the value of τ_k ranges from − 1 to 1. The copula parameter θ can be determined by solving the integral Eq. (7) once the Kendall rank correlation coefficient τ_k is known.

The dependence structure of the measured data is not easily observed, especially with limited observations. A useful approach is to transform the data into a standard uniform distribution $U_i = (U_{1i}, U_{2i})$ over the range of [0, 1]. The empirical distribution $U_i = (U_{1i}, U_{2i})$ can be defined by Eq. (8) using rank values. It ensures that $ank(x_{1i})$ and $rank(x_{2i})$ re less than or equal to the total number of observations N. Thus, the rank of x_{1i} and x_{2i} divided by $(N + 1)$ in Eq. (8) is a guarantee to ensure the generated value within the range of [0, 1]. Additionally, for datasets containing both positive and negative values, calculations should be performed separately for defined conditions. For instance, one condition can be defined as positive settlement and positive stress, while the other condition can depict negative value.

$$\begin{cases} u_{1i} = \frac{rank(x_{1i})}{N+1} \\ u_{2i} = \frac{rank(x_{2i})}{N+1} \end{cases} \quad i = 1, 2, \ldots, N \tag{8}$$

where $rank(x_{1i})$ and $rank(x_{2i})$ represent the ascending order in vector $\{x_{11}, \ldots, x_{1N}\}$ and $\{x_{21}, \ldots, x_{2N}\}$, respectively. x_{1i} and x_{2i} are random variables in the measured data, where u_{1i} and u_{2i} are standard uniform random variables.

The copula function plays a role in coupling the marginal distribution with the joint distribution. The most appropriate copula function can be chosen after the best-fitting marginal distribution has been determined. As the evaluation criteria, AIC and BIC are widely used to evaluate candidate copulas, which can be given as Eqs. (9)

and (10). In sum, the best-fitting copula function can be determined by (1) calculating the Kendall rank correlation coefficient τ_k; (2) solving integral for copula parameter θ underlying various candidate copulas; (3) obtaining empirical distribution $U_i = (U_{1i}, U_{2i})$ from the transformation of measured data; and (4) computing AIC and BIC to find out the minimum value, which represents the best-fitting copula.

$$\text{AIC} = -2\sum_{i=1}^{N} lnc(u_{1i}, u_{2i}) + 2k \tag{9}$$

$$\text{BIC} = -2\sum_{i=1}^{N} lnc(u_{1i}, u_{2i}) + klnN \tag{10}$$

where $c(u_{1i}, u_{2i})$ is the copula density function; k is the number of copula parameters and $k = 1$ in this research, as there is only one parameter in the four copula functions; N is the sample size of measured data. Values of AIC and BIC can be calculated by putting the value of (u_{1i}, u_{2i}) from Eq. (8) into Eqs. (9) and (10). The best-fitting copula function can be picked out where the value of AIC or BIC is the smallest.

3.3 Simulation-Based Reliability Analytics

In practice, the measured data from real projects in the construction industry are often limited, known as incomplete probability. Collecting high-quality data is challenging due to uncertainties in sensors and environmental factors. The Monte Carlo simulation (MCS) serves as a robust statistical tool, generating numerous random samples based on the characteristics of the measured data [12]. The failure probability in reliability analysis is determined by the ratio of failure sampling points L to the total number of sampling points N. MCS is widely used for probabilistic analysis due to its time-saving advantages and high accuracy under uncertainty. Probability density functions (PDFs) form the foundation of MCS, providing necessary probabilities for simulations [13]. Notably, MCS effectively simulates correlated variables even with insufficient data. For instance, Aladejare and Wang [14] employed Markov chain Monte Carlo (MCMC) simulation incorporated with joint PDF of cohesion (c) and friction angle (ϕ) of rock to generate a large number of sample pairs of c and ϕ with their correlation on reliability analysis of rock slope stability, where enough measured data is not available. Thus, this research applies MCS to generate a substantial dataset using various copula functions, aiming to model a bivariate distribution that better reflects reality and addresses the limitations of existing observations.

The considerable application of the copula function is to simulate dependent variables from bivariate distributions. The main steps for generating samples by the copula function are summarized as: (1) Simulate two independent standard normal random variables $K = (K_1, K_2)^T$; (2) Simulate corresponding standard uniform distribution between random variables U $= (U_1, U_2)^T$ which are associated with the prescribed copula function; and (3) Map U $= (U_1, U_2)^T$ into correlated non-normal variables X $= (X_1, X_2)^T$ by the usual CDF transform method by $(X_1, X_2) = (F_1^{-1}(U_1), F_2^{-1}(U_2))$, in which $F_1^{-1}(U_1), F_2^{-1}(U_2)$ are the inverse cumulative distribution function of X_1, X_2. Algorithms for U $= (U_1, U_2)^T$ could be varied from different types of the copula function, which are illustrated below in detail.

(1) Gaussian copula

(a) Do the Cholesky decomposition on the correlation coefficient matrix θ, making up by corresponding parameters θ, and then obtain the lower triangle matrix L_0
(b) Set Z $= L_0K$, then get the correlated standard normal distribution variables Z $= (Z_1, Z_2)^T$
(c) Set U $= \Phi$(Z), then get the correlated standard uniform distribution variable U $= (U_1, U_2)^T$

(2) Archimedean Copula

(a) Set V $= \Phi$(Z), then obtain independent standard uniform distribution variables V $= (V_1, V_2)^T$
(b) Set $U_1 = V_1$, and then solve a different nonlinear equation for U_2 to get U $= (U_1, U_2)^T$:
(1) No.16 Copula: Solve $V_2 = \frac{1}{2}\left(1 + \frac{\theta}{V_1^2}\right)\left[1 + S\left(S^2 + 4\theta\right)^{-1/2}\right]$ where $S = V_1 + U_2 - 1 - \theta\left(\frac{1}{V_1} + \frac{1}{U_2} - 1\right)$
(2) Frank Copula: $U_2 = -\frac{1}{\theta}\ln[\frac{1+V_2(e^{-\theta}-1)}{e^{-\theta V_1}-V_2(e^{-\theta V_1}-1)}]$, then get U $= (U_1, U_2)^T$

(3) Plackett Copula

(a) Set V $= \Phi$(Z), then obtain independent standard uniform distribution variables V $= (V_1, V_2)^T$
(b) Set a $= V_2(1 - V_2)$, b $= \theta + a(\theta - 1)^2$, c $= 2a\left(V_1\theta^2 + 1 - V_1\right) + \theta(1 - 2a)$
(c) Set d $= \sqrt{\theta}\sqrt{\theta + 4aV_1(1 - V_1)(1 - \theta)^2}$
(d) Set $U_1 = V_1$, $U_2 = \frac{[c-(1-2V_2)d]}{2b}$, then get U $= (U_1, U_2)^T$.

In general, the procedures for reliability analysis based on the MC-copula simulation are presented as follows: (1) Set the joint probability distribution on parameters X $= (X_1, X_2, \ldots, X_n)$ model and simulate enough samples of dependent random variables in accordance with their distributions and dependence structures with the copula approach which has been presented above; (2) Take sampling points X_i into the reliability analysis function and calculate the structure response $g_i = g(x)$; (3) Count the number of condition $g_i < 0$, which can be regarded as the number of failure samples L. The failure probability of the structure can be expressed as $p_f = L/N$.

4 Tunnel Face Reliability Modelling Framework

4.1 Identification of Key Influential Factors

Construction projects involving earth pressure balanced (EPB) shield tunnels face various risk factors. Accurately assessing these risks is crucial for ensuring safety in tunneling operations. The primary operating principle of the shield tunneling method is to apply supporting pressure to the tunnel face. Yang et al. [15] emphasized the significance of the supporting pressure, which can limit the environmental impact and ensure the stability of the tunnel face. Kasper and Meschke [16] studied the influence of face pressure with respect to the settlement of the ground surface for the tunnel design. To sum up, the supporting pressure plays an important role in controlling the magnitude of the ground settlement and affects the reliability of the excavation face.

However, maintaining supporting pressure at the initial ground stress level is challenging due to the dynamic nature of the excavation process. Variations in supporting pressure can lead to ground deformation and nearby foundation displacement. Excessive pressure may cause ground heave, while insufficient pressure can result in ground settlement. Both scenarios pose risks of excavation failure and potential hazards to surrounding structures. Therefore, it is essential to apply an appropriate magnitude of supporting pressure to prevent excavation face failure. Most shield tunnels in construction projects utilize the EPB method, where the supporting pressure often does not reach its upper limit, underscoring the need for careful management of these parameters.

4.2 Estimation of Tunnel Face Failure Probability

For structural systems in civil engineering, there are two cases to express the probability of the failure P_f, as seen in Eqs. (11) and (12). One is that $y(Q)$ denotes the minimum limit value and a failure condition is $\mathrm{y} - y(Q) < 0$, and thus, P_f can be calculated by Eq. (11). Another situation is that when $y(Q)$ stands for the maximum limit value, P_f can be estimated based on $\mathrm{y} - y(Q) > 0$ by using Eq. (12)

$$P_f = P\left[\mathrm{y} - y(Q) < 0\right] \tag{11}$$

where $y(Q)$ is the minimum limit value and y is the actual value.

$$P_f = P\left[\mathrm{y} - y(Q) > 0\right] \tag{12}$$

where $y(Q)$ is the maximum limit value and y is the actual value.

This research adopts the traditional Rankine's earth pressure theory along with the limit equilibrium theory for the soil to calculate the value of passive earth pressure in the limit of the supporting pressure condition, which is 70 kPa. Moreover, the ground settlement is another key factor that should be controlled strictly. According to construction requirements, the final settlement should be under 30 mm for construction safety [17]. Taking the supporting pressure and the ground settlement into consideration, the probability of the excavation face failure can be represented as:

$$P_f = P\left[g_1(X_1)\langle 0 \cup g_2(X_2)\rangle 0\right] \tag{13}$$

$$g_1(X_1) = X_1 - \overline{X_1}, g_2(X_2) = X_2 - \overline{X_2} \tag{14}$$

where X_1 and X_2 are the real value of the supporting pressure and the ground settlement, respectively. $\overline{X_1}$ are the minimum limit value of the supporting pressure, and $\overline{X_2}$ are the maximum limit value of the ground settlement. The shield tunnel construction will be broken, once X_1 is less than $\overline{X_1}$ or X_2 is greater than $\overline{X_2}$. Different kinds of copula functions with their dependence structures will produce a different probability of the excavation face failure.

5 Experimental Studies

5.1 Data and Resources

The excavation of a tunnel section, that is from Orient Lucky City station to Changfeng station (O-C section) in the Metro Line 7 in the Wuhan Metro System, China, is used as a case to test the applicability and effectiveness of the developed approach. Figure 3 illustrates the specific location information of the tunnel section. It is a twin-line tunnel with a left line of 1717 m in length and a right line of 1714 m in length. There are three curves with a radius of 450 m within the tunnel excavation routine. The thickness of soil cover is 9.74~21.4 m and the diameter of the inside tunnel is 5.5 m.

The O-C tunnel section is located within cohesive soil, with some sections passing through an artesian aquifer where the water level is 2.2–3.5 m below ground. Excavation of shield tunnels inevitably causes soil deformation. To maintain excavation face stability, equivalent stress must be applied to prevent significant soil deformation and minimize disturbances to surface structures. Given the complex geological conditions, studying the correlation between supporting pressure and ground settlement is crucial for enhancing the reliability evaluation of the excavation face.

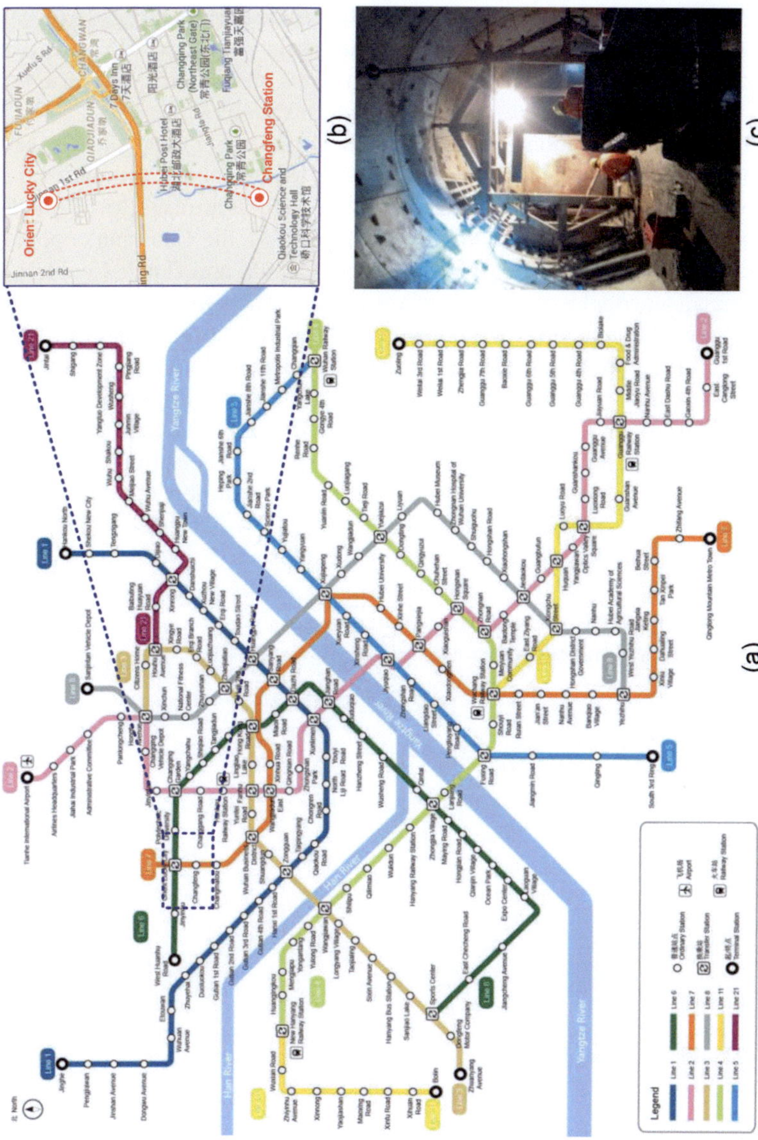

Fig. 3 Location of the tunnel case: **a** map of the Wuhan metro system; **b** layout of the O-C tunnel section; and **c** site picture of the construction of the O-C tunnel section

5.2 Copula Model Development

In the O-C tunnel section, 61 samples are collected measuring supporting pressure and ground settlement. The scatter plot in Fig. 4a shows a weak linear negative relationship. After applying Eq. (8) for standard uniform distribution, Fig. 4b reveals a stronger linear negative dependency, nearly symmetric along the 135° diagonal of a unit square. Therefore, candidate copula functions should be symmetric and effectively describe the negative correlation of random variables.

Four candidate copula functions, including Gaussian, No.16, Frank, and Plackett, can meet those requirements to fit the measured data [18], which are summarized in Table 2. The main reasons to choose the four candidate copulas are as follows: Firstly, they all belong to the classical copula family. Among them, the Gaussian copula is classified as an elliptical copula family, the No. 16 Copula and Frank Copula are from the Archimedean copulas, and the Plackett copula is the special type of copula. Secondly, they can all characterize the positive and negative correlations well. The absolute values of the correlation coefficient in the four copulas are all close to 1, and a negative correlation can be captured as well by just a simple conversion $-$ 1.

Methods for identifying the best-fitting marginal distribution function and copula function have been introduced in the methodology section. The summary statistic (e.g., mean, standard deviation, and others) of the measured data about the supporting pressure and the ground settlement serve as a theoretical basis for determining the marginal distribution function. Table 3 presents these statistics. It reveals a coefficient of variability (CoV) of 0.046 for ground settlement, significantly lower than the CoV of 0.097 for supporting pressure. This indicates greater variation in supporting pressure compared to ground settlement.

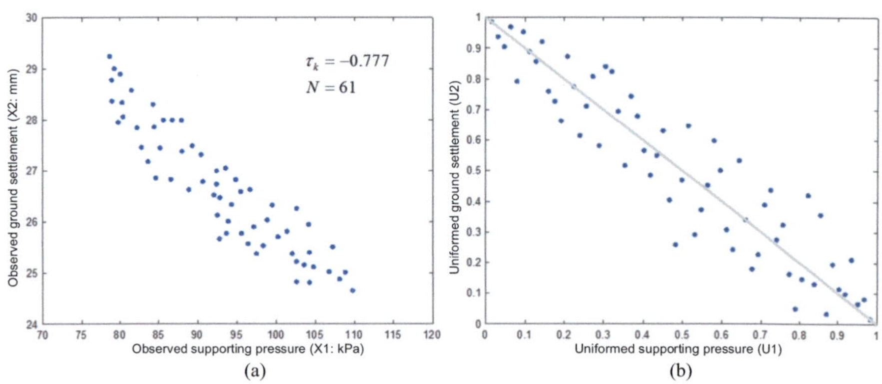

Fig. 4 Scatter plot of **a** the observed supporting pressure and ground settlement (X_1, X_2) and **b** the uniformed supporting pressure and ground settlement (U_1, U_2)

Table 2 A summary of candidate bivariate copula functions

Copula	Copula function $C(u_1, u_2; \theta)$	Generator function $\varphi(t, \theta)$	Range of θ
Gaussian	$\Phi_\rho(\Phi^{-1}(u_1), \Phi^{-1}(u_2))$	–	[−1,1]
No. 16	$\frac{1}{2}\left(S + \sqrt{S^2 + 4\theta}\right)$ $S = u_1 + u_2 - 1 - \theta(\frac{1}{u_1} + \frac{1}{u_2} - 1)$	$(\frac{\theta}{t} + 1)(1 - t)$	$[0, \infty]$
Frank	$-\frac{1}{\theta}\ln\left(1 + \frac{(e^{-\theta u_1} - 1)(e^{-\theta u_2} - 1)}{e^{-\theta} - 1}\right)$	$-\ln\frac{e^{-\theta t} - 1}{e^{-\theta} - 1}$	$(-\infty, \infty)$
Plackett	$\frac{S - \sqrt{S^2 - 4u_1u_2\theta(\theta - 1)}}{2(\theta - 1)}$ $S = 1 + (\theta - 1)(u_1 + u_2)$	–	$(0, \infty)$

Note Φ is the cumulative distribution function of the standard normal distribution; θ is the copula parameter to characterize the dependency in X_1 and X_2; u_1 and u_2 are the marginal distribution of X_1 and X_2

Table 3 Summary statistic of measured data of the supporting pressure and the ground settlement

Monitoring parameter	Mean	Standard deviation	Maximum	Minimum	Coefficient of variability	Skewness	Kurtosis
Supporting pressure (kPa)	93.053	8.982	109.770	78.791	0.097	0.044	1.911
Ground settlement (mm)	26.639	1.230	29.232	24.643	0.046	0.288	2.055

Note CoefficientofVariability(CoV) $= \frac{\sigma}{u}$, Skewness $= \frac{1}{n}\sum_{i=1}^{n}\left(\frac{x_i - u}{\sigma}\right)^3$, Kurtosis $= \frac{1}{n}\sum_{i=1}^{n}\left(\frac{x_i - u}{\sigma}\right)^4$, where μ is the mean value and σ is the standard deviation of measured data

5.3 *Marginal Distribution and Function Selection*

Four different kinds of marginal distributions, namely the log-normal distribution, truncated gumbel distribution, truncated normal distribution, and Weibull distribution, are compared to determine the best-fitting one based on AIC and BIC values, calculated using Eqs. (3) and (4). Table 4 indicates the calculated AIC and BIC values of the four marginal distribution functions, with the bested fitted values highlighted in bold. For the supporting pressure (X_1), the truncated normal distribution which results in the smallest AIC and BIC values is regarded as the best-fitting distribution. The goodness of fit in the log normal distribution is relatively close to the truncated normal distribution. For the ground settlement (X_2), the log-normal distribution is considered as the best-fitting marginal distribution. To visualize the fitting degree, histograms in Fig. 5 display the probability distribution functions of all four distributions against the measured data. The best-fitting distributions demonstrate superior alignment with the data characteristics compared to the others. These findings, corroborated by Table 4, highlight the effectiveness of AIC and BIC in selecting optimal marginal distributions.

To validate the marginal distributions, the Kolmogorov-Smirnov (K-S) test is applied, assessing the fit between candidate distributions and measured data. The K-S test calculates the distance (D) between the hypothetical probability distribution and the empirical data, and the p-value is derived from the K-S probability function. A higher p-value indicates a stronger fit, and if D is smaller than the p-value at significance level α, the distribution is accepted. Table 5 presents the K-S test results for the supporting pressure (X_1) and the ground settlement (X_2), with best fitted results highlighted in bold. None of the distributions are rejected at $\alpha = 0.05$. For the supporting pressure (X_1), the truncated normal distribution shows the smallest D and highest p-value (81.9%), confirming its suitability. Similarity, for ground settlement (X_2), the log-normal distribution yields the largest p-value (81.5%), making it the best fit.

To determine the best-fitting copula, the copula parameter is estimated using the Kendall rank correlation coefficient τ_k, calculated by Eq. (7). With τ_k at -0.777, a strong negative correlation is confirmed in the measured data, making it unreasonable to assume independence between variables. Figure 6 illustrates the differences between independent parameters and copula functions, showing that results based

Table 4 Values of AIC and BIC for four marginal distributions

Monitoring parameter		Log-normal	Trunc Gumbel	Trunc normal	Weibull
Supporting pressure (X_1)	AIC	443.931	450.199	**443.921**	447.554
	BIC	448.153	454.420	**448.143**	451.776
Ground settlement (X_2)	AIC	**200.602**	201.032	201.323	214.961
	BIC	**204.824**	205.254	205.544	219.183

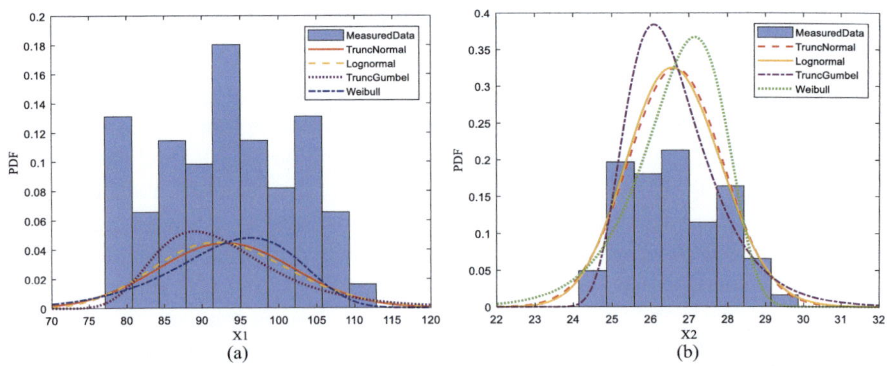

Fig. 5 Frequency histogram of the measured data and PDF of marginal distributions: **a** supporting pressure; and **b** ground settlement

Table 5 K-S test for four marginal distributions

Monitoring parameter	Marginal distribution	D value	P-value $\alpha = 0.05$
Supporting pressure (X_1)	Log-normal	0.091	0.710
	Trunc Gumbel	0.123	0.138
	Trunc normal	**0.082**	**0.819**
	Weibull	0.094	0.675
Ground Settlement (X_2)	**Log-normal**	**0.083**	**0.815**
	Trunc Gumbel	0.092	0.689
	Trunc normal	0.101	0.585
	Weibull	0.140	0.191

on the independence assumption may not accurately capture the real data characteristics. This highlights the necessity of using copula-based models to account for system dependence and better reflect the true behavior of the measured data.

The fitness results between simulated and measured data vary significantly depending on the chosen copula function. Selecting the appropriate copula should align with the characteristics of the measured data to accurately capture variable dependencies. Values of (u_{1i}, u_{2i}) come from Eq. (8), and then Eqs. (9) and (10) are used to calculate the AIC and BIC values of four candidate copulas, respectively. Table 5 summarizes the copula parameters and Kendall rank correlation coefficients for the four copulas, with the bested fitted values highlighted in bold. The best-fitting copula has the smallest AIC and BIC values, as seen in Table 6. The Frank copula performs best, suggesting that the commonly used Gaussian copula (Nataf distribution) is not always suitable for modeling correlations.

To validate the effectiveness of the Frank copula in modeling the measured data, Fig. 7 (the frequency histogram) is compared with Fig. 8a (the Frank copula probability density plot). The similarity in shape between these figures indicates that the

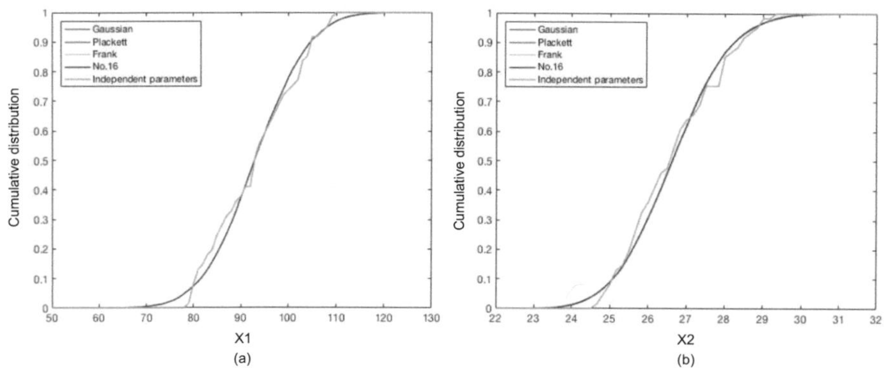

Fig. 6 Cumulative distribution of independent parameters and copula functions: **a** supporting pressure; and **b** Ground settlement

Table 6 Simulation results of four copula functions

Copula function	Gaussian	No. 16	Frank	Plackett
AIC	– 107.118	– 88.347	– **116.195**	– 96.199
BIC	– 105.007	– 86.236	– **114.084**	– 94.088
Parameter θ	– 0.939	0.002	– 16.097	0.011
τ_k between U_1, U_2	– 0.777	– 0.779	– 0.777	– 0.777
τ_k between X_1, X_2	– 0.777	– 0.779	– 0.777	– 0.777
τ_k between measured data	– 0.777			

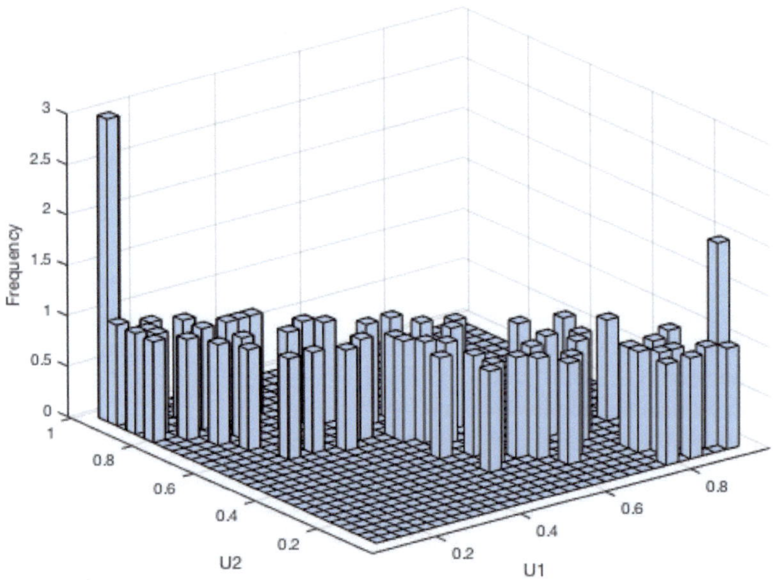

Fig. 7 Frequency histogram of measured data

Frank copula accurately captures the dependence structure between supporting pressure and ground settlement. Additionally, Fig. 8b shows a contour plot with clear symmetry along the diagonal, reflecting a strong negative correlation between the two variables. These results confirm the Frank copula's suitability for modeling the bivariate distribution, supporting the findings from Table 6.

5.4 *Tunnel Face Reliability Analysis*

In order to verify the superiority and necessity of the bivariate distribution modeled by the copula function, the reliability of the tunnel excavation face applied to this

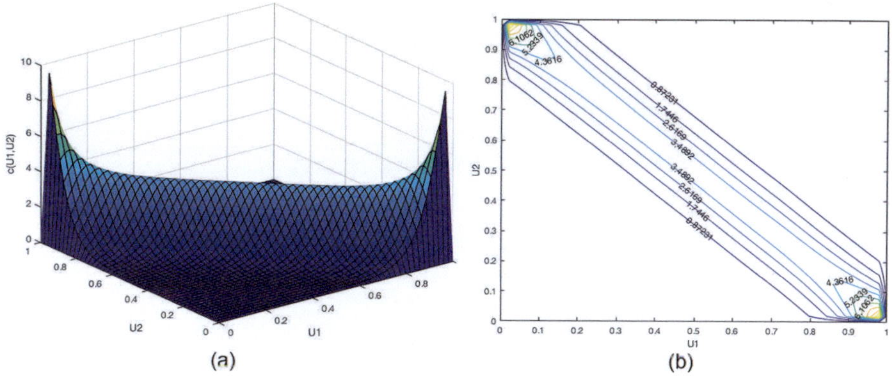

Fig. 8 Frank copula function plot of **a** probability density function; and **b** Contour

tunnel excavation case is estimated using the MC-copula simulation method. It is noted that the supporting pressure follows the truncated normal distribution and the ground settlement follows the log-normal distribution. This method is useful in evaluating the safety status of the excavation face precisely in tunneling construction, which will further guide the safety control in the construction progress.

Due to the limited size of the measured data, the MCS approach is used to simulate 10^5 samples. Without accounting for variable correlation, the supporting pressure (X_1) and the ground settlement (X_2) will follow the normal distribution, as displayed in Fig. 9. Obviously, when the supporting pressure and the ground settlement are independent, they distribute discretely and uniformly in the coordinate. Subsequently, the MCS approach is implemented to simulate 10^5 samples about the supporting pressure and the ground settlement underlying four candidate copula functions respectively, aiming to describe the dependency characteristics between variables. Figure 10 shows scatter plots of the measured (61 samples) and simulated data (10^5 samples), indicating that all four copula functions exhibit similar variation trends along the 135° diagonal. While it's difficult to identify the best-fitting copula function directly from the scatter plot due to limited measured data, the Frank copula, as identified in Table 6, best captures the data's characteristics.

Building on prior studies, this research sets two conditions for the excavation face failure: the minimum value of the supporting pressure is 70 kPa and the maximum value of the ground settlement is 30 mm. Neither one condition fails, the structure system will be regarded as a failure state. Figures 9 and 10 highlight the safety areas within the green shadows, while points outside represent failures. The probability of the face failure is the ratio of the number of points in the failure area to the total number of points, which can be calculated by Eqs. (13) and (14).

As a matter of fact, the selection of the copula function will be of great significance for the reliability analysis. Since the Frank copula has been identified as the best-fitting copula function in Sect. 2 based on AIC and BIC criteria, the failure probability p_f underlying the Frank copula is regarded as the reference in the reliability analysis. Table 7 lists the failure probability p_f using different copula functions and the relative

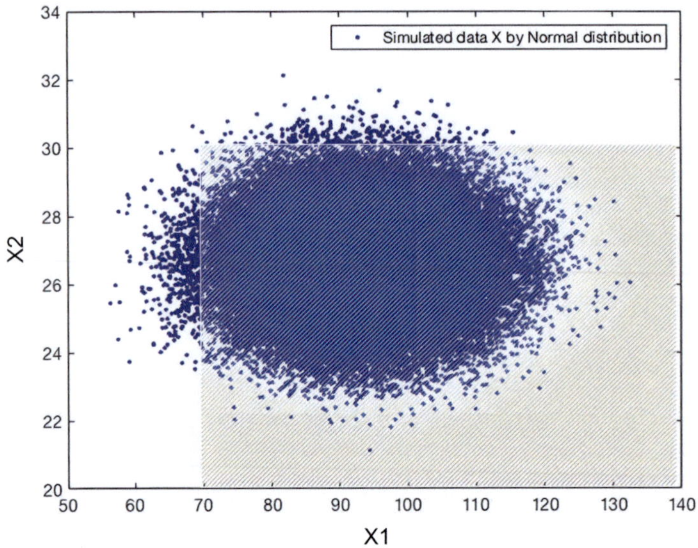

Fig. 9 Scatter plot of independent data and the failure area

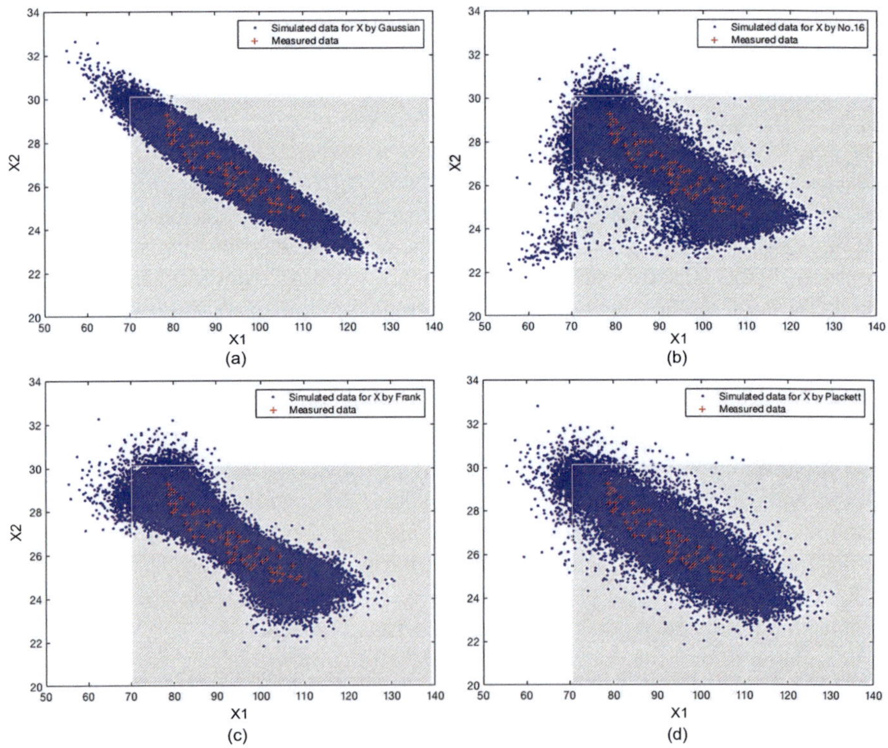

Fig. 10 Scatter plots of four copulas and failure areas: **a** Gaussian; **b** No. 16; **c** Frank; and **d** Plackett

Table 7 Probability of the excavation face failure under different copula models

Model	Plackett	Gaussian	No. 16	Frank	Independence
Failure probability p_f	8.48×10^{-3}	6.72×10^{-3}	9.56×10^{-3}	9.39×10^{-3}	1.98×10^{-2}
Relative error	0.10	0.29	0.02	**0**	1.11

Note The relative error is calculated by $\left|p_f - p_{fFrank}\right|/p_{fFrank}$, where p_f is the failure probability of candidate copula function. p_{fFrank} is the failure probability of Frank copula function

errors, with the smallest error highlighted in bold. Results of the reliability analysis for the excavation face are analyzed as follows:

(1) The results verify that ignoring the correlations in variables for simplicity is both unreasonable and unreliable. The measured data shows a strong negative dependency between supporting pressure and ground settlement, making traditional independent models inaccurate. The failure probability under the independence assumption is 0.0198, which is considerably larger than the expected value of 0.0094, and the error can even exceed 100%. Consequently, the failure probability based on the independent model will largely underestimate the reliability, making the further structural design overly conservative and inevitably increasing the project cost. One effective way to reduce the dispersion in the failure probability evaluation is to employ copula functions to characterize the dependence structure. As an illustration, the failure probability based on four copula functions is all smaller than the result calculated under the independence assumption, which is 0.0198.
(2) Simulating samples from different copula functions can lead to significant variations in failure probability estimation. As shown in Fig. 10, the choice of copula affects the scattering points' distribution, impacting reliability analysis results. Values of the excavation face failure probability p_f (smallest to the largest) are: Gaussian (6.72×10^{-3}), Plackett (8.48×10^{-3}), Frank (9.39×10^{-3}), and No.16 (9.56×10^{-3}). The Gaussian copula produces the largest error, underestimating the failure probability by more than 29% compared to the reference. This highlights its inadequacy in capturing non-linear correlations. Consequently, using the Gaussian copula can lead to less conservative reliability estimates. Selecting the right copula function is crucial to avoid inaccurate results and ensure system safety.
(3) The failure probability exhibits an inverse trend with increasing limit values for supporting pressure and ground settlement, set at 70 ± 5 kPa and 30 ± 5m, respectively. To simplify the problem, when one variable changes, the other one remains unchanged. As shown in Fig. 11, the failure probability rises with higher supporting pressure limits. While the trends for the four copulas are similar, notable differences exist in their numerical values. The Gaussian copula consistently yields lower failure probabilities than the others, indicating its tendency to

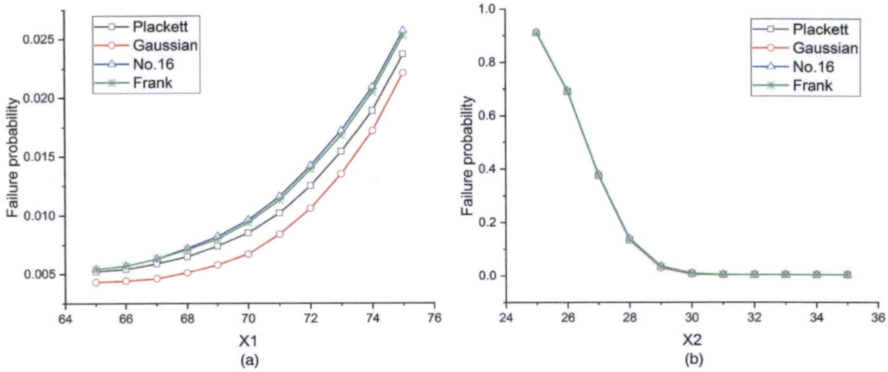

Fig. 11 Failure probability with various limit values of **a** the supporting pressure (unit: kPa); and **b** the ground settlement (unit: mm)

produce less conservative and potentially unrealistic results in reliability analysis. In contrast, the Frank and No. 16 copulas show minimal differences in their failure probability estimates.

Conversely, the failure probability shows a clear downward trend as the limit value of ground settlement increases, as illustrated in Fig. 11b. The significant variation in failure probabilities, ranging from 0.9128 to 0.0050, makes it challenging to directly compare the results from different copula functions. Notably, the Gaussian function yields the lowest failure probabilities when the ground settlement limit is below 31 mm, with values decreasing from 0.9110 to 0.0050 as the limit increases. In contrast, the Frank copula produces similar probabilities, slightly higher than those from the Gaussian function. When the ground settlement limit exceeds 32 mm, the failure probability approaches zero, indicating that such values are unlikely to occur. In contrast, when the ground settlement is less than 26 mm, more than half of the samples will fail, indicating a high risk. Therefore, it is crucial to dynamically adjust the limit values for failure criteria based on changing conditions. The findings from Fig. 11 enhance our understanding of how failure probability varies with these limit values.

6 Merits Analysis

There are two significant factors affecting the effectiveness of the reliability analysis: one is the Monte Carlo simulation (MCS) sampling size, and the other is the sample size of measured data, both of which are discussed as follows:

6.1 Impacts on the Monte Carlo Simulation

The reliability analysis results are based on 10^5 samples simulated using the MC-copula approach. Selecting an appropriate MCS sample size is crucial. Increasing the sample size enhances the precision of failure probability estimates, as supported by the law of large numbers and the central limit theorem. This is particularly important for very small failure probabilities, which require numerous samples for accurate predictions. However, a larger sample size also increases computation time, challenging efficiency. Conversely, a sample size that is too small yields limited information and results in significant errors, leading to unreliable conclusions. Balancing these factors is essential for effective reliability analysis.

A key factor in determining the MCS sampling size is comparing the quality of estimation results. Figure 12 illustrates how excavation face failure probability varies with different sampling sizes and four copula functions. For the sampling size below 10^4, the failure probability fluctuates significantly, resulting in high uncertainty in reliability estimates. In contrast, the sampling sizes above 10^4 show more stable results without sudden changes. The estimation results will tend to be more stable as the sample size increases. The values of failure probability in 10^5 samples and 2×10^5 samples do not have a significant difference. Thus, a setting size of 10^5 samples is reasonable in this case for achieving a balance between the required reliability of the results and the calculation speed. To estimate error, Lü and Low [19] employed Eq. (15) to calculate error ε of the failure probability estimation based on the MCS size *N*. In this case, using the Frank copula model, the estimation error with 105 samples is 0.06.

$$\varepsilon = k\sqrt{\frac{1-p_f}{Np_f}} \tag{15}$$

where p_f is the failure probability, N is the MCS sampling size and k is the confidence limit. For instance, when a 95% confidence interval is given, the value of k comes out to be 1.96.

Notably, the MCS sampling size depends on the failure probability level of interest. For the purpose of obtaining a satisfactory accuracy and convincing result, Wang [20] utilized a criterion to determine the minimum number of MCS samplings. Equation (16) gives the estimation method for minimum MCS sampling size, representing that the minimum number should be larger than the reciprocal of the probability level of interest by at least ten times.

$$n_{\min} = \frac{10 n_{x_1} n_{x_2}}{p_T} \tag{16}$$

where p_T is the target failure probability and $n_{x_1} n_{x_2}$ are the number of possible discrete values for X_1 and X_2. In this case, p_T is set to 0.01 and $n_{x_1} n_{x_2}$ is equal to 61, which satisfies the reliability requirement defined in [21].

The minimum MCS sampling size derived from Eq. (16) turns out to be 61,000. That is to say, a sampling size greater than 61,000 is able to ensure the accuracy

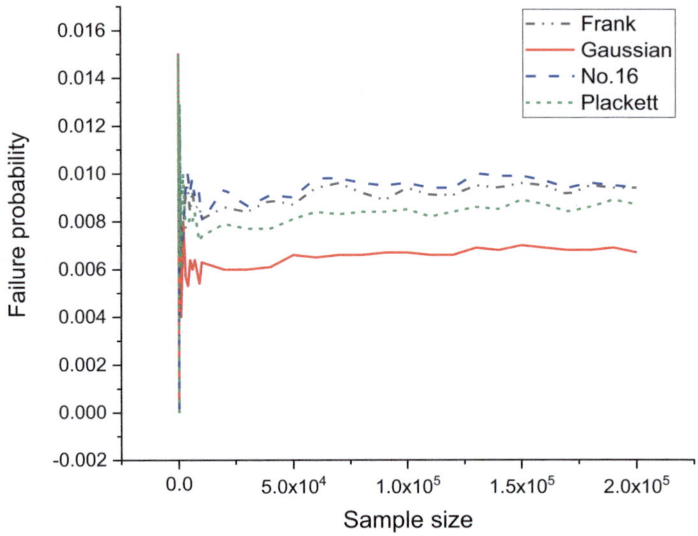

Fig. 12 Failure probability with different sampling sizes in four copula functions

of MCS. Besides, Eq. (15) enables estimating MCS sampling size in the sight of a given confidence level. Assuming that the failure probability is 0.01 and the expected estimation error is 0.1 in a 95% confidence interval, the desired MCS sampling size is 38,031. More specifically, the error of failure probability estimation can definitely be lower than 0.1 with a robust sampling size greater than 38,031 in a given 95% confidence level. Based upon the two approaches of total sampling size estimation above, it has been proved that 10^5 samples generated by MCS in this case are large enough to realize a credible reliability analysis without too long computing time.

6.2 Impacts on the Measured Sample Size

It is a fact that a small sample size is likely to generate inconclusive results, making the analysis pointless. The sample size could have an effect on the dependence structure and confidence regions. Nevertheless, measured data is usually limited from the real engineering project. To reach meaningful conclusions, it is of great importance to discuss the sample size of measured data.

The sample size of measured data exerts a direct impact on the shape of the scatter plot from MCS and the results of failure probability. To explore the influence of sample size, MCS is implemented to simulate 10^5 samplings based on best-fitting marginal distributions and copula functions using six different sizes of measured data, which are 11, 21, 31, 41, 51, and 61, respectively. By comparison of simulated data in Fig. 13 from a qualitative perspective, it is clear that the shape of simulated data from 11 measured data in Fig. 13a is distinctly different from Fig. 13f with 61 measured

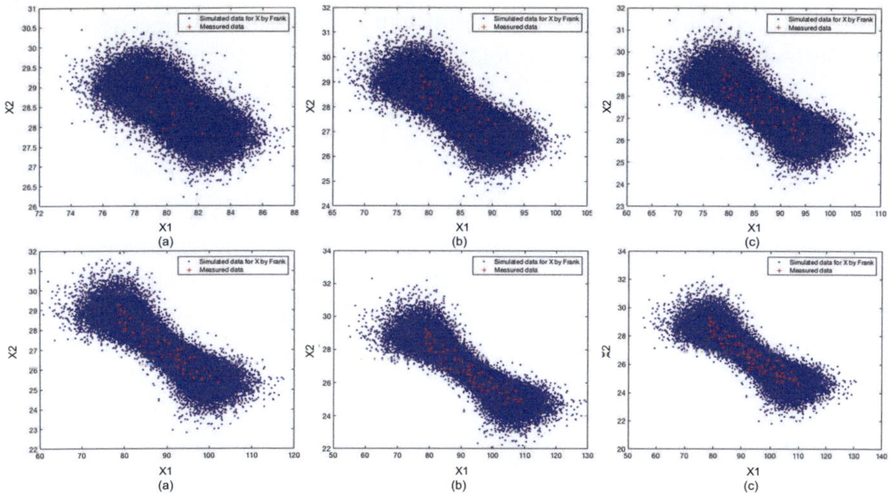

Fig. 13 Scatter plots under different sizes of measured data: **a** 11; **b** 21; **c** 31; **d** 41; **e** 51; and **f** 61

data. With the increase in data size, the shape of simulated data is more similar to Fig. 13f and the points are more concentrated in certain areas with a better shape. That is because a small size of the measurement data with no distinguished features inevitably adds more uncertainty to the sampling process. In the end, MCS will simulate samples more randomly and unreasonably, causing a reduction of statistical power.

To provide a deeper insight into the size of the measurement data, the failure probability is computed under different sample sizes. As seen in Table 8, the failure probabilities from 11, 21, and 31 measured data are experiencing relatively large fluctuations. Moreover, measured data with 11 and 31 samples come to incredible failure probability, resulting in a large deviation from the failure probability under 61 measured data. Thus, conclusions can be drawn that the sample size of measured data is not allowed to be less than 41 in this case. When the size of measured data

Table 8 Probability of the excavation face failure with different sample sizes

Size of measured data	11	21	31	41	51	61
Failure probability p_f	7.71×10^{-4}	7.13×10^{-3}	4.79×10^{-3}	7.41×10^{-3}	1.14×10^{-2}	$\mathbf{9.35 \times 10^{-3}}$
Relative error	0.92	0.24	0.49	0.21	0.22	**0**

Note The relative error is calculated by $|p_f - p_{f61}|/p_{f61}$, where p_f is the failure probability under different size of measured data increases, computations of failure probability will approach the reference value under 61 measured data as highlighted in bold, which is 9.35×10^{-3}.

7 Chapter Summary

This research aims to introduce a copula-based methodology for conducting robust probabilistic analyses of tunnel excavation face stability during construction. This approach serves as a valuable tool for accurately characterizing dependency structures among variables and enhancing reliability assessments, particularly in scenarios where probability information is incomplete. The analysis comprises two principal steps. Firstly, the bivariate joint probability distribution function is constructed and simulated associated with the copula function for modeling the dependence structure when both the best-fitting marginal distributions and copula functions are determined. Subsequently, reliability analysis is performed using the MC-copula method, which generates a substantial number of random samples reflective of the measured data characteristics, thereby increasing the accuracy and reliability of the failure probability estimates. The practicality and effectiveness of this reliability analysis approach are demonstrated through a case study within the Wuhan metro system.

Overall, the reliability analysis based on the copula approach offers a robust and credible framework for assessing the stability and safety of excavation faces in tunnel construction, particularly in contexts where comprehensive measured data are challenging to obtain. The developed approach contributes to (a) The state of the knowledge by presenting a novel method based on the copula theory to model the dependence structure of variables and calculate the failure probability under incomplete probabilities; and (b) The state of the practice by providing a more convincing estimation of the excavation face reliability in tunnels with limited observations.

References

1. Pan Y, Zhang L, Wu X, Qin W, Skibniewski MJ (2019) Modeling face reliability in tunneling: a copula approach. Comput Geotech 109:272–286
2. Zhang L, Wu X, Zhu H, AbouRizk SM (2017) Perceiving safety risk of buildings adjacent to tunneling excavation: an information fusion approach. Autom Constr 73:88–101
3. Anagnostou G, Kovári K (1994) The face stability of slurry-shield-driven tunnels. Tunn Undergr Space Technol 9:165–174
4. Paternesi A, Schweiger HF, Scarpelli G (2017) Numerical analyses of stability and deformation behavior of reinforced and unreinforced tunnel faces. Comput Geotech 88:256–266
5. Kim SH, Tonon F (2010) Face stability and required support pressure for TBM driven tunnels with ideal face membrane—drained case. Tunn Undergr Space Technol 25:526–542
6. Yang X-L, Qin C-B (2015) Limit analysis of supporting pressure in tunnels with regard to surface settlement. J Cent South Univ 22:303–309
7. Wang Y, Akeju OV (2016) Quantifying the cross-correlation between effective cohesion and friction angle of soil from limited site-specific data. Soils Found 56:1055–1070

8. Sklar A (1973) Random variables, joint distribution functions, and copulas. Kybernetika 9:449–460
9. Wang P, Lu Z, Zhang K, Xiao S, Yue Z (2018) Copula-based decomposition approach for the derivative-based sensitivity of variance contributions with dependent variables. Reliab Eng Syst Saf 169:437–450
10. Akaike H (1974) A new look at the statistical model identification. IEEE Trans Autom Control 19:716–723
11. Schwarz G (1978) Estimating the dimension of a model. Ann Stat 6:15–18
12. Jiang SH, Li DQ, Cao ZJ, Zhou CB, Phoon KK (2014) Efficient system reliability analysis of slope stability in spatially variable soils using Monte Carlo simulation. J Geotech Geoenvironmental Eng 141:04014096
13. Zhang L, Ding L, Wu X, Skibniewski MJ (2017) An improved Dempster-Shafer approach to construction safety risk perception. Knowl Based Syst 132:30–46
14. Aladejare AE, Wang Y (2018) Influence of rock property correlation on reliability analysis of rock slope stability: from property characterization to reliability analysis. Geosci Front 9:1639–1648
15. Yang Y, Zhou Q, Li H, Huang X, Tu X (2013) Analysis of face stability during excavation of double-O-tube shield tunnel. Sci World J, 781968
16. Kasper T, Meschke G (2006) On the influence of face pressure, grouting pressure and TBM design in soft ground tunnelling. TunnUndergr Space Technol Inc Trenchless Technol Res 21:160–171
17. Zhang L, Wu X, Ding L, Skibniewski MJ, Yan Y (2013) Decision support analysis for safety control in complex project environments based on Bayesian Networks. Expert Syst Appl 40:4273–4282
18. Nelsen RB (2006) An introduction to copulas (Springer series in statistics)
19. Lü Q, Low BK (2011) Probabilistic analysis of underground rock excavations using response surface method and SORM. Comput Geotech 38:1008–1021
20. Wang Y (2011) Reliability-based design of spread foundations by Monte Carlo simulations. Géotechnique 61:677–685
21. Wang Y (2009) Reliability-based economic design optimization of spread foundations. J Geotech Geoenvironmental Eng 135:954–959

Intelligent TBM Operations

Tunneling Equipment Selection

1 Introduction

With the advancement of underground construction technology, there are now various tunnel excavation equipment options available, such as hard-rock tunnel boring machine (TBM), shield TBM, and New Austrian tunneling method. Each type of equipment has distinct characteristics suited to specific scenarios. The selection of tunneling equipment is a complex, multi-faceted issue that involves several factors, including technical feasibility, environmental adaptability, economic considerations, and compatibility with the construction team [1]. The selection of tunneling equipment is essential for tunnel projects, since the inappropriate tunneling machine may reduce the efficiency, cause faults, and even may lead to casualties. Moreover, the modification of tunneling equipment is very time-consuming and costly. In order to maximize profits and mitigate risks in tunneling construction projects, the application of multi-criteria decision-making (MCDM) techniques that integrate and evaluate varied factors would be beneficial in the tunneling equipment selection.

MCDM is a branch of operational research methods to solve decision problems with multiple criteria or attributes. The technique of order preference similarity to the ideal solution (TOPSIS) is a popular MCDM method that is used frequently in decision-making problems with group decision-makers and multiple conflicting criteria. TOPSIS determines "distances" of alternatives to the best and worst cases and conducts trade-offs of the distances to the extreme cases. This trade-off aims to minimize the likelihood that the selected option is particularly weak in any aspect, thereby reducing the risk of accidents or project delays during tunnel construction due to weaknesses in specific areas of the tunneling equipment.

In the conventional TOPSIS method, the analysis is implemented based on the crisp numbers. However, for problems like tunneling equipment selection, due to the absence of quantitative evaluation indicators, experts tend to give evaluations in linguistic languages or subjective scores [2]. To solve the issues, the fuzzy set theory with the ability to represent linguistic concepts by fuzzy models with quantitative

L. Zhang et al., *Intelligent Construction in Tunnels*, Lecture Notes in Civil Engineering 679, https://doi.org/10.1007/978-981-96-8785-5_6

variables is introduced to TOPSIS. In previous research, many types of fuzzy models have been adopted to the TOPSIS method, for instance, triangular fuzzy model, trapezoidal fuzzy set, and interval-valued intuitionistic fuzzy set [3]. In these traditional fuzzy logic models, linguistic concepts will be transferred into pre-defined membership functions, and several calculation rules will be defined. In this case, only some simple calculation processes can be applied to the membership functions, which is not flexible. In addition, the belongingness of a value to a linguistic concept is deterministic. However, variance exists in the evaluation standards for different experts. A stochastic membership model is required to consider the divergence between experts.

Among various kinds of fuzzy models, the cloud model is an emerging approach that considers the stochastic fluctuation of membership. It introduces the probability theory to the Gaussian fuzzy set [4]. A new parameter called Hyperentropy (*He*) is defined to capture the variation of different interviewees with fluctuation. The cloud model produces a series of random numbers exhibiting a consistent tendency, effectively capturing the fuzziness and randomness of qualitative concepts when an element from the theoretical domain is inputted [5]. The introduction of the new parameter for randomness allows the decision-making process to more thoroughly consider the impact of uncertainty on the outcomes.

Inspired by the cloud model's effectiveness in addressing fuzziness and uncertainty, this chapter attempts to integrate the cloud model and TOPSIS to tackle the MCDM problem of tunneling equipment selection. This chapter presents an improved hybrid cloud-TOPSIS method under uncertainty. Cloud models are employed to transfer the fuzziness and randomness in the evaluation to quantitative values with random simulation. Unlike the traditional fuzzy logic sets, the cloud model can transfer the subjective evaluation into crisp values, rather than membership functions, which is more compatible with TOPSIS. The developed approach serves as a decision tool to provide insights into a better understanding of multi-source information fusion for alternative selection uncertainty with variations in the simulation and sensitivity analysis.

2 Multi-criteria Decision Making (MCDM)

For complex decision-making problems like tunneling equipment selection, alternatives usually have their pros and cons for certain evaluation criteria. When a decision-making task complies while assessing the performance of the considered alternatives against those criteria, it is termed multi-criteria decision-making (MCDM). MCDM techniques have been widely used for various purposes in the construction industry, for example, plan selection, site selection, and risk assessment. Challenges for MCDM problems in the construction industry include not only the numerous conflicting criteria but also the participation of multiple stakeholders with varying knowledge and interests. Therefore, the ability to balance opinions from diverse stakeholders is also required by MCDM techniques.

MCDM encompasses a variety of methods designed to aid decision-makers in evaluating and prioritizing alternatives. In a general MCDM method, information is usually collected and formatted into decision matrices with weights vectors as shown in Eq. (1). The goal of MCDM is to integrate evaluation values of all criteria for all alternatives and rank their favorability following certain algorithms. Traditional approaches, such as the Weighted Sum Model and Weighted Product Model, provide a straightforward means of aggregating criteria scores with given weights. However, they often fall short in handling uncertainty and non-linear relationships among criteria. Moreover, the simple weighted aggregation can sometimes obscure individual extreme deficiencies in options, making subsequent operations more susceptible to failure due to these critical flaws.

$$X = \begin{bmatrix} & C_1 & C_2 & \cdots & C_n \\ A_1 & x_{11} & x_{12} & \cdots & x_{1n} \\ A_2 & x_{21} & x_{22} & \cdots & x_{2n} \\ \vdots & \vdots & \vdots & \ddots & \vdots \\ A_m & x_{n1} & x_{n2} & \cdots & x_{mn} \end{bmatrix} \begin{bmatrix} W \\ w_1 \\ w_2 \\ \vdots \\ w_n \end{bmatrix} \quad (1)$$

where A_i $(i = 1, 2, \ldots, m)$ is the ith alternative options in an MCDM problem, C_j $(j = 1, 2, \ldots, n)$ is the jth selected criteria for the evaluation of options, w_j $(j = 1, 2, \ldots, n)$ is the assigned weight of the jth criterion, and x_{ij} is the score corresponding to the jth criterion of ith alternative.

To address the shortcomings of traditional MCDM, the Technique for Order Preference by Similarity to Ideal Solution (TOPSIS) has become one of the most popular techniques in recent years. This method ranks the preferences of options based on the relative distances of alternatives to the identified ideal solution. TOPSIS possesses the advantages of rationality, easy computation, sound logic, and clear visualization [6]. Compared with MCDM methods by human approach, such as the analytic hierarchy process and best–worst method, TOPSIS computes results by mathematical operations and shows higher objectivity and adequacy in the evaluation system [7]. Therefore, TOPSIS is selected as the MCDM technique to select the ultimate candidate.

A general TOPSIS analysis requires inputs of evaluation across selected attributes associated with different alternatives and weights based on the relative importance of attributes. The generic TOPSIS analysis procedures are presented as follows:

(1) Normalize the evaluation matrix.
(2) Assign the weights of criteria to the normalized matrix.
(3) Identify the positive ideal solution (PIS) and negative ideal solution (NIS) according to the weighted evaluation matrix.
(4) Determine the distances to PIS and NIS for each alternative.
(5) Determine the closeness coefficient with distances to PIS and NIS.
(6) Rank alternatives based on the closeness coefficients.

3 Tunneling Equipment Selection with Improved MCDM

In this research, a hybrid MCDM approach that integrates cloud model, TOPSIS, and Monte Carlo simulation is proposed to identify the optimal solution in decision making under uncertainty. Figure 1 illustrates the workflow of the proposed cloud-TOPSIS-based approach for tunneling equipment selection. Firstly, the evaluation framework of tunneling equipment is established by referring to literature and experiences. Then, evaluations from experts are collected and transferred to crisp values by the cloud model and simulation. TOPSIS is adopted to determine the ranking of various alternatives. Moreover, a sensitivity analysis is implemented to investigate the contributions of input factors. Those steps are elaborated as follows.

3.1 *Criteria Framework Formulation*

Initially, evaluation criteria are established for the tunneling equipment assessment problem. Typically, several common considerations inform the selection of appropriate excavation methods, including project cost, time constraints, site conditions, and local regulations. The decision-making for tunnel excavation projects usually possesses the characteristics of complex underground conditions, severe workspaces, and high impacts on the surrounding environment. For instance, Golestanifar et al. [8] introduced fuzzy-AHP and fuzzy-TOPSIS methods for evaluating excavation

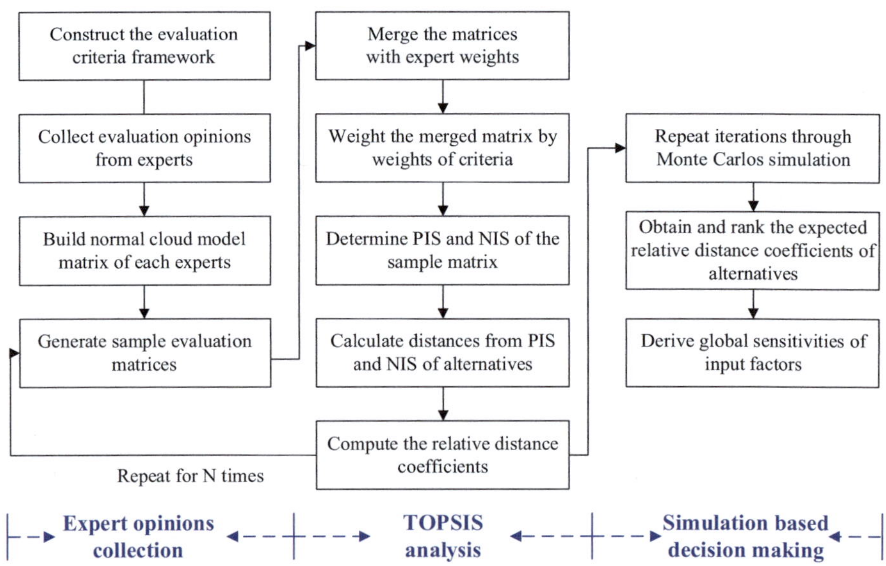

Fig. 1 Workflow of the proposed cloud-TOPSIS approach

technologies, addressing aspects such as tunnel design, geological and hydraulic conditions, operational management, safety concerns, and cost factors. Drawing on previous research and examining the project's nature, this study establishes a criteria framework, as detailed in Table 1. A total of 16 criteria are categorized into 4 groups which are technical adaptability, environmental adaptability, economical adaptability, and operational adaptability.

Table 1 List of evaluation criteria for tunneling equipment decision-making

Category	Criterion	Code	Cost/Benefit
Technical adaptability	Adaptability of excavation technology to tunnel shape	C_1	Benefit
	Adaptability of excavation technology to tunnel depth	C_2	Benefit
	Adaptability of excavation technology to working space	C_3	Benefit
	Adaptability of excavation technology to the construction schedule	C_4	Benefit
Environmental adaptability	Acceptance of subsoil condition to the excavation technology	C_5	Benefit
	Acceptance of seepage condition to the excavation technology	C_6	Benefit
	Impact of excavation technology on surface disturbance	C_7	Cost
	Impact of excavation technology on the existing railway	C_8	Cost
Economical adaptability	Acceptance of the construction company to the initial cost	C_9	Benefit
	Acceptance of the construction company to the operation cost	C_{10}	Benefit
	Fund utilization of excavation technology	C_{11}	Benefit
	Reliability of the excavation technology	C_{12}	Benefit
Operational adaptability	Supportability of excavation technology to the construction quality	C_{13}	Benefit
	Supportability of the excavation technology to the construction safety	C_{14}	Benefit
	Mastery of the construction company to the excavation technology	C_{15}	Benefit
	Acceptance of the policy to the excavation technology	C_{16}	Benefit

3.2 Conversion of Linguistic Concepts by the Cloud Model

It is difficult to objective indicators to measure the adaptability of equipment to tunneling projects. Therefore, usually, several experienced experts are invited to provide evaluations for criteria. Linguistic languages are usually used to express opinions by experts, which often involve several different levels, such as "poor", "fair", and "good". Fuzziness occurs due to the divergences in the individual perspectives of experts in each evaluation grade.

The cloud model, proposed by Li et al. [9], is a combination of probability theory and the fuzzy theory. Similar to other fuzzy sets, the cloud model is able to transform concepts between qualitative and quantitative forms. In the cloud model theory, given a domain of discourse U, there is a qualitative concept T corresponding to U. For $x \in U$ as a numerical instantiation (also called cloud drop) of T, its degree of belongingness to T is $y = (Ex, En, He)$, which is the standard form of the cloud model. Ex (Expectation) is the expected value of the numerical instantiation of the qualitative conception T. En (Entropy) is the uncertainty measurement of the cloud model, which is determined by the randomness and the fuzziness of the qualitative concept. He (Hyperentropy) is the uncertain degree of En, which quantifies the uncertainty that is not considered by En.

Considering the wide applicability of normal distribution when describing the phenomenon, the normal distribution function and the Gaussian fuzzy set are adopted frequently to the cloud model to form the normal cloud model. A normal cloud model can be considered as a bunch of Gaussian membership functions that can be presented as Eq. (2)

$$y = e^{-\frac{(x-Ex)^2}{2En'^2}} \tag{2}$$

where x is a random instantiation of the subjective linguistic concept, Ex is the expectation of x, and En' is a set of random samples generated from the normal distribution $N(En, He^2)$.

The cloud model transfers the qualitative concepts into the quantitative values by generating random instantiation x with the forward normal cloud generator. The detailed process of the forward normal cloud generator is presented in Algorithm 1.

Algorithm 1 Forward normal cloud generator

Input: Characteristics values of the cloud model (Ex, En, He), and the number of droplets
Output: Cloud model with N drops and their memberships
Steps:
1. Generate a sample En_i' from the normal distribution with the expectation En and standard deviation He, i.e. $En_i' \sim N(En, He^2)$;
2. Generate a sample cloud drop x_i from the normal distribution with expectation Ex and standard deviation En_i', i.e. $x_i \sim N(Ex, En_i'^2)$;

(continued)

(continued)

3. Calculate the membership of x_i by Eq. 1, i.e., $y_i = e^{-\frac{(x_i-Ex)^2}{2En_i'^2}}$
4. Repeat steps 1-3 to obtain N cloud drops

A 5-level evaluation scale is adopted in this chapter, given by $L = (l_1 =$ very low, $l_2 =$ low, $l_3 =$ medium, $l_4 =$ high, $l_5 =$ very high). For each evaluation level, a corresponding cloud model will be induced. As a result, the five cloud models corresponding to the five evaluation levels can be expressed as (Ex_1, En_1, He_1), (Ex_2, En_2, He_2), (Ex_3, En_3, He_3), (Ex_4, En_4, He_4), and (Ex_5, En_5, He_5), respectively.

Referring to [10], the principle of the golden section ratio is commonly used to define the upper and lower bounds of the standard score interval. The membership functions of the 5-level cloud models can be derived by applying the principle of the golden section ratio, which is given by Eq. (3).

$$\begin{cases} Ex_1 = x_{\min} \\ Ex_2 = (1 - 0.382)(x_{\min} + x_{\max})/2 \\ Ex_3 = (x_{\min} + x_{\max})/2 \\ Ex_4 = (1 + 0.382)(x_{\min} + x_{\max})/2 \\ Ex_5 = x_{\max} \\ En_2 = En_4 = 0.382(x_{\max} - x_{\min})/6 \\ En_3 = En_2/0.618 \\ En_1 = En_5 = En_2/0.618 \\ He_3 = He' \\ He_2 = He_4 = He_3/0.618 \\ he_1 = He_5 = He_2/0.618 \end{cases} \tag{3}$$

where $x_{\min}$ and $x_{\max}$ are the globally pre-defined lower and upper bounds of the evaluation criteria, respectively, and He' is the pre-defined hyperentropy of l_3.

Generally, the values of the criteria would be standardized before the evaluation. The lower and upper bounds of each evaluation criterion could be standardized to be 0 and 1, respectively. The 5-level evaluation scale can be obtained by using Eq. (3). Table 2 presents the parameters of cloud models for the 5-level evaluation scale. During the transformation process from the linguistic terms to the cloud models, special treatments should be given to the start and end cloud models. To control the representation of linguistic terms within an interval [0, 1], the start cloud model l_1 (corresponding to the linguistic term "Very low") belongs to a half-up cloud model, while the end cloud model l_5 (corresponding to the linguistic term "Very high") both belong to the half-down cloud model. Figure 2 illustrates the corresponding plots of the 5 cloud models.

Table 2 The 5-level evaluation scale and corresponding cloud models

Code	Linguistic term	Cloud model
l_1	Very low	(0, 0.103, 0.013)
l_1	Low	(0.309, 0.064, 0.008)
l_1	Medium	(0.5, 0.039, 0.005)
l_1	High	(0.691, 0.064, 0.008)
l_1	Very high	(1, 0.103, 0.013)

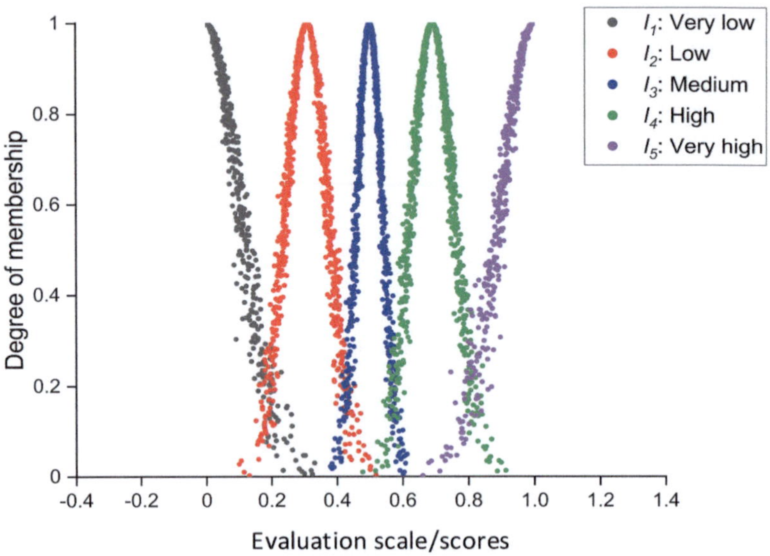

Fig. 2 Cloud model plots of the 5-level evaluation scales/scores

In a decision-making problem, supposing there are m alternatives A_i ($i = 1, 2, \ldots, m$) and n criteria C_j ($j = 1, 2, \ldots, n$), l experts E_k ($k = 1, 2, \ldots, l$) are involved in the assessment committee. Denote the linguistic evaluation matrix of the kth expert as $L^k = \left(l_{ij}^k\right)_{m \times n}$, where l_{ij}^k represents his/her grading opinion towards the jth criteria of the ith alternative. These linguistic matrixes can be transformed into the cloud model matrix by converting all the elements l_{ij}^k to the corresponding cloud models $\left(Ex_{ij}^k, En_{ij}^k, He_{ij}^k\right)$ following Eq. (3). For example, the evaluation of the kth expert to the jth criterion of the ith alternative is "high". Then the corresponding cloud model of the element l_{ij}^k is (0.691, 0.064, 0.008). The converted cloud model matrix of the kth expert is expressed in Eq. (4). After the conversion from qualitative concepts to quantitative models is completed, the forward cloud generator will be implemented through random sampling, to obtain sample matrices formed by pure numbers.

$$L^k = \begin{bmatrix} \left(Ex_{11}^k, En_{11}^k, He_{11}^k\right) & \left(Ex_{12}^k, En_{12}^k, He_{12}^k\right) & \cdots & \left(Ex_{1n}^k, En_{ij}^k, He_{ij}^k\right) \\ \left(Ex_{21}^k, En_{21}^k, He_{21}^k\right) & \left(Ex_{22}^k, En_{22}^k, He_{22}^k\right) & \ldots & \left(Ex_{2n}^k, En_{2n}^k, He_{2n}^k\right) \\ \vdots & \vdots & \ddots & \vdots \\ \left(Ex_{m1}^k, En_{m1}^k, He_{m1}^k\right) & \left(Ex_{m2}^k, En_{m2}^k, He_{m2}^k\right) & \cdots & \left(Ex_{mn}^k, En_{mn}^k, He_{mn}^k\right) \end{bmatrix} \quad (4)$$

3.3 Improved TOPSIS for Alterative Evaluation

The selection process in TOPSIS is based on the principle that the optimal alternative has the relatively shortest distance from the PIS and the largest distance from the NIS, where positive solutions maximize the beneficial factors while negative ones maximize the cost criterion. Traditional TOPSIS methods, however, do not account for uncertainty. This study addresses this gap by incorporating the cloud model to incorporate uncertainty in evaluations, utilizing a forward cloud model generator to process expert opinions. This algorithm will produce sample deviations *En'* for each cloud model in the evaluation matrices as shown in Eq. (4). Sample ratings corresponding to cloud models are generated from the normal distribution $N\,(Ex, En')$. Sample evaluation matrices formed by sample ratings are established, which are in the form of a crisp number and suitable for TOPSIS analysis. The subsequent steps of the TOPSIS process, based on the sample evaluation matrices, are outlined in detail below.

(1) Aggregating the matrices of the evaluation group

The group decision-making matrices of the kth experts will be aggregated together firstly for the subsequent operation. Considering the variance in the characteristics, such as professional title and work experiences, the weights index will be assigned to evaluation matrices to reflect the relative importance of experts. A scoring table based on professional title and work experience for expert weight allocation is shown in Table 3. Referring to the scoring system, the scores of kth assessment members can be calculated as H^k. The weighted index of the kth expert λ^k can be determined by Eq. (5). The sample evaluation matrix of the kth expert is denoted as $X^k = \left(x_{ij}^k\right)_{m\times n}$. The aggregated sample matrix of the evaluation group X is the summation of the individual evaluation matrices with the corresponding expert weights as given by Eq. (6).

$$\lambda^k = \frac{H^k}{\sum_{k=1}^{l} H^k} \quad (5)$$

$$X = \sum_{k=1}^{l} \lambda^k X^k \quad (6)$$

Table 3 Expert weightage allocation system

Aspect	Class/Level	Score
Professional title	Senior	5
	Sub-senior	4
	Intermediate	3
	Associate	2
	Junior	1
Working experience	Over 20 years	4
	10–19 years	3
	5–9 years	2
	Under 5 years	1

(2) Assigning the weight of criteria

The design of criterion weights is an important step in TOPSIS. Subjective weight methods, such as the Delphi method, expert survey method, and AHP, are usually used to determine the weights of attributes. However, these methods can result in divergent attribute weights due to individual biases. Furthermore, they necessitate additional evaluative input from experts, and the inherent fuzziness and randomness in weight assessments may compromise the analysis of uncertainty in criterion evaluations. In contrast, objective weighting methods derive weights from the inherent characteristics of the attributes, effectively minimizing subjective influences. Therefore, the entropy weight method as an objective weight method is applied in the proposed approach. The entropy weight approach as a data-driven weighting method is appropriate for decision-making problems with qualitative judgments. This method assigns attribute weights based on the information entropy of attributes, where the information entropy is a concept to measure the disorder degree of the information. The principle of this method assumes that a smaller disorder degree of the system will lead to a smaller entropy. Thus, criteria with lower entropy values are considered to convey more significant information, resulting in higher assigned weights. The detailed procedure for applying the entropy-weighted method within the TOPSIS framework is outlined as follows:

The standardization of the elements is firstly conducted to all of the evaluation scores into one dimension, which is implemented by calculating the proportion of the ith alternative in the summation scores of all the alternatives for the jth criterion, denoted as p_{ij}, as shown by Eq. (7). The entropy of the jth criterion is then calculated with these proportions as shown in Eq. (8). Finally, the weight of entropy of the ith indicator can be defined based on the entropy, as given by Eq. (9). In this way, the criteria with a higher discrimination degree will be assigned to a larger weight.

$$p_{ij} = \frac{x_{ij}}{\sum_{i=1}^{m} x_{ij}} \tag{7}$$

$$e_j = -\frac{1}{\ln m}\sum_{i=1}^{m} p_{ij}\ln\left(p_{ij}\right) \tag{8}$$

$$w_j = \frac{1-e_j}{\sum_{j=1}^{n}\left(1-e_j\right)} \tag{9}$$

where x_{ij} is the aggregated sample score of the jth criterion of the ith alternative, m is the total number of alternatives, and n is the total number of criteria.

The weight vector of the criteria is given by $w = \{w_j|j = 1, 2, \ldots, n\}$. The weighted evaluation matrix is denoted as $V = \left(v_{ij}\right)_{m\times n}$, where V is calculated by assigning the corresponding weight to the elements as presented in Eq. (10).

$$V = \begin{bmatrix} v_{11} & v_{12} & \cdots & v_{1n} \\ v_{21} & v_{22} & \cdots & v_{2n} \\ \vdots & \vdots & \ddots & \vdots \\ v_{m1} & v_{m2} & \cdots & v_{mn} \end{bmatrix} = \begin{bmatrix} w_1x_{11} & w_2x_{12} & \cdots & w_nx_{1n} \\ w_1x_{21} & w_2x_{22} & \cdots & w_nx_{2n} \\ \vdots & \vdots & \ddots & \vdots \\ w_1x_{m1} & w_2x_{m2} & \cdots & w_nx_{mn} \end{bmatrix} \tag{10}$$

where w_j is the weight of the jth criterion and x_{ij} is the aggregated sample score of the jth criterion of the ith alternative.

(3) Identifying PIS and NIS of alternatives

PIS and NIS will be determined based on the weighted matrix. In this step, the criteria should be classified into two categories: benefit criteria J_1 and cost criteria J_2. The PIS is supposed to contain the highest values of the benefit criteria and lowest values of the cost criteria, and conversely for NIS. PIS and NIS are represented by Eqs. (11) and (12), respectively.

$$\begin{aligned} PIS = A^+ &= \left\{(\max_i v_{ij}|j \in J_1), (\min_i v_{ij}|j \in J_2)\right\} \\ &= v_1^+, v_2^+, \ldots v_j^+ \ldots, v_n^+ \end{aligned} \tag{11}$$

$$\begin{aligned} NIS = A^- &= \left\{(\min_i v_{ij}|j \in J_1), (\max_i v_{ij}|j \in J_2)\right\} \\ &= v_1^-, v_2^-, \ldots, v_j^-, \ldots, v_n^- \end{aligned} \tag{12}$$

where A^+ and A^- are the positive and negative ideal solutions of alternatives, respectively, and v_{ij} is the weighted sample score of the jth criterion of the ith alternative. v_j^+ and v_j^- are the PIS and NIS of the jth criterion, respectively.

(4) Calculating the expected closeness coefficients

The alternatives are ranked based on the relative distance to PIS and NIS. The distance of the ith alternative to PIS is denoted as d_i^+, and the distance to NIS is denoted as d_i^-, which are given by Eqs. (13) and (14), respectively. The closeness coefficient CC_i of the alternative A_i respect to PIS is calculated by Eq. (15).

$$d_i^+ = \sqrt{\sum_{j=1}^{n}\left(v_{ij} - v_j^+\right)^2} \tag{13}$$

$$d_i^- = \sqrt{\sum_{j=1}^{n}\left(v_{ij} - v_j^-\right)^2} \tag{14}$$

$$CC_i = \frac{d_i^-}{d_i^+ + d_i^-} \tag{15}$$

where v_{ij} is the simulation weighted sample score of the jth criterion of the ith alternative v_j^+ and v_j^- are the PIS and NIS of the jth criterion, respectively.

3.4 Decision-Making Under Uncertainty

Subjectivity in human judgment introduces inherent fuzziness and randomness that are often unavoidable. The uncertainties present within the linguistic constructs of an evaluation system can be effectively characterized using qualitative values through the cloud model. By employing characteristic values such as expectation, entropy, and hyperentropy, these concepts can be quantitatively understood and manipulated. To fully leverage the uncertainty parameters of cloud models and integrate them with TOPSIS, Monte Carlo simulation is utilized to sample data from the cloud model evaluation matrices. Given the high adaptability of Gaussian membership [4] and the well-developed normal cloud model theory, a normal cloud model based on Gaussian membership is applied in this research. The probability distribution function of the normal distribution is given by Eq. (16).

$$f(x, \mu, \sigma) = \frac{1}{\sigma\sqrt{2\pi}} e^{-\frac{(x-\mu)^2}{2\sigma^2}} \tag{16}$$

where μ is the mean of the distribution and σ is the standard deviation.

In a cloud model, *Ex* is the mean of the instantiation for the fuzzy membership function, and *En* is the expectation uncertainty measurement of the model. The value representing fuzziness and randomness is assumed to satisfy the normal distribution, with the mean of *En* and the deviation of *He*. The cloud drops can be obtained by the forward cloud generator through Algorithm 1. The conditional probability distribution function of a normal cloud model is obtained by Eq. (17), and the probability distribution function of En' is represented by Eq. (18).

$$f\left(x|En'\right) = \frac{1}{|En'|\sqrt{2\pi}}e^{-\frac{(x-Ex)^2}{2En'^2}} \tag{17}$$

$$f\left(En'\right) = \frac{1}{He\sqrt{2\pi}}e^{-\frac{(x-En)^2}{2He^2}} \tag{18}$$

where Ex is the expectation of the cloud model, En is the entropy which is the expectation of the deviation of the cloud model, and He is the hyperentropy of the cloud model.

The cloud drops can be regarded as the sample scores of expert evaluations. The converted numerical evaluation of the cloud model $y_{ij}^k\left(Ex_{ij}^k, Ex_{ij}^k, Ex_{ij}^k\right)$ is denoted by x_{ij}^k, where x_{ij}^k is sampled by generating a normally distributed random number $En_{ij}^k\prime$ with expectation En_{ij}^k and variance He_{ij}^2, which follows $\mathrm{En}_{ij}^{k'} \sim N\left(En_{ij}^k, He_{ij}^k\right)$. Then, a normally distributed random number x_{ij}^k is generated with expectation Ex and variance $En_{ij}^{k'}$, which follows $x_{ij}^k \sim N\left(Ex_{ij}^k, En_{ij}^k\right)$. The generated sample matrix of the kth expert can be expressed by $X^k = \left(x_{ij}^k\right)_{n\times m}$.

For each simulation, a set of sample matrices $\left(X^1, X^2, \ldots, X^l\right)_h$ ($h = 1, 2, \ldots, H$) will be generated, where H is the total number of simulations, and there is a corresponding output closeness coefficient CC_i^h for each simulation. By repeating operations, thousands of closeness coefficients corresponding to the sample evaluation matrices will be obtained. In this way, the mean value and the distribution of these coefficients can be obtained. The finalized result is given by Eq. (19).

$$CC_i^f = \frac{1}{H}\sum_{h=1}^{H} CC_i^h \tag{19}$$

where $CC_i{}^h$ is the closeness coefficient of the ith alternative in the hth iteration. Finally, all the alternatives will be ranked by comparing the coefficients, where the one with the largest $CC_i{}^f$ will be considered as the optimal solution.

Upon completion of the TOPSIS decision-making process, a sensitivity analysis is conducted. The primary objective of this sensitivity analysis is to examine the effects of minor variations in model inputs on the outputs. Additionally, the contribution of criteria to the final result will be revealed by comparing the measurement of sensitivities, which can help the weight assignment and identify critical criteria. In the proposed method, the global sensitivities of the optimal alternative determined in the last step to changes of the sample evaluation scores of criteria are calculated. Spearman's rank correlation coefficient is adopted to measure sensitivities. The sensitivity of the jth criterion to the closeness coefficient of the ith alternative, denoted by $GSA(C_j)$, is given by Eq. (20).

$$GSA(C_j) = \frac{\sum_{h=1}^{H}\left(R\left(x_{ij}^h\right) - \overline{R}\left(x_{ij}^h\right)\right)\left(R\left(CC_i^h\right) - \overline{R}\left(CC_i^h\right)\right)}{\sqrt{\sum_{h=1}^{H}\left(R\left(x_{ij}^h\right) - \overline{R}\left(x_{ij}^h\right)\right)^2}\sqrt{\sum_{h=1}^{H}\left(R\left(CC_i^h\right) - \overline{R}\left(CC_i^h\right)\right)^2}} \tag{20}$$

where H is the total number of iterations, h stands for the hth iteration, $R\left(x_{ij}^h\right)$ is the rank of the x_{ij}^h among all the sample evaluations of the jth criterion, $\overline{R}\left(x_{ij}^h\right)$ is the mean value of $R\left(x_{ij}^h\right)$, $R\left(CC_i^h\right)$ is the rank of CC_i^h among all the closeness coefficients of the jth criterion, and $\overline{R}\left(CC_i^h\right)$ is the mean value of $R\left(CC_i^h\right)$.

4 Experimental Studies

4.1 Data and Resources

Part of the underground project located in the Wuhan Metro Line 2A tunnel project in China is selected for demonstration and validation in this study. Specifically, data from the Zhongshan Park-Xunlimen (Z-X) tunnel section is collected. Details of the studied zone are presented in Fig. 3. The Z-X tunnel section is a double-track tunnel going under a crowded residential area, with the left line covering 1019 m, and the right line covering 999 m. The spacing between the two lines varies from 9 to 16 m and the minimum railway curve radius reaches 300 m. The varying gap and sharp curvature of the tunnel significantly increase the difficulty of the construction. Additionally, the ground elevation in this area ranges from 19.5 to 31.5 m with a maximum longitudinal gradient of 28.3°. The soil in this area consists of artificial fill, clays, dense fine sands, dense coarse sands, and gravels, distributing from top to bottom. Such complex underground condition also causes hazards to the construction process. Therefore, the Z-X tunnel section is particularly studied as a critical case.

According to the site condition and the design requirement of this project, a total number of 5 alternative excavation methods are identified, including Hard-rock TBM (A_1), New Austrian tunneling method (A_2), the mine tunneling method (A_3), Shield TBM (A_4), and Soft-rock TBM (A_5). The information and characteristics of the 5 alternatives are listed in Table 4.

4.2 Model Development

Five experts in the field of tunnel excavation were consulted to evaluate these 16 criteria based on 16 criteria in Table 1. The linguistic evaluations of the 5 experts are first converted to a cloud model. The cloud model evaluation matrices of the experts

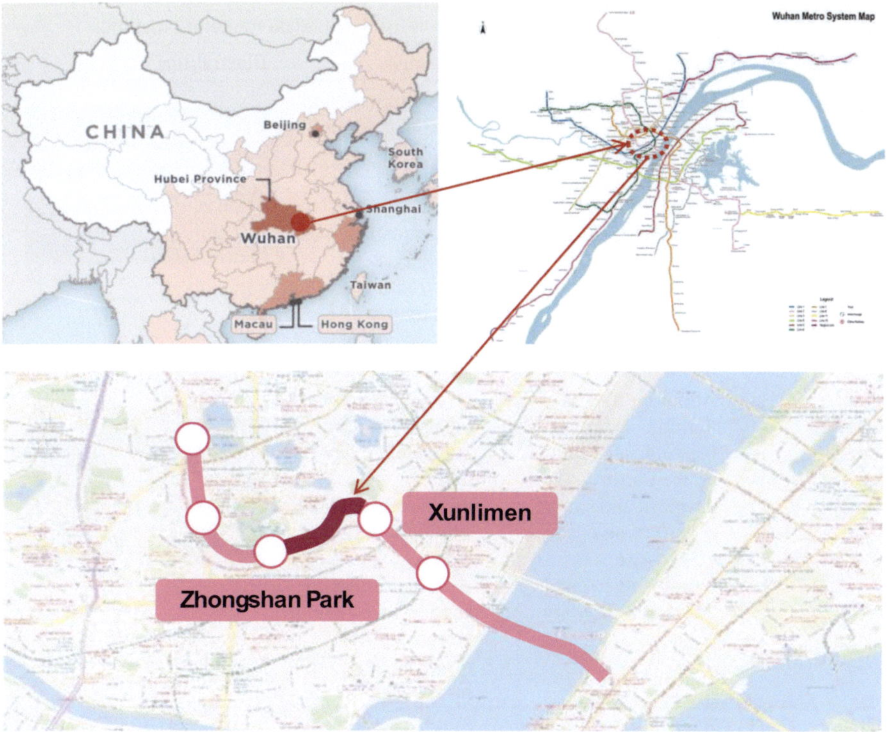

Fig. 3 Map of the Zhongshan Park-Xunlimen section of the Wuhan Metro Line 2

for the 5 alternatives are constructed as $Y^k = \left(y_{ij}^k\right)_{5\times 16}$. Then, sample data will be generated from each cloud model by the forward cloud generator to establish the sample evaluation matrix. This process is repeated 1000 times for each evaluation by experts to model the uncertainty carried by the linguistic languages.

After the sample matrices of one iteration are prepared, TOPSIS analysis is implemented. Since the TOPSIS method is conducted on a single evaluation matrix, the matrices of different experts are required to be aggregated. The weights are allocated to the members of the evaluation team for the aggregation according to Table 3. The aggregation results (x_{ij}) are computed by Eqs. (5) and (6). Considering those criteria in the same evaluation framework will carry unequal weightage in the decision-making process due to the importance of reliability issues, the entropy method is implemented to determine the weight of each criterion following Eqs. (7–9). The entropy-based weights obtained from one iteration are presented in Table 5.

Later on, the entropy-based weights are applied to the criteria matrix. Then, PIS and NIS can be obtained following Eqs. (11) and (12). The distances of each alternative to PIS ($d_i{}^+$) and NIS ($d_i{}^-$) are calculated by Eqs. (13) and (14), respectively, and the closeness coefficients (CC_i) are determined by combing $d_i{}^+$ and $d_i{}^-$ using Eq. (15), and the results of one iteration are listed in Table 6. Generally, the higher

Table 4 Advantages, disadvantages, and illustrations of 5 candidate tunneling methods

Alternatives	Advantages	Disadvantages	Illustrations
Hard-rock TBM (A_1)	(1) High excavation rate in hard rocks; (2) Environment-friendly; (3) Energy saving and high efficiency; (4) High security and reliability	(1) High purchase and maintenance cost; (2) sensitive stratum; (3) long equipment procurement cycle, auxiliary construction preparation; (4) poor adaptability of TBM face	
New Austrian tunneling method (A_2)	(1) Adopting flexible lining to control blasting, speeding up the construction progress and saving construction costs; (2) On-site monitoring to ensure construction safety	(1) Over-excavation or under-excavation may happen due to the poor measurement and irrational design of blasting; (2) Artificial initial support takes a long period of time, resulting in a long process interval	
Mine tunneling method (A_3)	(1) Excavating with segments to ensure safety; (2) High economic efficiency	(1) Surface settlement is difficult to control; (2) Low waterproof; (3) Higher risk probability than shield TBM	
Shield TBM (A_4)	(1) High safety and reliability; (2) High degree of mechanization; (3) Good working environment; (4) Less earthwork, fast progress, low construction costs, etc.	(1) Difficult to guarantee the stability of the excavation face in the section with the uneven stratum, hard rocks, and rich groundwater; (2) Low utilization rate of the equipment; (3) High risk of construction delay; (4) High additional cost	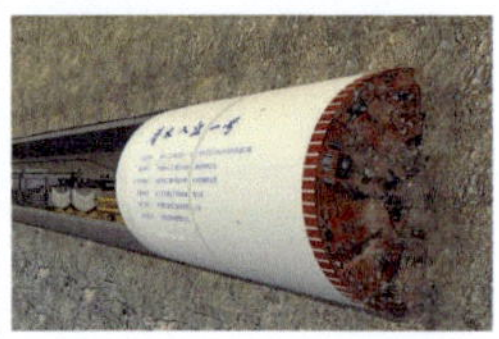

(continued)

Table 4 (continued)

Alternatives	Advantages	Disadvantages	Illustrations
Soft-rock TBM (A_5)	(1) Complete equipment for crushing and transporting rocks; (2) Long service life and high continuous construction ability	(1) High purchase and maintenance cost; (2) sensitive stratum; (3) long equipment procurement cycle, auxiliary construction preparation	

Table 5 Weights of criteria determined by the entropy method

Criterion	C_1	C_2	C_3	C_4	C_5	C_6	C_7	C_8
Weight	0.062	0.062	0.062	0.062	0.069	0.065	0.066	0.062
Criterion	C_9	C_{10}	C_{11}	C_{12}	C_{13}	C_{14}	C_{15}	C_{16}
Weight	0.061	0.061	0.062	0.061	0.061	0.061	0.062	0.062

closeness coefficient value implies that the alternative is closer to the PIS, which is more favorable in decision making. The preferences of alternatives in one iteration are ranked in a descending order of the coefficients.

4.3 Analysis of Results

In this case study, a 1000-iteration simulation is implemented to model the possible MCDM process. Finally, the mean value of the 1000 iterations is selected as the expected result, where the mean values of processed evaluation scores, PIS, and NIS are plotted in Fig. 4. The finalized closeness coefficients of each alternative are calculated by Eq. (19), which has been presented in Table 7. In the real situation, due to the uncertainty in the qualitative evaluations, the true ranking result may vary from the expected results in the extreme case. Therefore, this research explores the variations between simulation iterations. The distances to PIS and NIS of alternatives for the 1000-step run are plotted in Fig. 5. The counts of rankings for alternatives according to the simulation results are plotted in Fig. 6. Hypothesis test is implemented to validate the reliability of the expected result, as shown in Table 8. To examine the contributions of the criteria to the outcomes, sensitivity analysis based on Spearman's rank correlation coefficient of A_4 (CC_4) is performed, and the results are shown in Fig. 7. The detailed analysis of the results in the aspects of expected rankings, uncertainty modeling, and sensitivity analysis is presented below.

(1) The preferences of the alternatives are ranked as $A_4 > A_3 > A_5 > A_1 > A_2$, where A4, the shield TBM method is considered to be the optimal option. This is

Table 6 Weighted sample evaluation matrix $V = (v_{ij})_{m \times n}$ of one iteration

Alternative	Technical adaptability				Environmental adaptability			
	C_1	C_2	C_3	C_4	C_5	C_6	C_7	C_8
A_1	0.036	0.033	0.035	0.030	0.017	0.021	0.017	0.040
A_2	0.024	0.018	0.020	0.020	0.014	0.015	0.019	0.031
A_3	0.040	0.036	0.040	0.031	0.058	0.044	0.050	0.026
A_4	0.036	0.035	0.035	0.039	0.067	0.055	0.047	0.029
A_5	0.032	0.034	0.032	0.033	0.037	0.036	0.044	0.025
PIS	0.040	0.036	0.040	0.039	0.067	0.055	0.017	0.025
NIS	0.024	0.018	0.020	0.020	0.014	0.015	0.050	0.040
Alternative	*Economical adaptability*				*Social adaptability*			
	C_9	C_{10}	C_{11}	C_{12}	C_{13}	C_{14}	C_{15}	C_{16}
A_1	0.046	0.043	0.038	0.036	0.034	0.041	0.036	0.033
A_2	0.038	0.044	0.037	0.037	0.041	0.043	0.029	0.026
A_3	0.046	0.036	0.045	0.046	0.028	0.044	0.046	0.031
A_4	0.052	0.044	0.033	0.039	0.039	0.028	0.041	0.038
A_5	0.038	0.035	0.028	0.051	0.042	0.037	0.032	0.031
PIS	0.052	0.044	0.045	0.051	0.042	0.044	0.046	0.038
NIS	0.038	0.035	0.028	0.036	0.028	0.028	0.029	0.026

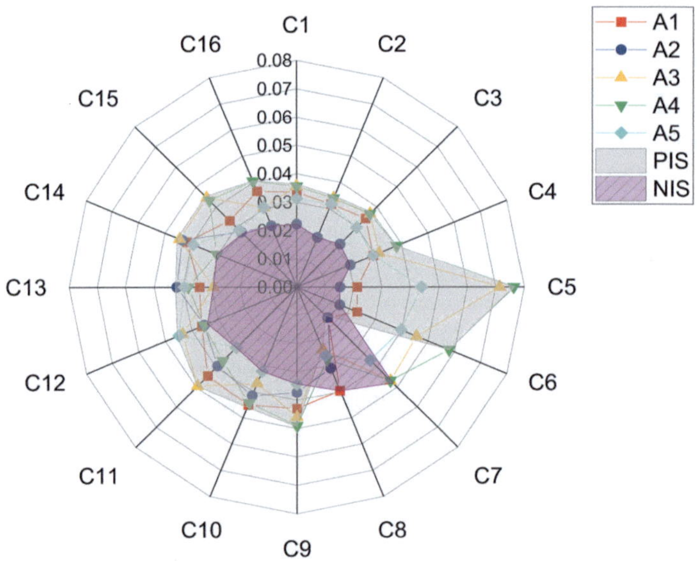

Fig. 4 Radar plot of the mean value of the mean weighted evaluation matrix, PIS, and NIS

Table 7 Finalized results of the case study throughout 1000 iterations

Alternative	CC_i	Standard deviation	Rank
A1	0.416	0.0245	4
A2	0.320	0.0207	5
A3	0.656	0.0239	2
A4	0.684	0.0206	1
A5	0.463	0.0258	3

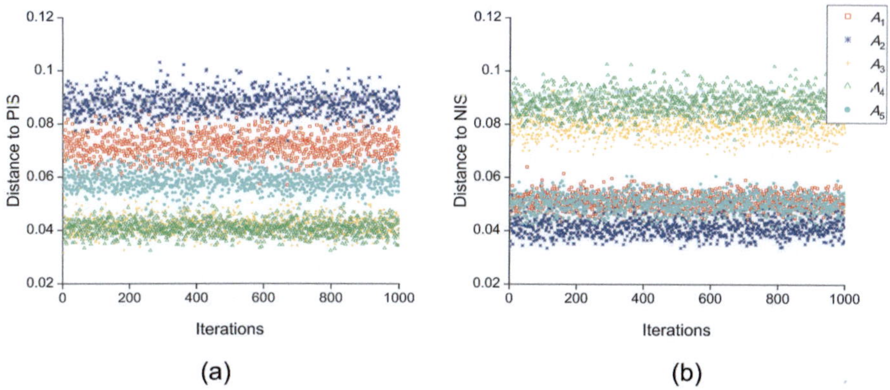

Fig. 5 Scatter plot of distances of alternatives to the **a** positive ideal solutions and **b** negative ideal solutions in the 1000-step simulations

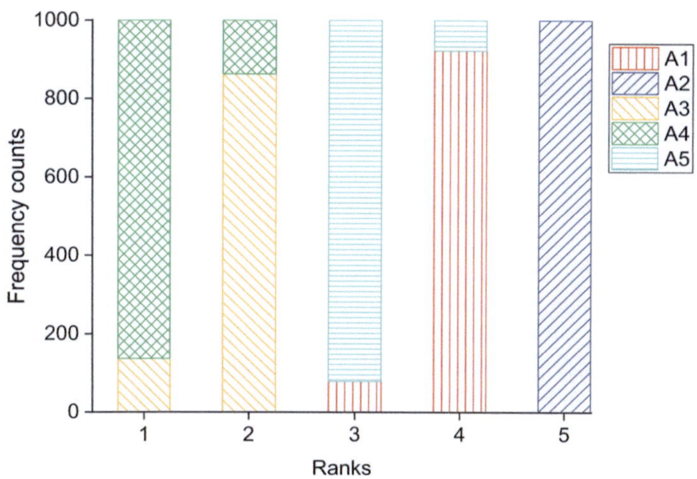

Fig. 6 Counts of rankings of alternatives in the simulations

Table 8 Paired *t*-test of closeness coefficients of A_3 and A_4

Alternative	N	Mean	Standard deviation	Standard error
A3	1000	0.656	0.024	7.575×10^{-4}
A4	1000	0.684	0.021	6.520×10^{-4}
Difference	1000	− 0.028	0.025	7.905×10^{-4}
t statistics	Degree of freedom	*p*		
35.37	999	0		

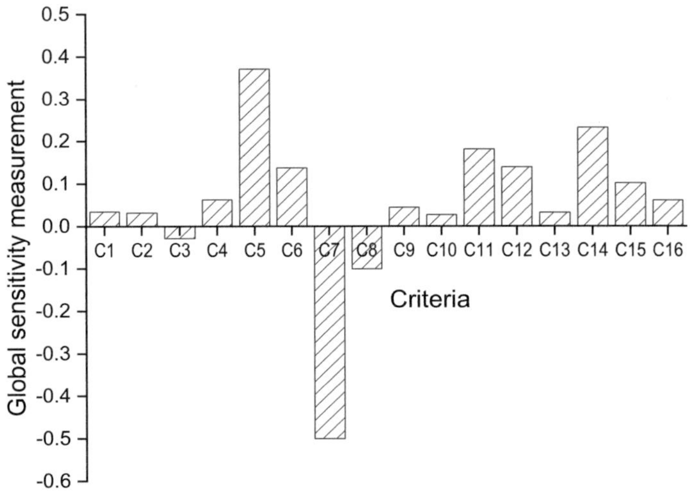

Fig. 7 Global sensitivity measurement of criteria to the closeness coefficient of A_4

consistent with the real situation that shield TBM is selected as the excavation method. As presented in Table 7, the closeness coefficient of A_4 is 0.684, which is slightly higher than A_3 in the second place. The result indicates that A_4 is the optimal solution. As observed from Fig. 4, for half of the criteria, A_4 possesses the most favorable evaluations among all the alternatives. Meanwhile, except C_7, C_8, and C_{10}, the performance of A_4 is much better than the worst scheme. This is also reflected in the closeness of this alternative to PIS (d_i^+) and NIS (d_i^-) as shown in Fig. 7. Overall, d_4^+ is the smallest among all the d_i^+, and d_4^- is the largest among all the d_i^-.

(2) The uncertainty in the MCDM process can be captured by the simulation, which results in variations in ranking results. As seen in Fig. 5, d_i^+ and d_i^- of the alternatives are not consistent during the simulation. The boundaries of these distances are not distinct, and there are overlaps between A_3 and A_4, as well as A_1 and A_5. The randomness in d_i^+ and d_i^- leads to the unstable closeness coefficients, as illustrated by Fig. 5. Finally, the ranking result may also vary with the simulation. It can be observed from Fig. 6, in 137 out of 1000 iterations, A_3

becomes the optimal scheme, rather than A_4. To further explore the uncertainty modeling of the simulation, the normality test at the 5% level is conducted on the distribution of closeness coefficients. The test results show that closeness coefficients are significantly drawn from the normal distribution, and standard deviations of the 5 alternatives are close, ranging from 0.0206 to 0.0258 as listed in Table 8. To identify the certainty of the true situation conforms with the final results, a paired-sample hypothesis test is implemented to determine whether the closeness coefficient of A_4 is significantly higher than A_3 with a 5% significance level. The test result is shown in Table 13, where the t value is 35.37 and the p value is 0. As a result, the null hypothesis that CC_4 is significantly higher than CC_3 at 5% level is not rejected. This result indicates that it can be believed that the preference for A_4 is higher than A_3.

(3) C_5 and C_{14} are discovered to be the highest positive sensitive factors, and C_7 is discovered to be the highest negative sensitive factor that contributes to the outstanding performance of the selected optimal alternative A_4. The correlation coefficient of C_5 (Acceptance of subsoil condition to the excavation technology) with the closeness coefficients of A_4 is 0.371, and 0.232 for C_{14} (Supportability of the excavation technology to the construction safety), which ranks first and second in the top positive sensitivity factors. The factor C_7 (Impact of exaction technology to surface disturbance) possesses the sensitivity factor with the largest magnitude to be 0.501. The magnitude of sensitivities of criteria is related to their importance in practice, and stakeholders should pay great attention to the soil conditions and safety control [11]. Meanwhile, as the railway section, in this case, is under the crowded residential area, the influence of the tunnel excavation is one of the domain considerations in the project. Additionally, only the sensitivities of C_7 and C_8 are significantly negative as observed by Fig. 7. This is consistent with that the higher value in cost attributes will induce the lower preference, which validates the effectiveness of the proposed method.

5 Merits Analysis

In the proposed approach, the same degree of uncertainty of the evaluation team is assumed. To further examine the impact of uncertainty in the qualitative assessment on the decision-making results, uncertainty factors $\alpha = \{0.8, 0.9, 1.1, 1.2\}$ are introduced to analyze the dispersions in results. Two sets of tests where different uncertainty factors are multiplied to *En* and *He* of the cloud models respectively. The tests with smaller uncertainty factors are regarded as situations with lower uncertainty. The frequency counts of the ranks of A_4 through simulations are presented in Fig. 8. Affected by the uncertainty, multiple potential ranking results occur for the case study, as presented by Fig. 9. The comparison between potential results is discussed. To further demonstrate the capability of the proposed cloud-TOPSIS, a comparison between fuzzy-TOPSIS and cloud-TOPSIS is conducted, and the results are listed in Table 6. The detailed discussions are presented below.

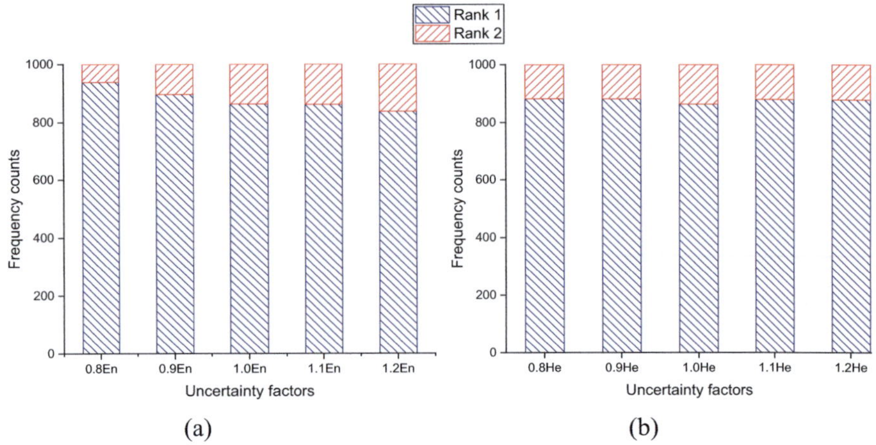

Fig. 8 Frequency counts of the ranks of A_4 by controlling **a** *En* and **b** *He*

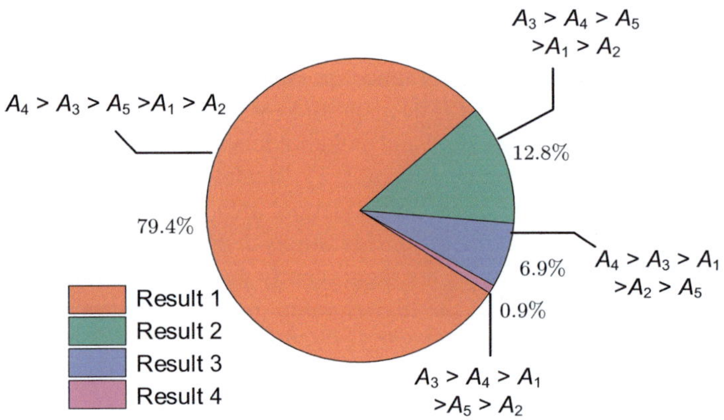

Fig. 9 Potential ranking results of the case study

(1) The simulation tests show that the inconsistency of final results grows as *En* increases, and the impact of the changes in *He* is not significant. It can be observed from Fig. 8a that the possibility that A_4 is ranked secondly raises gradually from 0.8*En* to 1.2*En*, from 0.6 to 1.6%. The degree of inconsistency in the rankings increases with the rise of the factor for *En*. The inconsistency should be addressed when the evaluation of two alternatives is close, where a further comparison may be required. The influence of *He* on the variation is not significant. There is no significant tendency that the inconsistency level of results changes with *He*, as shown in Fig. 8b. This observation is also reflected

by plots in Fig. 5b that the cloud models are hardly distinguished. Therefore, compared with *He*, adjusting *En* is a more appropriate and effective method to simulate the various degrees of uncertainty of experts. However, the precision and accuracy of experts in evaluations may be different according to the factors, such as the diversity of their proficiency, the project information provided by decision-makers, and the definition of each description level.

(2) With the Monte Carlo simulation for uncertainty, multiple possible ranking results with probabilities can be obtained. The 4 potential ranking results of the case study are presented in Fig. 9. The variations of the evaluation results model both normal and extreme scenarios of the evaluation. When there is a certain bias in the evaluation standards between experts, an extreme scenario may occur. As observed by Fig. 9, the dominant result $A_4 > A_3 > A_5 > A_1 > A_2$ takes up 79.4% of the simulation results. In the second result, A_3 the mine tunneling method surpasses A_4 the shield TBM method becoming the ultimate choice. A_3 possesses higher overall assessment results of C_7, C_8, and C_{11} compared with A_4, and the performance on C_{14}, C_{15}, and C_{16} is close. Therefore, it is rational that A_3 becomes the ultimate result in some cases. Overall, the portion of the first result is significantly larger than other potential results, which validates the robustness of the dominant result. The suggestion of selecting A_4 is reliable in most cases for the case study.

(3) The proposed cloud-TOPSIS approach addresses the criteria where significant differences exist between candidates and provides more comprehensive evaluation results. The computed closeness coefficient results by the fuzzy-TOPSIS and cloud-TOPSIS approaches are shown in Table 9. The ranking result of the two approaches is the same, which demonstrates the validation of the proposed method. There are slight differences in the closeness coefficient between the fuzzy-TOPSIS and the mean values obtained by the cloud-TOPSIS. The fuzzy-TOPSIS approach is implemented without weightage learning, which assumes criteria share the same contributions. The weightage learning of the cloud-TOPSIS assigns a higher weight to the criterion with a large deviation assessment for alternatives. With the weightage learning, the alternative with significant weakness on several criteria will be unfavorable. Furthermore, since the simulation process is implemented, the cloud-TOPSIS approach is able to provide a set of simulation results, rather than a single value of the closeness coefficient. Table 14 presents the mean, minimum, and maximum of the simulated closeness coefficients of alternatives. The final rankings can be decided not only according to the mean values, but also other parameters. In this way, a more comprehensive evaluation can be achieved.

Table 9 Closeness coefficients of fuzzy-TOPSIS and cloud-TOPSIS

Methods		Closeness coefficients				
		A_1	A_2	A_3	A_4	A_5
Fuzzy-TOPSIS		0.431	0.318	0.633	0.674	0.467
Cloud-TOPSIS	Mean	0.416	0.320	0.656	0.684	0.463
	Min	0.347	0.257	0.581	0.616	0.376
	Max	0.489	0.387	0.727	0.744	0.557

6 Chapter Summary

TOPSIS is a popular MCDM method to analyze the numeric decision-making matrix provided by decision-makers. Since the evaluations provided by experts in this field are usually in linguistic languages, the cloud model is utilized to describe the fuzziness and randomness in the assessment. In this research, a novel soft computing approach, cloud-TOPSIS, is proposed to solve the problems in the selection of the optimal tunnel excavation method under uncertainty. A realistic case study of the excavation project of the Z-X tunnel section located at the Wuhan Metro Line 2 in China is used to testify the applicability and effectiveness of the developed approach.

In light of the case study, some findings are obtained. A4 (shield TBM) is identified to be the optimal method that can be used in the excavation of the Z-X tunnel section. The sensitivities of criteria vary, and the attribute with higher sensitivity also shows higher importance. The proposed method has the advantage of simulating the uncertainty by random sampling and learning the criteria weightage from evaluation matrices. Sample evaluation matrices will be generated by random sampling, which will affect the consistency of closeness coefficients and expected rankings. Additionally, in the proposed method, the criterion weights can be learned from the evaluation matrix, without additional assessments on the relative importance of experts.

The contribution of this research lies in two aspects. From the theoretical perspective, the proposed cloud model-TOPSIS approach addresses the uncertainty in the MCDM problem by simulation. With the simulation process, rather than the deterministic ranking result computed by the traditional fuzzy-TOPSIS approach, a probabilistic ranking result that represents the probabilities of all possible ranking results will be provided. From the practical perspective, the realistic tunnel case study in the city of Wuhan provides insights into a better understanding of selecting the optimal tunneling excavation method in large-scale infrastructure projects.

References

1. Zhang L, Chen W (2022) Multi-criteria group decision-making with cloud model and TOPSIS for alternative selection under uncertainty. Soft Comput 26:12509–12529
2. Zhang J, Chang W, Zhou S (2015) An improved MCDM model with cloud TOPSIS method. In: The 27th Chinese control and decision conference (2015 CCDC), pp 873–878
3. Liu DS, Zhang J, Wang CL, Ci WL, Wu BX, Quan HF (2024) Integrated triangular fuzzy KE-GRA-TOPSIS method for dynamic ranking of products of customers' fuzzy Kansei preferences. J Intell Fuzzy Syst 46:19–40
4. Li D, Liu C, Gan W (2009) A new cognitive model: cloud model. Int J Intell Syst 24:357–375
5. Sun P, Zhang RZ, Qiu XW (2023) A survey on cloud model. J Internet Technol 24:1159–1167
6. Pandey V, Komal, Dincer H (2023) A review on TOPSIS method and its extensions for different applications with recent development. Soft Comput 27:18011–18039
7. Salih MM, Zaidan BB, Zaidan AA, Ahmed MA (2019) Survey on fuzzy TOPSIS state-of-the-art between 2007 and 2017. Comput Oper Res 104:207–227
8. Golestanifar M, Goshtasbi K, Jafarian M, Adnani S (2011) A multi-dimensional approach to the assessment of tunnel excavation methods. Int J Rock Mech Min Sci 48:1077–1085
9. Li D, Cheung D, Shi X, Ng V (1998) Uncertainty reasoning based on cloud models in controllers. Comput Math Appl 35:99–123
10. Liu H, Wang L, Li Z, Hu Y (2019) Improving risk evaluation in FMEA with cloud model and hierarchical TOPSIS method. IEEE Trans Fuzzy Syst 27:84–95
11. Kines P, Andersen LPS, Spangenberg S, Mikkelsen KL, Dyreborg J, Zohar D (2010) Improving construction site safety through leader-based verbal safety communication. J Safety Res 41:399–406

TBM Performance Prediction

1 Introduction

TBM is a widely-used method for tunnel construction due to its safety and efficiency. Controlling the TBM to follow the designed route is crucial during tunneling operations. Although there has been extensive research on geological and physical characteristics, studies on TBM attitudes are limited. Currently, the laser positioning system experiences significant delays in capturing the TBM attitude. Operators adjust the TBM attitude manually, processing very limited data and selecting TBM parameters based on experience, which prevents them from anticipating future TBM attitudes. That makes operations short-sighted. Consequently, the wealth of time series data from TBM sensors underscores the urgent need to develop an accurate TBM attitude prediction approach [1].

Traditional TBM attitude prediction mainly relies on theoretical analysis, and the position of the TBM is calculated using limited TBM and geological parameters. For instance, Shen et al. [2] examined shield-soil interaction for attitude calculation. Zhang et al. [3] investigated the attitudinal change behavior of earth pressure balanced (EPB) shield machines during the tunneling process. The nonlinear nature of TBM attitude prediction, with its multiple inputs and outputs and vast sample size, renders traditional methods poorly adaptable to varying geological conditions and inadequate for real-time prediction.

With the advancement of artificial intelligence in TBM applications, many problems are being addressed, making machine learning a focal point in TBM performance prediction. Intelligent models can discern intricate structures within datasets and solve engineering problems, even when feature relationships are unknown. For example, Adoko et al. [4] used Bayesian inference to select the best model for estimating penetration rates among eight candidates. Jakubowski et al. [5] introduced a model combining multivariate linear regression and classification and regression trees (CART) to link TBM penetration rate, penetration per revolution, and field penetration index with TBM operational and geotechnical features. Wei et al. [6] utilized

L. Zhang et al., *Intelligent Construction in Tunnels*, Lecture Notes in Civil Engineering 679, https://doi.org/10.1007/978-981-96-8785-5_7

Monte Carlo simulation and back propagation neural network to predict penetration rate by inputting the features. These machine learning methods typically preprocess data through cleaning, filtering, and denoising, which can result in information loss. Real-time predicting and adjustment of TBM attitude are crucial during tunneling, necessitating accurate prediction from noisy raw data. In addition, as excavation proceeds, the increasing volume of TBM data and the limited nonlinear representation capabilities of traditional machine learning methods hinder high prediction accuracy. Therefore, this research adopts a deep learning approach for TBM attitude prediction.

Multi-parameter models have higher computational complexity and require more time for training. By extracting as many effective features as possible for input to a deep learning model, prediction accuracy can be achieved without sacrificing computational efficiency. Convolutional neural networks (CNNs) are effective for extracting features from unprocessed input data. The convolution operation is crucial in feature extraction and has been widely applied in civil engineering fields such as crack detection, and seismic response prediction. Although CNNs are predominantly used in classification tasks, their applications in regression problems, particularly TBM attitude prediction, remain underexplored. Given that TBM data features are one-dimensional temporal data, this study employs a one-dimensional convolution neural network (1D CNN) to extract high-level features.

It is important to note that TBM attitude data is time-dependent and not isolated, making attitude prediction a time series data problem. The shield attitude changes continuously, and each time step's attitude is influenced by multiple previous time steps. Predicting based on time series data leverages these correlations, using past TBM attitude data as input features to enhance prediction accuracy. Recurrent neural networks (RNNs) are effective for processing time series data and widely used in TBM parameter prediction. There are three main types of RNN layers: traditional RNN, long-short-term memory (LSTM), and gated recurrent unit (GRU). Traditional RNN layers often struggle with vanishing and exploding gradient issues, making them difficult to train effectively. LSTM layers address these problems and are used in various projects. For example, Fu and Zhang [7] utilized multiple LSTM layers for real-time TBM operating parameter prediction. Zhou et al. [8] applied LSTM after a CNN network for TBM attitude and position prediction. However, as time series research progresses, LSTM limitations become apparent. For lower-dimensional input data, LSTM layers may overfit, making them less effective than GRU layers. Additionally, GRU layers have only two gates, which not only improve calculational efficiency but also maintain prediction accuracy when there is one less gate than LSTM. Therefore, this research combines GRU layers with 1D CNN to build the prediction model.

The core technical challenges of this research are: (1) How to predict future TBM attitude based on current TBM parameters, enabling timely adjustments and providing sufficient data for attitude warnings. (2) How to verify the accuracy and robustness of the TBM attitude prediction method using TBM time series data. (3) How to quantitatively analyze the impact of input features multiple steps ahead on multi-step prediction outcomes? To address these challenges, this chapter presents a robust multi-step prediction method using the proposed hybrid C-GRU model for TBM attitude and analyzes the global sensitivity of prediction results to determine the influence of each variable. The model has strong multi-step prediction robustness while maintaining accuracy, while the interpretability of the relationship between input and output sequences in the process of time-series-based multi-step prediction is enhanced.

2 Time Series Prediction

Effective TBM attitude control is crucial for safe TBM operation, and accurate attitude prediction facilitates this control. Extensive research has been conducted on using deep learning methods for TBM attitude prediction. Generally, time series data methods fall into two categories: single-step prediction and multi-step prediction.

(1) **Single-step prediction**. A Single-step prediction uses time-series features as input to output the value for a single subsequent time step. As the interval between the input sequence and the output time step increases, prediction accuracy typically decreases. Cachim and Bezuijen [9] utilized a multilayer feedforward artificial neural network to predict cutter head torque based on previously measured torque and the volume of injected foam. Lin et al. [10] developed an LSTM neural network, optimized using the particle swarm method, to predict main thrust during tunnel excavation. Zhang et al. [11] proposed a hybrid PCA-GRU model for real-time TBM movement prediction. In summary, TBM attitude measurements lag, often leading to adjustments only after significant deviations occur. Besides, current TBM attitude predictions based on time series data are usually single-step, which provides insufficient adjustment time for practical applications due to the short duration each time step represents.
(2) **Multi-step prediction**. A multi-step prediction employs a sequence-to-sequence approach, inputting time series features to predict future time series outputs. While multi-step prediction in TBM control has received limited attention, related studies exist in similar fields. For instance, Wei et al. [12] used adaptive empirical wavelet transform and bi-LSTM for ship roll motion forecasting, optimizing hyperparameters with PSOGSA. Lu et al. [13] introduced a temporal attention encoder-decoder network to predict building heat load using LSTM for temporal data relationship. Gao et al. [14] developed a multi-step prediction method for ship trajectory incorporating TPNet and LSTM. However, current multi-step predictions primarily rely on LSTM, which lacks ideal computational efficiency, hindering effective shield attitude adjustments.

GRU features a more concise structure, which has been shown in past studies to improve computational efficiency while maintaining satisfactory performance. In TBM applications, GRU has been used sparingly and primarily for single-step predictions. A challenge for the multi-step prediction is that forecast accuracy typically decreases as the output sequence lengthens. How to output as many time steps as possible in the process of TBM attitude prediction and ensure the high accuracy of the prediction results is a difficult problem.

Additionally, input features often contain unavoidable disturbances, and the multi-step prediction model must maintain accuracy under such conditions, which is essential for engineering applications. This requires the model to be robust. 1D CNN can effectively extract high-dimensional features from time series data without preprocessing, enhancing the model's resistance to disturbances. In addition, Extracting features from "healthy" TBM data and training a "normality" model can support decision-making in similar scenarios and help infer abnormal operations and states. Therefore, this study proposes a multi-step TBM attitude prediction method by merging 1D CNN and GRU.

3 TBM Performance Prediction Framework

To achieve precise control of TBM parameters, this research proposes a robust multi-step prediction method for TBM attitude by combining 1D CNN and GRU. The results of the method are analyzed using the random balance design Fourier amplitude sensitivity test for sensitivity analysis, further examining the influence of each TBM parameter on attitude. This approach consists of three main steps, detailed in the following section. The overall approach is illustrated in Fig. 1.

3.1 Time Series Reconstruction

The data acquisition process involves exporting and storing TBM data in a CSV file. This data comprises two main categories: active data and passive data. Active data includes parameters that the operator can adjust during shield tunneling, such as articulation displacement, CHD RPM, and thrust force. Passive data includes parameters returned by sensors, such as earth pressure, torque, and penetration. Data recorded during TBM shutdown periods is not valuable for multi-step TBM attitude prediction, so it is removed from the dataset first. Thrust values are 0 during shutdown, and the features are sequentially cleaned using the following equation.

$$f_i(n) = f_i(n) \times st,\ st = \begin{cases} 1,\ thrust \neq 0 \\ 0,\ thrust = 0 \end{cases} \tag{1}$$

where $f_i(n)$ is the nth value in the time series of the ith feature in the original data. st is the operating status of TBM. When the thrust is 0, indicating a shutdown status, $st = 0$. Conversely, when the thrust is not 0, indicating an active status, $st = 1$.

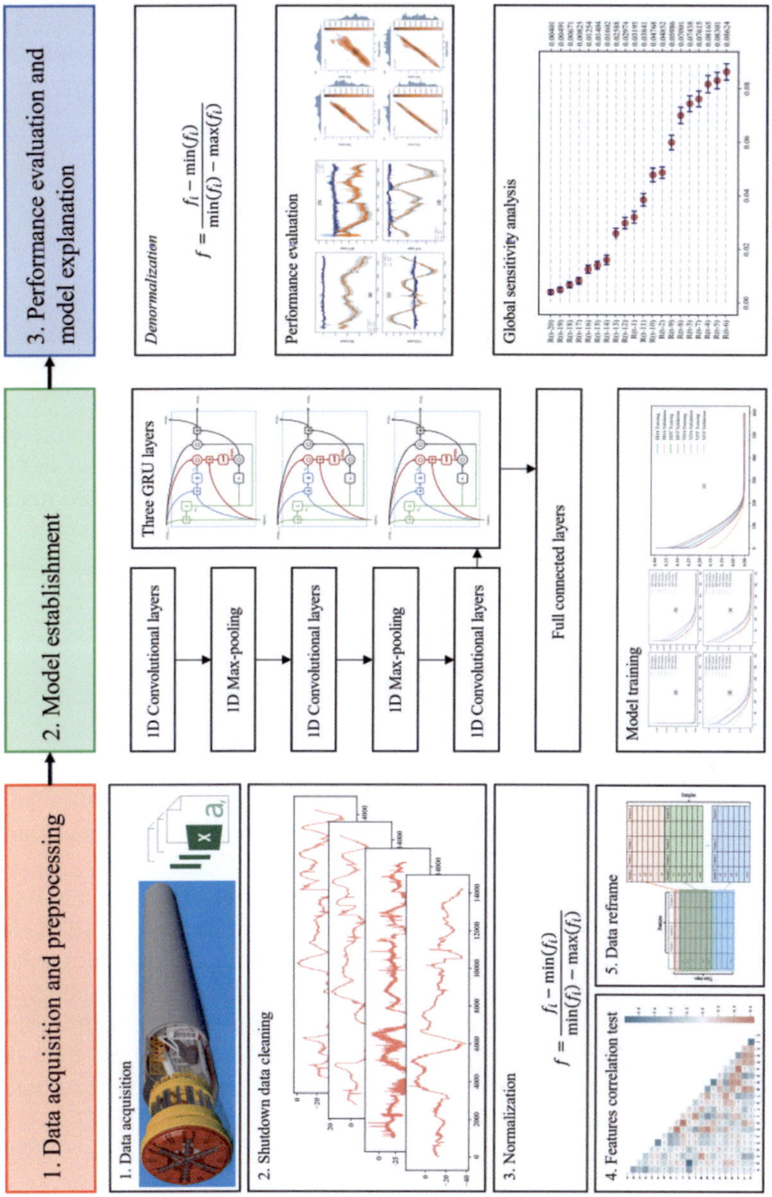

Fig. 1 Overall flowchart for TBM attitude multi-step prediction

After acquiring the excavation data, it is essential to check for high linear correlations between selected input and output features. A complex prediction model is necessary only if there is no clear linear relationship between TBM attitude parameters and input features. The Pearson correlation coefficient, recognized for assessing linear relationships between variables, is calculated as shown in (Eq. 2).

$$corr\left(f_x, f_y\right) = \frac{\mathrm{cov}\left(f_x, f_y\right)}{\sigma_{f_x}\sigma_{f_y}} = \frac{\sum_{i=1}^{n}\left(f_{xi} - \overline{f_{xi}}\right)\left(f_{yi} - \overline{f_{yi}}\right)}{\sqrt{\sum_{i=1}^{n}\left(f_{xi} - \overline{f_{xi}}\right)^2}\sqrt{\sum_{i=1}^{n}\left(f_{yi} - \overline{f_{yi}}\right)^2}} \tag{2}$$

where $corr\left(f_x, f_y\right)$ is the Pearson correlation coefficient of different features f_x, f_y. $\sigma_{f_x}, \sigma_{f_y}$ are the standard deviation of f_x, f_y, respectively. f_{xi}, f_{yi} are the ith value of f_x, f_y, respectively. $\overline{f_{xi}}, \overline{f_{yi}}$ are the average of f_x, f_y, respectively.

The different dimensions and value ranges of each selected input feature can increase errors in model training and TBM attitude prediction. Therefore, before training the model, normalization is applied (Eq. 3) to prevent issues related to data scale and units. Since the prediction model uses the normalized dataset, its results are not the actual TBM attitude parameters. To obtain the real values, denormalization is necessary (Eq. 4).

$$Fn_i = \frac{f_i - \min(f_i)}{\min(f_i) - \max(f_i)} \tag{3}$$

$$Fd = Fn_{output}\left[\min(fn) - \max(fn)\right] + \min(fn) \tag{4}$$

where f_i and Fn_i is the ith TBM feature before and after normalization. fn is the feature before normalization corresponding to Fn_{output}. Fd is the result of denormalization corresponding to Fn_{output}.

Normalized time series data cannot be directly used as input for the multi-step prediction model but needs to be reframed into an appropriate data structure. The normalized data is a two-dimensional matrix with time steps and variables, where each time step is a sample. To use past TBM attitude data for predictions, this matrix must be converted into a three-dimensional matrix of sample numbers, time steps, and TBM features. Specifically, the ith sample corresponds to the current time step t and includes the most recent n past time steps. The other samples can be reframed accordingly. The reframing process is illustrated in Fig. 2.

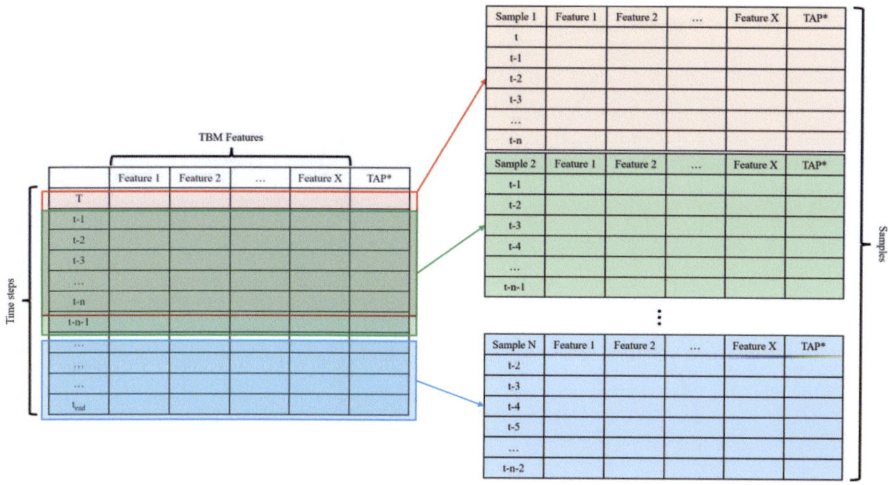

Fig. 2 Schematic diagram of data structure and reconstruction process. *TAP denotes TBM attitude parameters

3.2 *Model Establishment*

The Gated recurrent unit (GRU) is an improved RNN introduced in 2014. Compared to LSTM, GRU has a simpler structure, faster training speed, and higher computational efficiency. The GRU structure, primarily consists of two components: the update gate and the reset gate. The update gate determines how much past information is passed to the future, that is, how much information from the current and previous time steps is passed to the next. The reset gate determines how much information is to be forgotten. After processing through these two gates, the final memory for the current time step is output. The internal parameter transfer of the GRU unit is mathematically described by Eqs. (5–9).

$$u_t = S\left(w_1^{(u)} input_t + w_2^{(u)} step_{t-1}\right) \tag{5}$$

$$r_t = S\left(w_1^{(r)} input_t + w_2^{(r)} step_{t-1}\right) \tag{6}$$

$$step_t\prime = tanh\left(w' input_t + r_t \odot step_{t-1}\right) \tag{7}$$

$$step_t = (1 - u_t) \odot step_t\prime + input_t \odot step_{t-1} \tag{8}$$

$$\tanh(x) = \frac{\sinh(x)}{\cosh(x)} = \frac{e^x - e^{-x}}{e^x + e^{-x}} \tag{9}$$

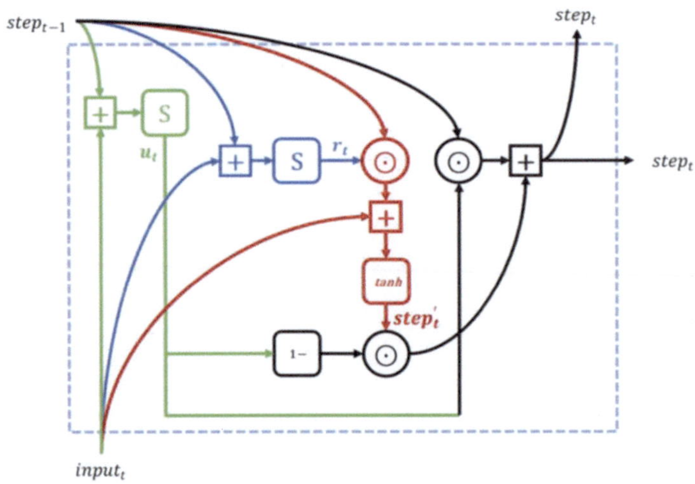

Fig. 3 Structure of the GRU cell

where $input_t$ is the input at time step t, and $step_{t-1}$ is the status of the last time step $t-1$. $w_1^{(u)}$, $w_2^{(u)}$, $w_1^{(r)}$, and $w_2^{(r)}$ are the weights of $input_t$ and $step_{t-1}$ in the update and reset gates, respectively. S is the sigmoid function, $\odot$ is the Hadamard product. w' is the weight of $input_t$ when calculating $step_t\prime$. The function $\tanh(x)$ serves as the activation function of the GRU cell (Fig. 3).

Inputting all raw data into the GRU can be inefficient and may not yield optimal performance. Given that the time series data of each feature is 1D and independent, GRU and 1D CNN are fused to form a C-GRU network, as illustrated in Fig. 4.

The network comprises 12 layers: 3 1D convolutional layers for high-level feature extraction, 2 max-pooling layers for expanding the receptive field, 3 GRU layers, and 4 fully connected layers for mapping features to target intervals. The detailed structure and parameters of the C-GRU model are outlined in Table 1. The activation functions of the convolutional and fully connected layers use exponential linear units (ELU, Eq. 10) to enhance the nonlinear expression and prediction accuracy of the network. The dimensions of the input layer vary with the number of selected TBM features and past time steps, while the dimensions of the output layer adjust based on TBM attitude parameters and the length of the predicted sequence.

$$ELU(x) = \begin{cases} x, & if\ x > 0 \\ \alpha(e^x - 1), & if\ x \le 0 \end{cases} \tag{10}$$

After the model is built, the training parameters need to be defined. Mean squared error (MSE, Eq. 11) is employed during training to evaluate model performance more comprehensively by reflecting the dispersion of deviations between predicted and true values. Root mean square propagation (RMSprop) is used as the model's optimizer to identify the global minimum of the loss function.

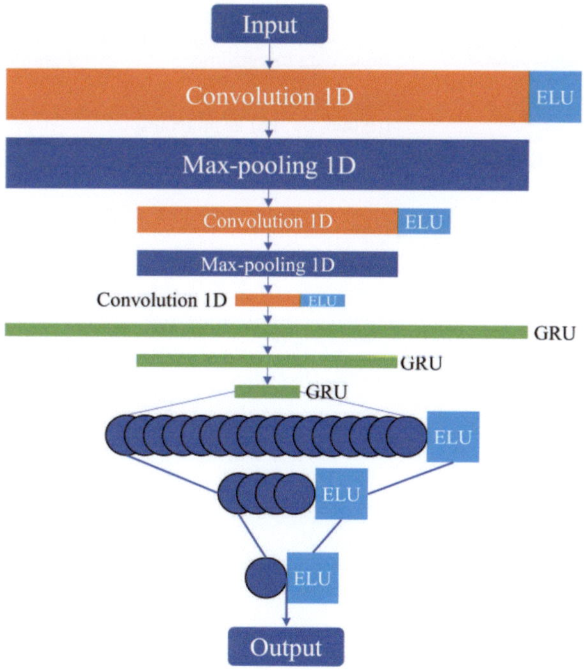

Fig. 4 Structure of the C-GRU model

$$MSE = \frac{1}{N}\sum_{i=1}^{N}(y_i - \hat{y}_i)^2 \tag{11}$$

where y_i is the prediction result of the proposed model, and $\hat{y}_i$ is the true value of the output features.

3.3 Model Evaluation and Explanation

To further evaluate the performance of the proposed method, mean absolute error (MAE, Eq. 12) and R-squared (R^2, Eq. 13) are used as indicators. MAE reflects the average error between predicted and true values, intuitively indicating prediction quality. R^2 standardizes prediction results to 0–1, allowing for performance comparison across different models within the same range.

$$MAE = \frac{1}{N}\sum_{i=1}^{N}|(y_i - \hat{y}_i)| \tag{12}$$

$$R^2 = 1 - \frac{\sum_{i=1}^{N}(y_i - \hat{y}_i)^2}{\sum_{i=1}^{N}(y_i - \overline{y}_i)^2} \tag{13}$$

Table 1 Details of the C-GRU model

Operation	Input data dimensions	Parameters	Proportion of parameters
Input	Changeable ((None, 64, 18) as an example)	0	0
Convolution 1D	(None, 64, 128)	16,256	10.2%
ELU			
Max-pooling 1D	(None, 32, 128)	0	0
Convolution 1D	(None, 32, 64)	41,024	25.7%
ELU			
Max-pooling 1D	(None, 16, 64)	0	0
Convolution 1D	(None, 16, 16)	3088	1.9%
ELU			
GRU	(None, 16, 128)	56,064	35.1%
tanh			
GRU	(None, 16, 64)	37,248	23.3%
tanh			
GRU	(None, 16)	3936	2.5%
Tanh			
Dense	(None, 64)	1088	0.7%
ELU			
Dense	(None, 16)	1040	0.7%
ELU			
Dense	(None, 4)	68	0.0%
ELU			
Dense	Changeable ((None, 21) as an example)	105	0.1%

where y_i is the prediction result of the proposed model, $\overline{y}_i$ is the mean of y_i, and $\hat{y}_i$ is the true value of the output features.

To further assess the accuracy of the multi-step prediction model, the impact of each input feature on the final output is analyzed to enhance the model's applicability in engineering. Sensitivity analysis is employed to elucidate the relationship between input and output indicators. The random balance design Fourier amplitude sensitivity test (RBD-FAST) enhances computational efficiency and robustness over FAST and is ideal for complex input-output scenarios. The RBD-FAST method has been applied in civil engineering for tasks such as heat transfer in roller-compacted concrete dam construction, hygrothermal simulation of brick buildings, and building energy models.

The RBD-FAST method for calculating first-order sensitivity indices involves four steps: (1) Select N design points along a periodic curve in the factor space; (2) Randomly permute the coordinates to generate scrambled points covering the factor space; (3) Input these points into the model; (4) Sort the model outputs to align with ascending design points and calculate the sensitivity index (Eqs. 14–18).

$$S1 = \frac{\sum_{n=1}^{M} \Lambda_{nw_i}}{\sum_{n=1}^{\frac{N}{2}} \Lambda_n} \tag{14}$$

$$\Lambda_p = A_k^2 + B_k^2 \tag{15}$$

$$A_k = \frac{1}{N} \sum_{j=1}^{N} \cos(ks_j) F(s_j) \tag{16}$$

$$B_k = \frac{1}{N} \sum_{j=1}^{N} \sin(ks_j) F(s_j) \tag{17}$$

$$F(s) = \frac{2}{\pi} \arcsin(\sin(w_i s_j)) \tag{18}$$

where $s_j = -\pi + \frac{2\pi}{N} j, \forall j = 1, 2, \ldots, N$. M is the number of harmonics to sum in the Fourier series decomposition. w_i is the frequency corresponding to each index. N is the number of samples to generate. S1 is the first-order sensitivity.

4 Experimental Studies

To verify the effectiveness of the proposed approach, the C-GRU model is applied to the TBM tunneling process of the T308 tunnel construction in Singapore for a multi-step prediction of the TBM attitude. Details on TBM data processing, model training, performance evaluation, and model explanation are elaborated in the following sections.

4.1 Data and Resources

The T308 tunnel connects Marine Terrace Station and Siglap Station from underground, spanning 1480 m and comprising 968 rings. Each ring is circular with an inner diameter of 6.8 m and consists of 1.4 m wide segments. The TBM has a main length of 9.81 m and a cutter head radius of 6.67 m. It is equipped with numerous sensors for real-time data acquisition. A project overview is shown in Fig. 5.

To verify the validity of the model, a suitable experimental environment is crucial. Sudden changes in soil conditions can significantly affect the model's performance in predicting TBM parameters, necessitating further adjustments. To mitigate geological influence, a region with stable and uniform soil conditions (from ring No.500 to No.699) composed of Marine clay is selected. While not entirely static, this region's

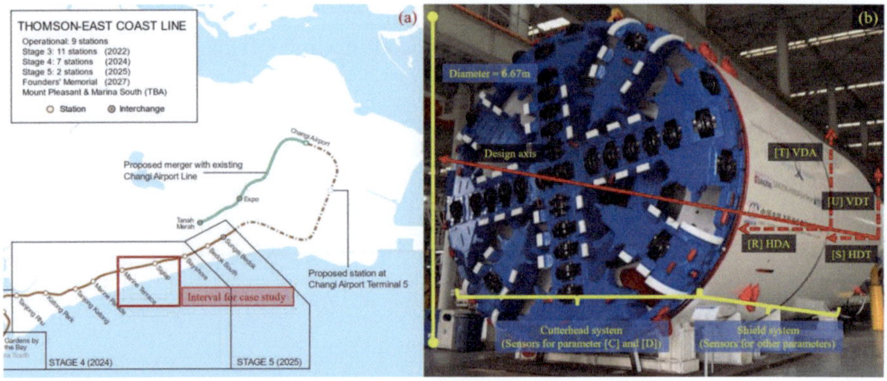

Fig. 5 A project overview: **a** interval for the case study; **b** TBM information

soil shows no sudden change, making it suitable for analysis. Unexpected TBM attitude changes due to sudden soil conditions are also unpredictable. Deep learning methods can predict TBM attitude as uniform soil variations are reflected in TBM parameters. Additionally, this region is chosen because it exhibits complex TBM attitude data, with varying degrees of disturbances, providing a robust test for the proposed method.

The originally recorded TBM data includes 767 features and 14,247 data points collected during a 200-ring tunneling process. After eliminating data from shutdown periods (as per Eq. 1), 13,855 data points remain, each representing a time step. Given the uniformly distributed soil conditions in the tunneling area, soil parameters are excluded from feature selection. Using essential TBM parameters enhances prediction accuracy and computational speed. Based on engineering experience, a total of 18 TBM parameters and 4 TBM attitude parameters are selected and then normalized according to Eq. 3.

The distribution characteristics of these features are illustrated in Figs. 6 and 7, with detailed information provided in Table 2. Most TBM features display a concentrated distribution, with the interquartile range of "Foam flow rate" at approximately 0.6 post-normalization, while other features do not exceed 0.4. Among them, the "Cutter head rotation speed per minute" has the smallest interquartile range, indicating the most concentrated data distribution. TBM attitude parameters are represented by horizontal deviation of articulation (HDA), horizontal deviation of the tail (HDT), vertical deviation of articulation (VDA), and vertical deviation of the tail (VDT). Notably, the articulation and tail deviations differ, especially in the horizontal direction where tail deviations change frequently, complicating multi-step predictions. In the vertical direction, while deviations of articulation and tail show similar trends, their values diverge significantly.

To test the correlation between selected TBM features and TBM attitude parameters, a correlation coefficient heatmap (Fig. 8) is generated based on Eq. 2. The highest negative correlation is between "Bearing deviation" and "HDT" at -0.63,

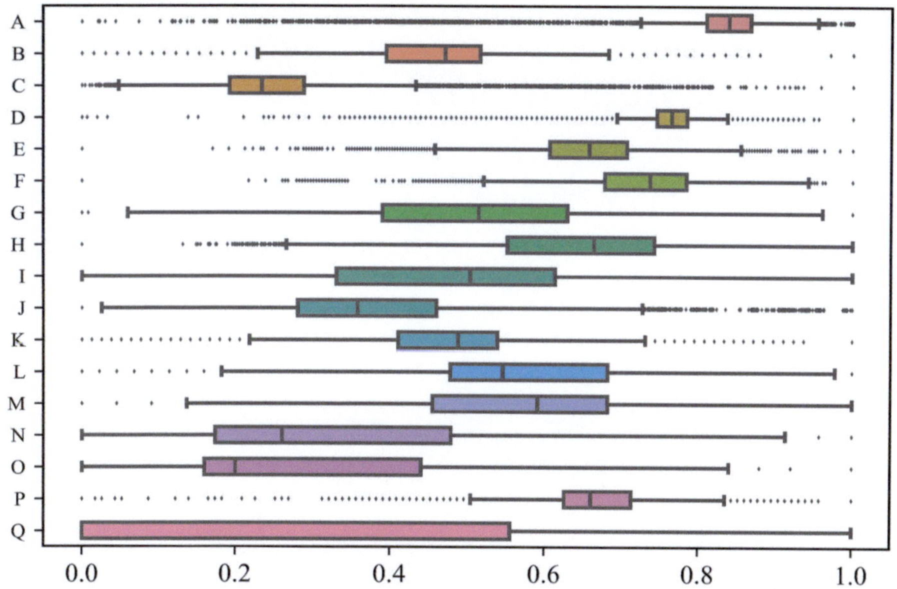

Fig. 6 Distribution characteristics of input features. The vertical line within each box represents the median of the group of data. The left and right boundaries of the box indicate the lower and the upper quartiles, while the endpoints of the line extend to (*lower quartile* $-$ 1.5 $*$ *interquartile range*) and (*upper quartile* $+$ 1.5 $*$ *interquartile range*), respectively

while the highest positive correlation is between "Articulation Displacement D" and "VDA" at 0.56. The absolute values of the correlation coefficients between other TBM features and TBM attitude parameters are all below 0.5. Thus, there is no evident linear relationship between TBM features and TBM attitude parameters.

4.2 Training Implementation Details

The proposed model in this study is developed using Python. The core hardware includes an AMD 3500X CPU and an NVIDIA GEFORCE GTX1650 4G GPU. After reframing the time series data, each group of 256 samples is input into the model. The learning rate is set to 2×10^{-5}, with a total of 600 epochs trained. The dataset is split in a 4:1 ratio for training and testing. To align with engineering practice, the datasets are partitioned sequentially. That is, the first 80% of the reframed data is used for training, while the remaining 20% serves as the test set. Additionally, 25% of the training set is used for validation to prevent over-fitting and under-fitting.

Figures 9 and 10 display the training process for 1, 11, 21, 31, and 41-step prediction models. Figure 9 shows MSE curves for the training and validation sets. Fewer prediction steps result in faster convergence, but as steps increase, convergence speed

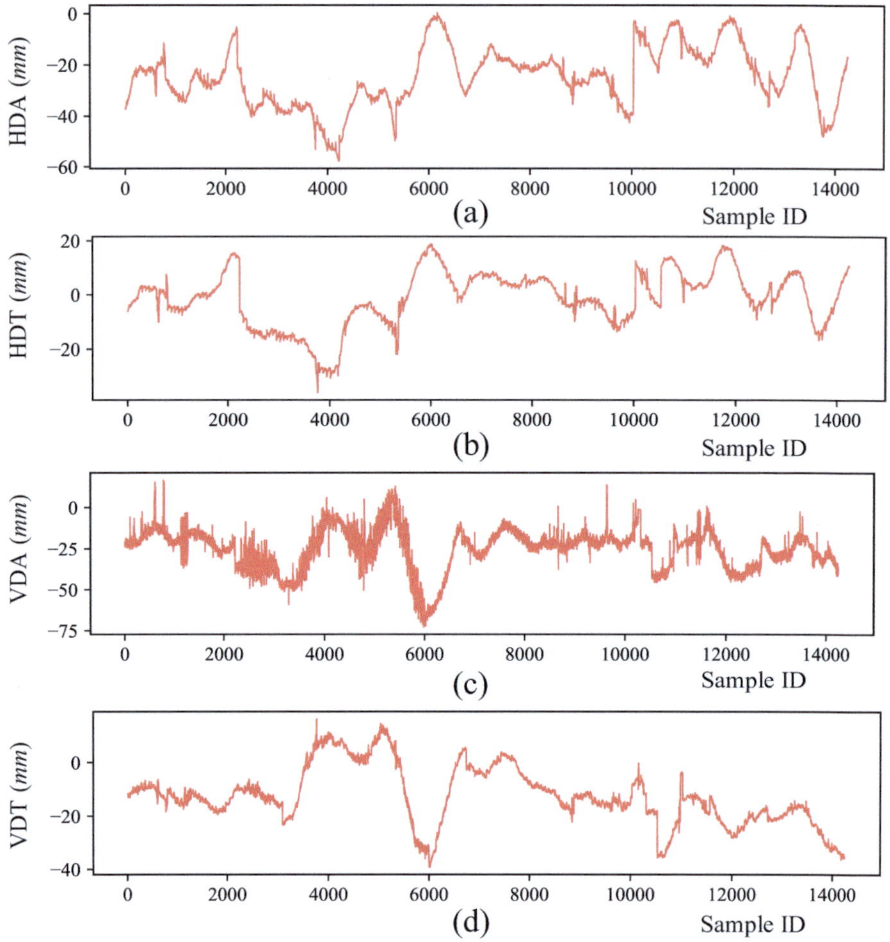

Fig. 7 Data of output features: **a** HDA; **b** HDT; **c** VDA; **d** VDT

stabilizes, completing around the 400th epoch. Meanwhile, the MSE for the training and validation sets remain similar. Figure 10 depicts the MAE curves, following a pattern similar to the MSE curve, though with more fluctuation in MAE changes.

4.3 Analysis of Results

The well-trained C-GRU model uncovers the nonlinear relationship between the TBM feature time series and TBM attitude parameters, enabling multi-step predictions of TBM attitude. Each reframed input sample contains 18 TBM features and 4 TBM attitude parameters over the past 64 time steps. The output sequence predicts

Table 2 Detailed information on each time series feature

Feature name	Mean	Std[a]	Min	Lower quartile	Median	Upper quartile	Max
[A] Thrust force (kN)	22,749.21	2953.78	226	22,482.5	23,285	24,068	27,729
[B] Penetration (mm/min)	29.38	7.74	0	26	31	34	66
[C] Cutterhead torque (kN·m)	851.29	234.38	281	713	808	931	2541
[D] Cutterhead rotation speed (rpm)	1.23	0.10	0.06	1.2	1.23	1.26	1.59
[E] Earth pressure (left) (bar)	3.81	0.24	1.92	3.67	3.82	3.96	4.81
[F] Earth pressure (right) (bar)	3.78	0.23	1.79	3.64	3.8	3.93	4.52
[G] Thrust cylinders (right) (kN)	3951.32	608.66	1864	3458	3967	4438.5	5957
[H] Thrust cylinders (down) (kN)	7600.45	793.81	3835	7067	7728	8186	9698
[I] Thrust cylinders (left) (kN)	2658.28	688.82	746	2039	2715	3145	4660
[J] Thrust cylinders (top) (kN)	1568.06	550.27	44	1163	1471	1876	4034
[K] Thrust speed average (mm/min)	36.26	9.34	0	32	38	42	78
[L] Articulation displacement A (mm)	61.52	8.15	37	58	61	67	81
[M] Articulation displacement B (mm)	57.11	4.64	45	55	58	60	67
[N] Articulation displacement C (mm)	46.82	5.48	39	43	45	50	62
[O] Articulation displacement D (mm)	43.01	5.43	35	39	40	46	60
[P] Bearing deviation (mm)	0.23	0.11	− 0.53	0.19	0.23	0.29	0.62
[Q] Foam flow rate (m^3/min)	4.47	4.88	0	0	0	10	18

(continued)

Table 2 (continued)

Feature name	Mean	Std[a]	Min	Lower quartile	Median	Upper quartile	Max
[R] HDA (mm)	– 12.50	10.83	– 39.2	– 18.6	– 13.3	– 6.1	16.2
[S] HDT (mm)	– 25.06	12.69	– 72.8	– 32	– 22.3	– 17.6	16.6
[T] VDA (mm)	– 1.20	10.07	– 36	– 6.7	– 0.5	5.75	18.8
[U] VDT (mm)	– 24.74	11.58	– 57.8	– 32.5	– 24.5	– 17.5	0.4

[a] Std denotes the standard deviation

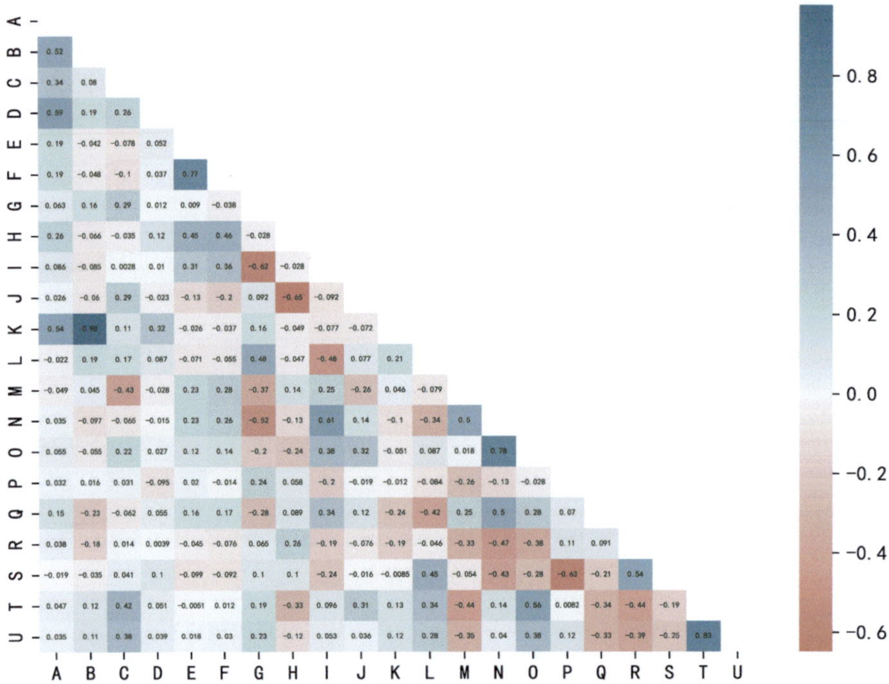

Fig. 8 Pearson correlation coefficient heatmap of input and output sequence

one of the 4 TBM attitude parameters several steps ahead. Figure 11 presents the TBM attitude sequence from the 21st step prediction model and the absolute error of each output compared to actual TBM attitude parameters in the test set. To assess the predictive ability of the C-GRU model, the overall R^2 for the 21-step prediction is calculated. The length of the output sequence is adjusted by changing the output layer dimensions to find the predictive limit. Figure 12 displays the R^2 variation from 1-step to 64-step predictions for the 4 TBM attitude parameters, while Fig. 13 shows the R^2 for the 21st step prediction. The longest prediction sequence is then evaluated, with the evaluation indices for each step in the multi-step prediction shown in Fig. 14.

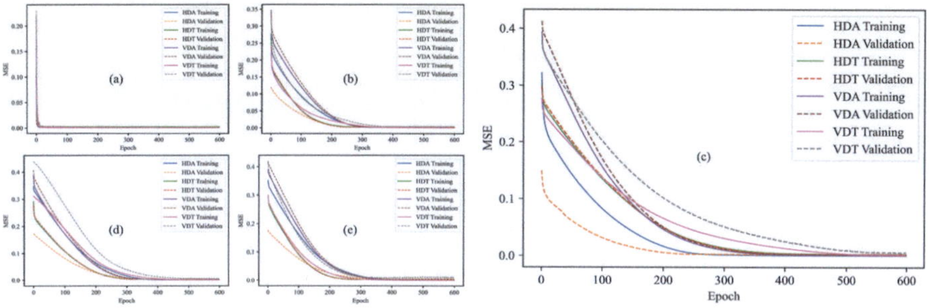

Fig. 9 MSE of the training set and validation set of various-step prediction model: **a** 1-step; **b** 11-step; **c** 21-step; **d** 31-step; **e** 41-step

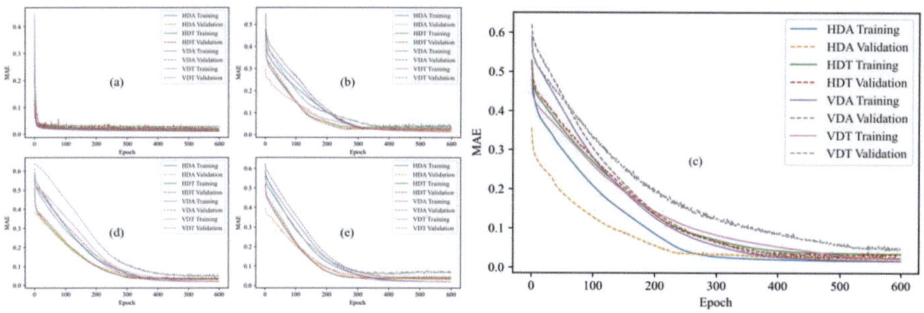

Fig. 10 MAE of the training set and validation set of various-step prediction model: **a** 1-step; **b** 11-step; **c** 21-step; **d** 31-step; **e** 41-step

The RBD-FAST method analyzes the contribution of the input data to the C-GRU model output. A set of 50,000 randomly generated normalized data points (range 0 to 1) are used as input. Similarly, the data is reframed into the trained model to obtain the output sequence, and the first-order sensitivity for each indicator is calculated according to Eqs. 13–17. The 20 most sensitive features are shown in Fig. 15. Figure 16 illustrates the first-order sensitivity excluding historical TBM attitude parameters to further explore the influence of TBM features on TBM attitude. The detailed analysis follows these findings.

(1) The prediction stability and accuracy remain satisfactory for output sequences of fewer than 21 time steps. According to Fig. 12, the single-step prediction is the most precise, yet its application value in engineering is limited. As the output time step increases, the prediction accuracy fluctuates. Due to the lack of a unified robustness evaluation standard in multi-step forecasting, this study proposes a 10% threshold for robustness. If the R^2 value for a time step changes by more than 10% compared to the previous time step, robustness is considered insufficient. Using this criterion, the VDT change at the 22nd time step exceeds the threshold at 13.99%. It is important to note that exceeding the 10% threshold does not render the model invalid. In engineering applications, the appropriate

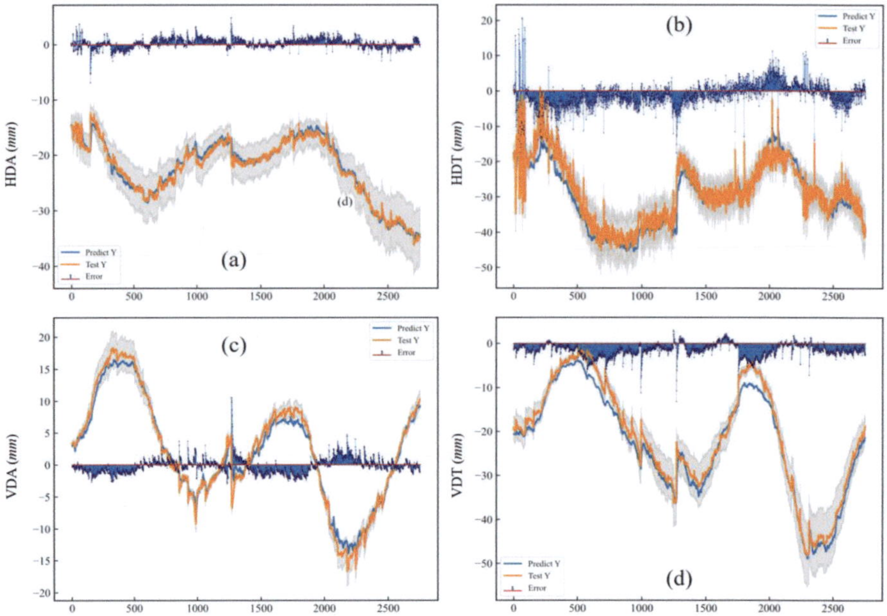

Fig. 11 Results of 21-step TBM attitude prediction: **a** HDA; **b** HDT; **c** VDA; **d** VDT

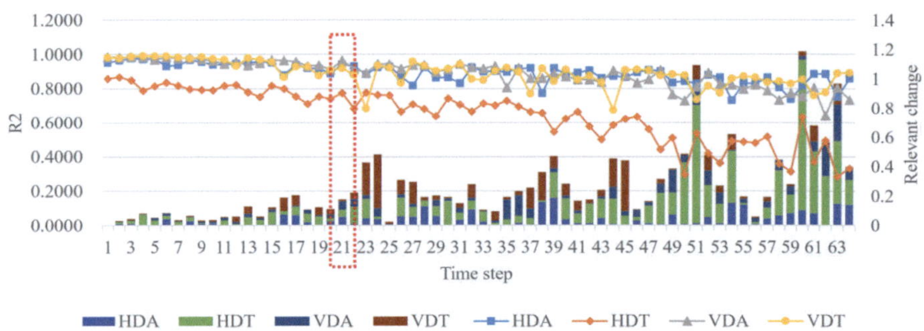

Fig. 12 R^2 of 4 different TBM attitude parameters by various-step prediction

output sequence length should be chosen based on accuracy requirements, taking into account the associated variability.

(2) The C-GRU model demonstrates high performance in multi-step prediction. As shown in Fig. 13, the R^2 of the 21-step prediction reaches a peak of 0.9652, and even with severe jitter in the horizontal deviation of the TBM tail, it maintains an R^2 of 0.7758. The R^2 values for the other two parameters are 0.9193 and 0.9412, both indicating satisfactory performance. The proposed deep learning model exhibits strong robustness. A key advantage of the multi-step prediction over the single-step prediction is the visibility of changes across the output sequence

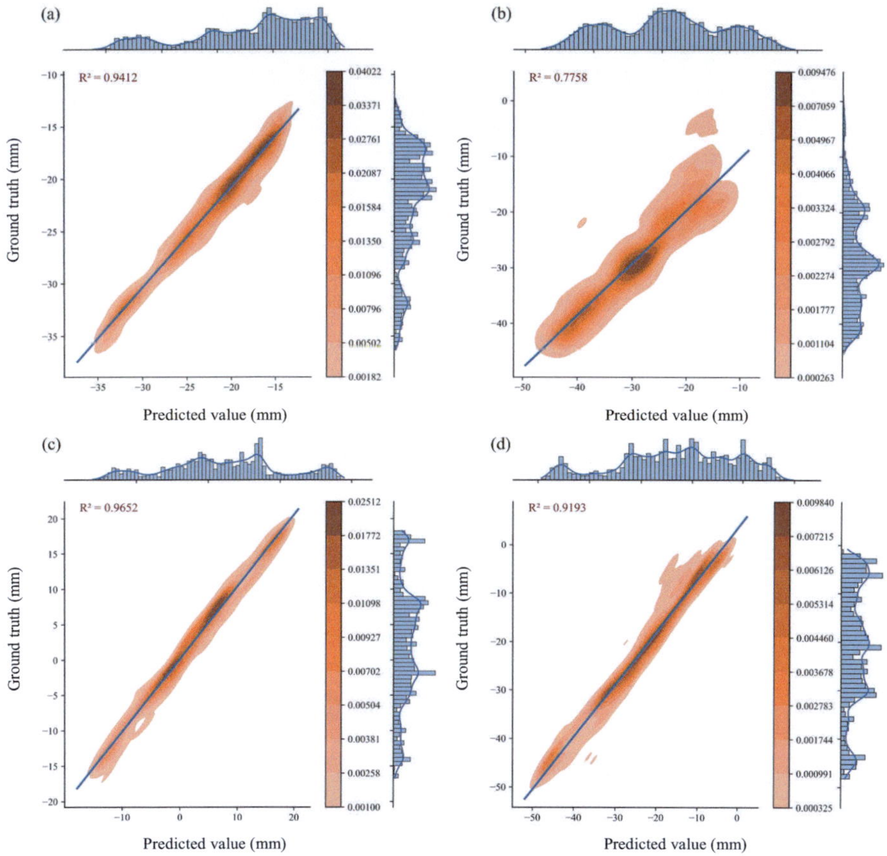

Fig. 13 R^2 of 4 TBM attitude parameters by 21-step prediction: **a** HDA; **b** HDT; **c** VDA; **d** VDT

of different TBM attitude parameters. The final prediction is robust only if the prediction of each step in the output sequence is stable, with evaluation indices fluctuating around the overall sequence index. According to Fig. 14, only the 8th and 14th steps in the 21-step prediction of HDT show a relative change greater than 5%. The R^2 values for each step in the output sequence of all parameters float around the overall R^2. Therefore, the robustness of the proposed model is confirmed.

(3) Historical TBM attitude data significantly impact the 21-step TBM attitude prediction, but closeness to the current time step doesn't necessarily increase its influence. As shown in Fig. 15, the top 20 sensitive indicators for the 4 TBM attitude parameters include the most recent 11 time steps of historical data, with the top 11 indicators for each parameter corresponding to these steps, though their order varies. This suggests that adjustments to achieve a predetermined TBM attitude should be made at least 11 time steps in advance. The most

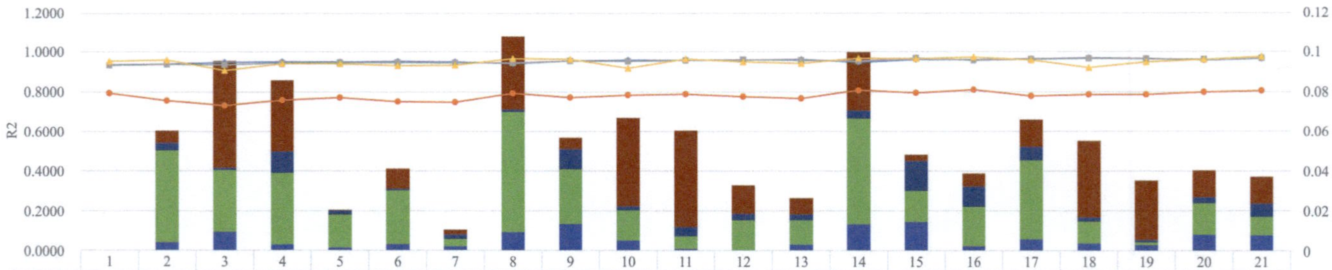

	1	2	3	4	5	6	7	8	9	10	11	12	13	14	15	16	17	18	19	20	21
VDT		0.63%	5.39%	3.60%	0.02%	1.00%	0.24%	3.67%	0.58%	4.46%	4.86%	1.45%	0.82%	2.91%	0.29%	0.66%	1.40%	3.82%	2.99%	1.35%	1.35%
VDA		0.39%	0.12%	1.06%	0.21%	0.07%	0.22%	0.11%	0.98%	0.20%	0.46%	0.30%	0.28%	0.41%	1.49%	1.03%	0.68%	0.20%	0.11%	0.30%	0.65%
HDT		4.62%	3.11%	3.61%	1.67%	2.73%	0.36%	6.08%	2.77%	1.52%	0.63%	1.52%	1.22%	5.34%	1.57%	1.97%	3.96%	1.12%	0.14%	1.59%	0.93%
HDA		0.41%	0.94%	0.31%	0.15%	0.32%	0.22%	0.93%	1.34%	0.51%	0.09%	0.03%	0.31%	1.33%	1.44%	0.23%	0.57%	0.37%	0.30%	0.80%	0.78%
HDA	0.9348	0.9387	0.9475	0.9505	0.9491	0.9521	0.9500	0.9412	0.9538	0.9587	0.9578	0.9581	0.9611	0.9484	0.9620	0.9598	0.9653	0.9688	0.9660	0.9582	0.9657
HDT	0.7937	0.7570	0.7334	0.7599	0.7726	0.7515	0.7488	0.7943	0.7723	0.7840	0.7890	0.7770	0.7676	0.8086	0.7958	0.8115	0.7793	0.7880	0.7870	0.7995	0.8070
VDA	0.9325	0.9362	0.9351	0.9450	0.9470	0.9463	0.9442	0.9453	0.9546	0.9526	0.9570	0.9599	0.9572	0.9533	0.9675	0.9575	0.9640	0.9659	0.9649	0.9620	0.9682
VDT	0.9524	0.9584	0.9067	0.9394	0.9396	0.9302	0.9324	0.9667	0.9611	0.9183	0.9629	0.9489	0.9411	0.9684	0.9656	0.9719	0.9583	0.9217	0.9492	0.9620	0.9750

Time steps

Fig. 14 R^2 and relative change of each step in 21-step prediction

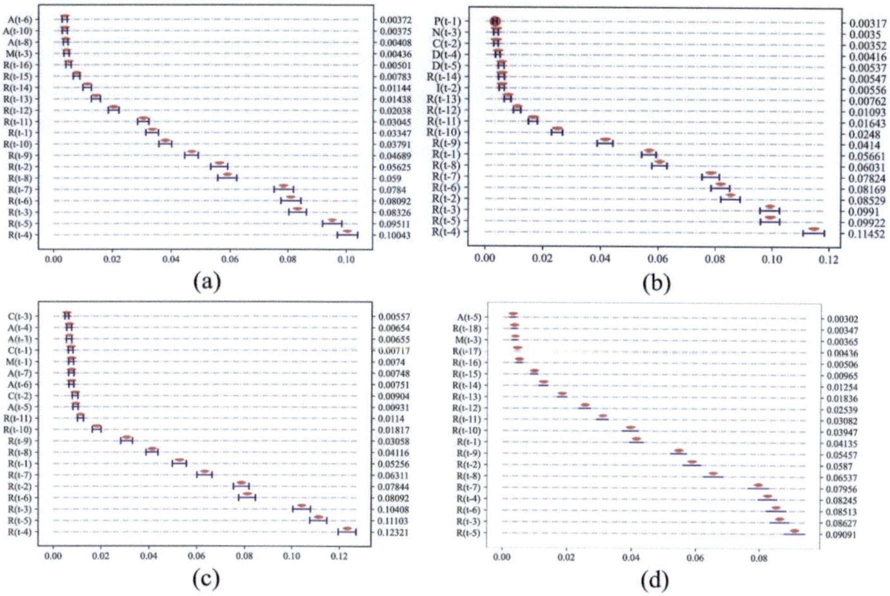

Fig. 15 First-order sensitivity of all TBM features and TBM attitude parameters of all time steps: **a** HAD; **b** HDT; **c** VDA; **d** VDT

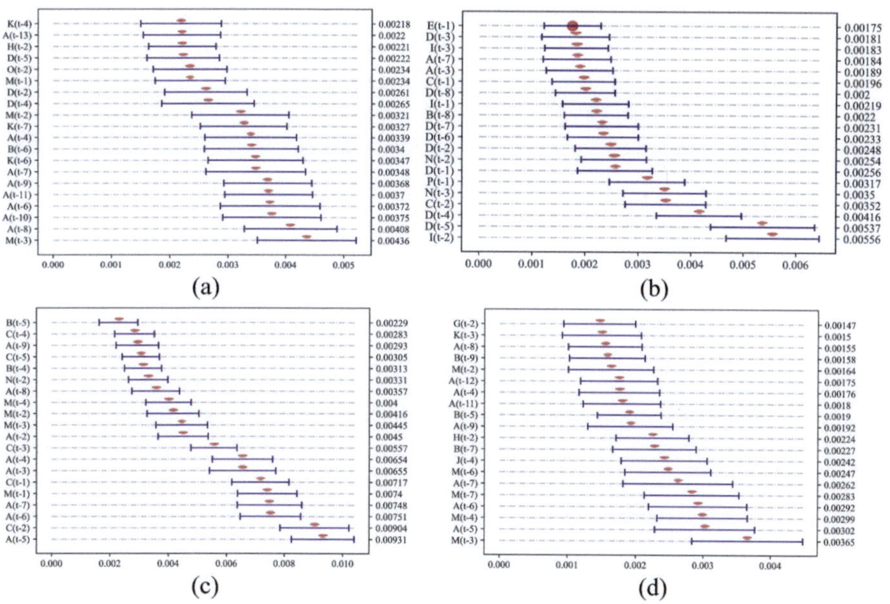

Fig. 16 First-order sensitivity of all TBM features of all time steps: **a** HAD; **b** HDT; **c** VDA; **d** VDT

sensitive data is from the past 4th time step of the historical data. In engineering practice, setting an attitude threshold and using the attitude at the time step (t − 4) as a key reference can help ensure proper adjustment. While manual control often overlooks negative impacts unless the threshold is exceeded, the C-GRU multi-step prediction emphasizes the importance of considering the impact of the current attitude on future states 4 steps ahead, allowing for parameter adjustments to prevent exceeding the control threshold.

(4) After removing its historical data, TBM features remain increasingly sensitive to attitude parameters as the time step approaches, with different TBM features affecting each parameter. As shown in Fig. 16, HDA, VDA, and VDT are all sensitive to "Thrust Force", though at different time steps. Historical data beyond 10 time steps are less impactful. Notably, one or two first-order sensitivities for each TBM attitude parameter are significantly higher than others. These parameters should receive extra attention during construction adjustments.

5 Merits Analysis

To further demonstrate the superiority of the proposed multi-step prediction approach, a comparison with the state-of-the-art methods (i.e. LSTM, GRU, and C-LSTM) based on time series data is necessary. LSTM, C-LSTM, and GRU methods are selected for this comparative study, with the hyperparameters of the GRU model set to match those of the C-GRU method. For a fair comparison, the structures and hyperparameters of C-LSTM and LSTM align with the reference [7], except that the 2D convolutional and max-pooling layers are replaced with corresponding 1D layers to fit our dataset. Figures 17 and 18 present the multi-step prediction results and R^2 values for the LSTM and GRU methods, respectively. A comparison between the C-GRU method and these traditional methods allows for the following discussion.

(1) The multi-step prediction accuracy of the C-GRU model surpasses that of the traditional LSTM and GRU methods. Figures 17 and 18 showcase the 21st step prediction results for the LSTM and GRU methods, respectively. While these methods generally align with the test data trends, a significant amount of data falls outside the confidence interval. Figures 19 and 20 display the R^2 values and distribution of prediction results for both methods. Except for some TBM attitude parameters matching the test set distribution, many predictions are invalid. The highest R^2 for LSTM is 0.9443, with the lowest at 0.1315, whereas the highest R^2 for GRU is only 0.8334, and the lowest is 0.4767. Both methods show a substantial gap compared to the C-GRU results. The mean R^2 for the C-GRU method over four TBM attitude parameters in 21-step prediction is 62.16% and 30.90% higher than the LSTM and GRU methods, respectively. Meanwhile, GRU demonstrates a slight accuracy advantage over LSTM, but both present less centrally distributed prediction results. The fitted line for these methods deviates from the diagonal, indicating poor symmetry. Here it can be concluded that for parameters with a stable trend, the shallow network performs adequately, with

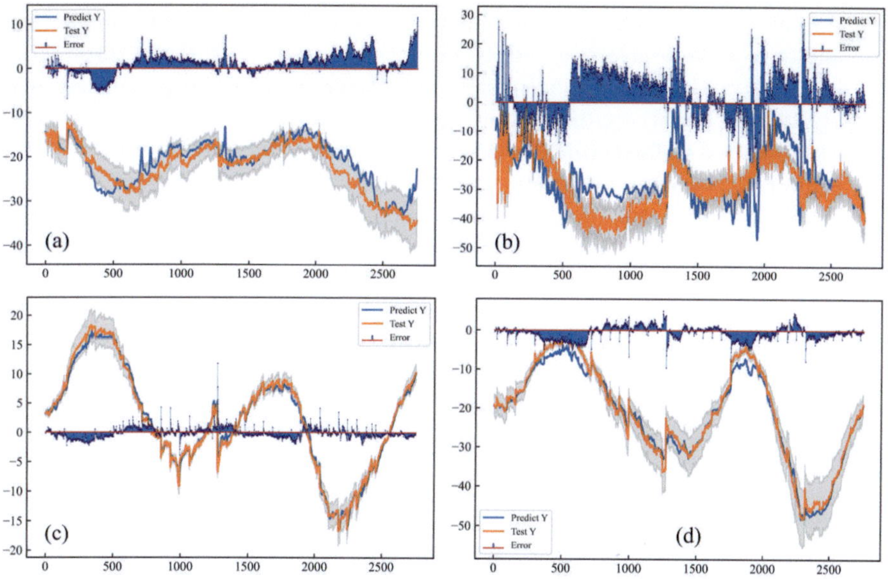

Fig. 17 Results of 21-step prediction of TBM attitude parameters with the LSTM method: **a** HDA; **b** HDT; **c** VDA; **d** VDT

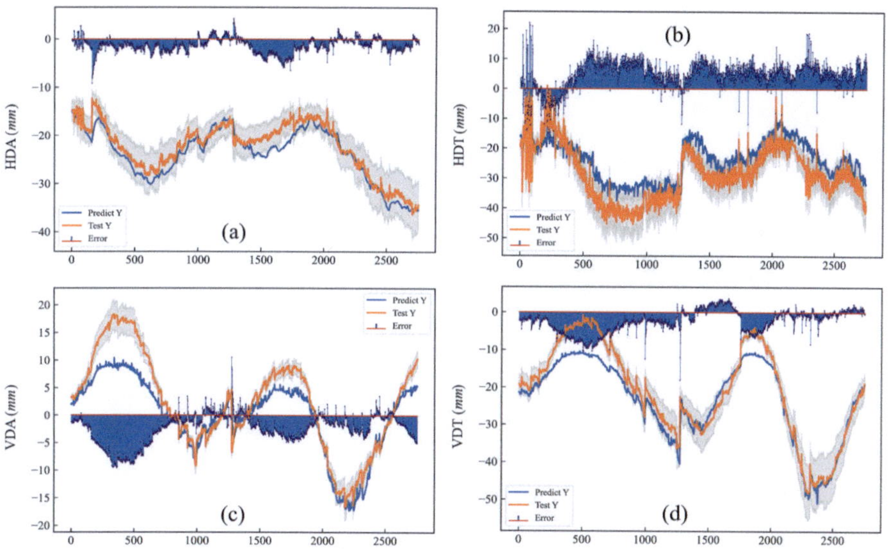

Fig. 18 Results of 21-step prediction of TBM attitude parameters with the GRU method: **a** HDA; **b** HDT; **c** VDA; **d** VDT

the deep network offering limited improvement. Conversely, 1D convolutional and max-pooling layers in C-GRU effectively extract meaningful information from the raw data, resulting in significant improvements for parameters with high variability, where the shallow network fails.

(2) The proposed method is more robust than both the LSTM and GRU methods. The overall accuracy of multi-step prediction evaluates the entire output sequence, which is crucial in engineering applications where the accuracy and stability of each step are paramount. According to Fig. 21a, LSTM shows robustness in the vertical direction, with no VDA and VDT data varying more than 10% compared to the previous step, and the highest relative change being 8.37%. However, the robustness of the LSTM for horizontal deviation is poorer. For HDA, while the first three prediction steps maintain high accuracy, it drops by nearly 20% after the fourth step and continues to decline. For VDT, the

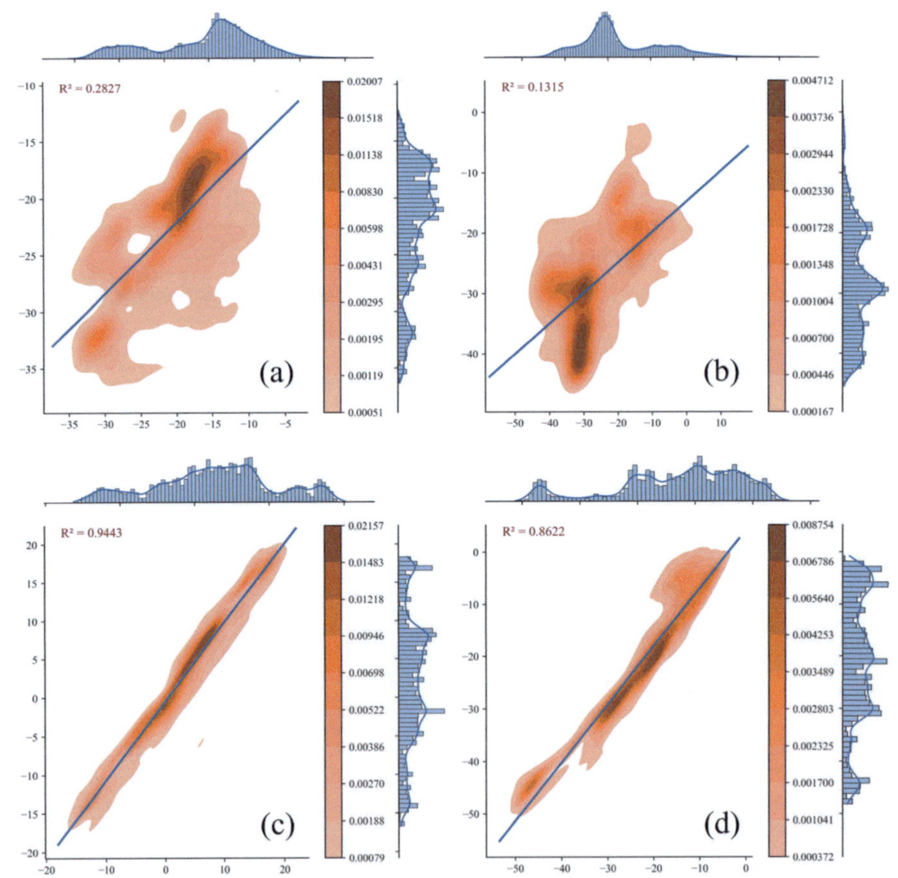

Fig. 19 R^2 of 21-step prediction of TBM attitude parameters with the LSTM method: **a** HDA; **b** HDT; **c** VDA; **d** VDT

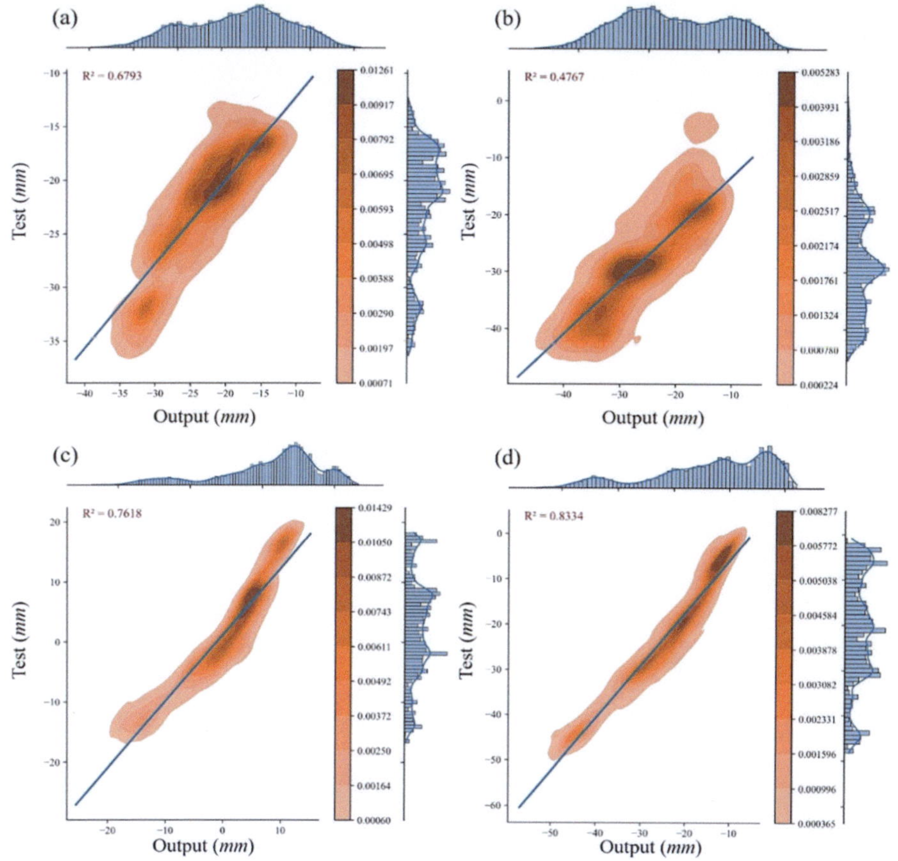

Fig. 20 R^2 of 21-step prediction of TBM attitude parameters with the GRU method: **a** HDA; **b** HDT; **c** VDA; **d** VDT

accuracy is not only lower, but the majority of relative changes exceed 20%, with a maximum variation reaching 93.79%. The GRU method performs even worse in robustness (Fig. 21b), particularly for HDT, where the relative change exceeds 17 times its previous step on the 2nd step.

(3) The proposed C-GRU method demonstrates superior accuracy and robustness compared to state-of-the-art deep learning methods. After comparing it with traditional time series methods, the C-GRU method is further evaluated by altering its structure. Table 3 summarizes the performance of all compared models. The detailed structure and hyperparameters of the C-LSTM training process are identical to those in reference [7]. The results indicate that the C-LSTM achieves higher horizontal prediction accuracy, significantly outperforming GRU and LSTM, with only a slight advantage in vertical accuracy. Once again, the effectiveness of the 1D convolutional and max-pooling layers in

Fig. 21 Robustness test of two comparison methods: **a** LSTM; **b** GRU

Table 3 R^2 score of comparison among deep learning methods

Method	R^2				
	HDA	HDT	VDA	VDT	Mean
C-GRU	0.9412	0.7758	0.9652	0.9193	0.9003
C-LSTM	0.8358	0.3245	0.9583	0.8816	0.7501
LSTM	0.2827	0.1315	0.9443	0.8622	0.5552
GRU	0.6793	0.4767	0.7618	0.8334	0.6878

Table 4 Significance of R^2 enhancement of C-GRU compared to different methods

Item	C-LSTM	LSTM	GRU
HDA	1.366×10^{-8}	2.746×10^{-10}	2.902×10^{-11}
HDT	2.916×10^{-20}	1.932×10^{-21}	2.006×10^{-6}
VDA	7.767×10^{-4}	2.886×10^{-2}	3.132×10^{-11}
VDT	1.030×10^{-5}	4.285×10^{-7}	2.311×10^{-8}
Total	1.279×10^{-11}	2.621×10^{-15}	3.218×10^{-22}

enhancing model accuracy, especially for fluctuating data, is reaffirmed. Overall, the C-GRU method remains optimal for the multi-step prediction of TBM attitude, illustrating that a well-constructed model can significantly enhance the prediction accuracy in multi-step predictions.

(4) The C-GRU method significantly improves multi-step prediction accuracy over state-of-the-art time-series deep learning models. To validate this improvement, a paired student's t-test is conducted for each output feature series after applying different models to the same TBM dataset (test set data). This test effectively assesses whether the results from different treatments of the same dataset are statistically significant. Table 4 lists the p-values obtained from the paired t-test for the three models. A p-value less than 0.05 indicates that the accuracy improvement of the C-GRU model is significant in the respective metrics. By calculating the p-values of the four attitude parameters (HDA, HDT, VDA, VDT) and their combined overall values, it is evident that the C-GRU significantly enhances TBM attitude multi-step prediction accuracy.

6 Chapter Summary

To address the limitations of the single-step TBM attitude prediction, the C-GRU method is proposed for the multi-step prediction. This approach effectively retains all features of the original tunneling process data and achieves high multi-step prediction accuracy. The method comprises three key steps: data acquisition and preprocessing, model establishment, and performance evaluation and model explanation. Preprocessing includes tunneling data extraction, correlation testing, data normalization, and data reframing, ensuring that only preprocessed data are used as the input dataset. The constructed model features three 1D convolutional layers, two 1D max-pooling layers, three GRU layers, and four fully connected layers. The input and output layer dimensions are adjustable based on specific scenarios and needs. RMSprop and MSE are employed as the optimizer and loss function, respectively. MAE and R^2 are selected as the evaluation indicators, and the RBD-FAST method is used to quantitatively analyze the first-order sensitivity between the input and output of the trained model.

The C-GRU model was applied to a portion of the Singapore T308 tunnel project to validate its effectiveness for the multi-step prediction. The main conclusions are as follows: (1) When the disturbance of the excavation data is minimal, the R^2 values of the multi-step prediction results are around 0.9, peaking at 0.9652. Even with significant fluctuations in the excavation data, the multi-step prediction results remain robust, with R^2 values being up to 0.7758. (2) With an input time step of 64, the output sequence for the multi-step prediction can extend to 21 steps, demonstrating strong robustness. (3) The predicted TBM attitude at the current time step is more sensitive to the TBM attitude parameters from the past 21 time steps, especially the most recent 11. The most sensitive parameter is the TBM attitude at the fourth past time step. After removing the sensitivity of the historical TBM attitude parameters, the sensitivity of each TBM feature varies. These sensitivity insights provide valuable reference points for adjusting TBM attitude and operation in engineering applications. (4) Compared to state-of-the-art prediction methods based on time series data (e.g., LSTM, GRU, and C-LSTM), the C-GRU method demonstrates significantly higher accuracy and robustness when applied in engineering contexts.

References

1. Wang K, Wu X, Zhang L, Song X (2023) Data-driven multi-step robust prediction of TBM attitude using a hybrid deep learning approach. Adv Eng Inform 55:101854
2. Shen X, Lu M, Chen W (2011) Computing three-axis orientations of a tunnel-boring machine through surveying observation points. J Comput Civ Eng 25:232–241
3. Zhang M, Zhang X, Wu H, Javadi AA, Dai Z (2024) Theoretical study of mechanical behavior of tunnels considering soil-machine interactions. KSCE J Civ Eng 28:4473–4486
4. Adoko AC, Gokceoglu C, Yagiz S (2017) Bayesian prediction of TBM penetration rate in rock mass. Eng Geol 226:245–256
5. Jakubowski J, Stypulkowski J, Bernardeau F (2017) Multivariate linear regression and CART regression analysis of TBM performance at Abu Hamour phase-I tunnel. Arch Min Sci 62:825–841
6. Wei M, Wang Z, Wang X, Peng J, Song Y (2021) Prediction of TBM penetration rate based on Monte Carlo-BP neural network. Neural Comput Appl 33:603–611
7. Fu X, Zhang L (2021) Spatio-temporal feature fusion for real-time prediction of TBM operating parameters: a deep learning approach. Autom Constr 132:103937
8. Zhou C, Xu H, Ding L, Wei L, Zhou Y (2019) Dynamic prediction for attitude and position in shield tunneling: a deep learning method. Autom Constr 105:102840
9. Cachim P, Bezuijen A (2019) Modelling the torque with artificial neural networks on a tunnel boring machine. KSCE J Civ Eng 23:4529–4537

10. Lin S-S, Zhang N, Zhou A, Shen S-L (2022) Time-series prediction of shield movement performance during tunneling based on hybrid model. Tunn Undergr Space Technol 119:104245
11. Zhang N, Zhang N, Zheng Q, Xu Y-S (2022) Real-time prediction of shield moving trajectory during tunnelling using GRU deep neural network. Acta Geotech 17:1167–1182
12. Wei Y, Chen Z, Zhao C, Tu Y, Chen X, Yang R (2021) A BiLSTM hybrid model for ship roll multi-step forecasting based on decomposition and hyperparameter optimization. Ocean Eng 242:110138
13. Lu Y, Tian Z, Zhou R, Liu W (2021) Multi-step-ahead prediction of thermal load in regional energy system using deep learning method. Energy Build 233:110658
14. Gao D-W, Zhu Y-S, Zhang J-F, He Y-K, Yan K, Yan B-R (2021) A novel MP-LSTM method for ship trajectory prediction based on AIS data. Ocean Eng 228:108956

Advanced TBM Attitude Control

1 Introduction

The growth of underground engineering projects has been witnessed around the world to facilitate regional communication and support economic development and relieve traffic pressure [1]. Tunnel boring machines (TBM) have been extensively accepted and used in underground projects due to their fast speed, high safety, low cost, and easy operation [2]. In the operation, the deviation of the position may interfere with the tunneling route and disturb the surrounding underground environment, resulting in increased safety risks, thus affecting the construction schedule and cost with increasing operation time. Therefore, the TBM tunneling position is critical to the successful construction of underground engineering projects.

Practically, TBM operators and site managers often encounter unexpected out-of-tolerance errors that significantly impact the progress of underground engineering projects under manual operation [3]. On one hand, operators need to constantly monitor TBM performance and make real-time adjustments, which is time-consuming and can lead to delays. On the other hand, human operators are prone to fatigue, which can reduce their efficiency and increase the likelihood of errors. As for reliability, TBM operations involve complex spatial–temporal dynamics that are challenging for human operators to manage accurately, where the consistency of decisions made by human operators is hard to guarantee. Varying levels of the operator's experience, expertise, and subjective judgment can affect the reliability of TBM operations. Additionally, consistently accurate judgment under dynamic environment conditions is very unlikely to happen even for an experienced operator. Therefore, establishing a more effective and reliable automated TBM operation method is urgently needed in the industry, which significantly motivates this study.

With these concerns, the deep reinforcement learning (DRL) method stands out as a cutting-edge approach, which supports processing large-scale data with uncertainty and complexity to mine the patterns of the position control of the TBM. The DRL method merges the advantages of deep learning and reinforcement learning,

L. Zhang et al., *Intelligent Construction in Tunnels*, Lecture Notes in Civil Engineering 679, https://doi.org/10.1007/978-981-96-8785-5_8

facilitating the integration of deep learning's feature extraction capabilities with the decision-making abilities of reinforcement learning [3]. This allows the direct learning of control policies from high-dimensional raw data. Specifically, this method enables the agent to dynamically interact with the environment and uses the environment information as input to obtain experience to update the parameters of the model [4]. Thanks to modern sensor technologies, the on-site operation data from the TBM provides valuable prerequisites for successfully implementing the machine learning method and examining its performance. Notably, the collected data exhibit significant variability, encompassing spatially diverse geological parameters, and temporally fluctuating operational parameters. However, current studies employing DRL in the domain of TBM control mostly use a numerical database or model to simulate the environment [5], lacking the simulation of the spatial-temporal dynamics along with the TBM operation.

In this chapter, a novel method for intelligent TBM position operation by employing DRL for automatic control with its spatial-temporal interaction with the environment simulated by a time-series forecasting model. Specifically, a feature selection and data processing algorithm are proposed to tackle large-scale data for providing detailed and valuable information as input. Afterward, the long-short-term memory (LSTM) is introduced to model the DRL environment, which considers multiple historical steps to simulate the spatial-temporal dynamics of the TBM. The policy network is then built to model the position control by combining actors and critics derived from TD3 for presenting the optimal strategy. Several suitable loss functions, such as environment loss, actor loss, and critic loss, are designed to assist the proposed method in learning from the environment. Ultimately, the insights gained from the results can lead to an optimal strategy that may serve to guide the TBM position control to smoothly advance the tunneling process.

2 Automated TBM Operation

The core technique for automated TBM operation is to replace manual operation by intelligent decision-making techniques. Traditional optimization algorithms include Genetic Algorithms (GA) [6], Particle Swarm Optimization (PSO) [7] and their variations have been widely used. Particularly, the optimization of TBM operation aims to optimize the safety and efficiency of the excavation process. For instance, Liu et al. [8] employed the nondominated sorting genetic algorithm (NSGA-II) to optimize TBM operational parameters for urban subway construction, improving decision-making under varying geological conditions. Qin et al. [9] introduced an adaptive decision-making method for shield machine operation parameters using a D-ResNet network and improved inertia weight adaptive constrained particle swarm optimization algorithm (IWACPSO), significantly enhancing tunneling performance by effectively considering geological conditions. These optimization strategies are all proven to be effective. However, as optimization focuses on finding the best solution to a static problem with a known objective function and constraints, it is unsuitable to be applied in a dynamic and contentious problem.

Recently, DRL has garnered considerable interest due to its combined advantages of deep learning and reinforcement learning with powerful functionality to fit strategies to solve state-action space problems by leveraging the powerful representational capabilities of neural networks [9]. With years of development, there are some applications in TBM operation. For instance, Zhang et al. [10] proposed a hybrid DRL model by integrating deep Q network (DQN) and PSO to improve the extreme learning machine (ELM) model for real-time prediction of tunneling-induced ground responses. Similarly, Erharter et al. [11] developed a novel DRL-based framework containing a virtual environment with tunnel breakthrough as a goal and a DQN as an agent for providing strategic guidance to perform tunneling in an economical and safe manner. Those methods have returned promising results with a strong impression, which opens the way to an efficient method with a low workload applied in the TBM field. However, all these methods are limited in the simulation of the environment by failing to consider the temporal characteristics of the excavation process.

In summary, most of the studies have provided promising results and proven applicability in specific scenarios but have limitations or potential for improvement. Firstly, the current application of DRL in tunnel construction is limited to the machine-soil interaction to simulate the environment, failing to consider TBM parameters in the environment. Additionally and most importantly, the application of DRL in TBM position control is still rare and the application is significantly held up by the difficulties in simulating the working environment. As the excavation of TBM is a dynamic process in both space and time, simultaneous consideration of the special-temporal dynamics of the TBM is of high importance. Therefore, this chapter is devoted to developing an intelligent TBM position forecasting and control by developing a time-series forecasting model to simulate the space-time dynamic interaction of TBM and the environment in DRL for automatic control. By filling these research gaps, this chapter is expected to further motivate the automation of TBM operation in the field of tunnel construction.

3 TBM Attitude Control Framework

To achieve autonomous control of the TBM position, a research framework is developed from four key modules: (1) feature selection and data processing, (2) construction of a simulated environment, (3) agent-enabled position control, and (4) utilization of loss functions and evaluation metrics. The details of the DRL method are depicted in Fig. 1.

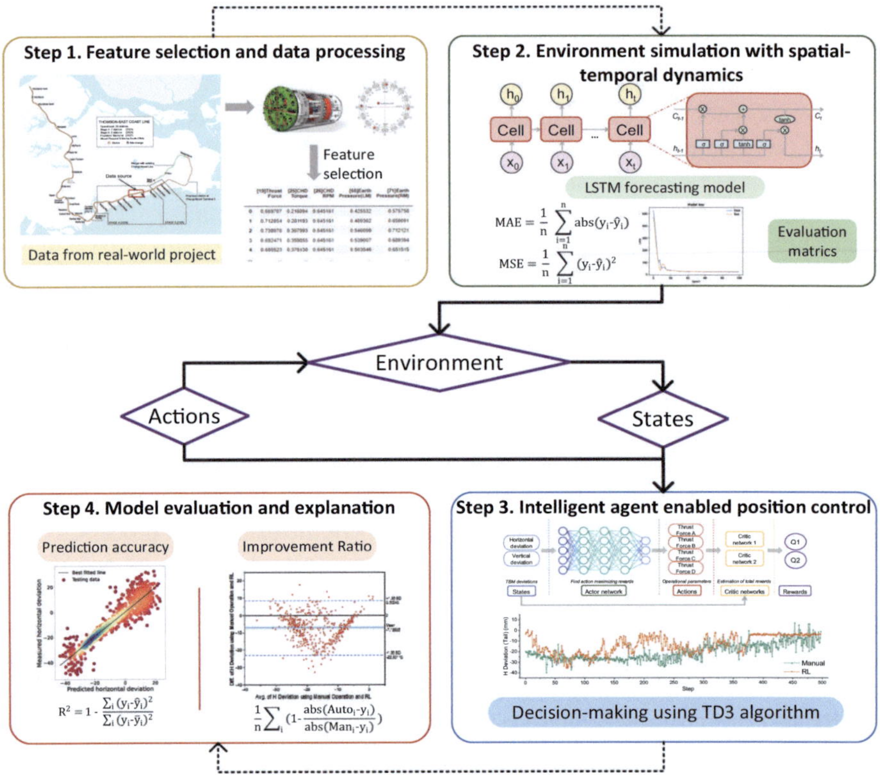

Fig. 1 Architecture of the proposed DRL approach

3.1 Feature Selection and Data Processing

Feature selection, as a data pre-processing strategy, effectively addresses data analysis challenges by efficiently preparing data—particularly high-dimensional data—through the removal of irrelevant and redundant information, resulting in a cleaner and more interpretable dataset. Feature selection techniques aim to build simpler and more capable models for data mining with the benefits of reducing computation time and improving learning accuracy. In this chapter, the original dataset contains hundreds of variables for TBM operation monitoring, such as the ring number, excavation time, TBM geometry, earth pressure, cutterhead torque, depth, thrust force, etc. If working directly on those variables, the computational cost could become prohibitively computationally demanding and complex without favoring the model performance. Therefore, this chapter adopts feature selection to reduce the data size to facilitate the model performance for data mining.

As shown in Fig. 2, the TBM has numerous components to produce the thrust and control the position of the TBM. Those thrust components can be divided into four groups, namely the cylinders right (A), lower (B), left (C), and top (D), generating

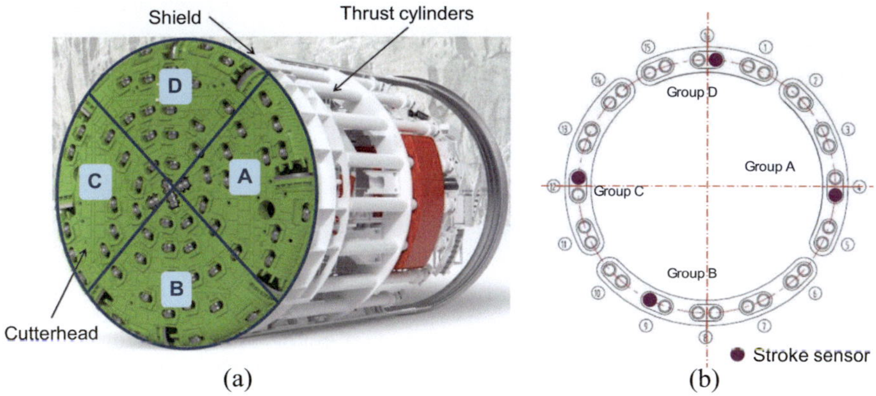

Fig. 2 Thrust cylinders with stroke sensor in TBM: **a** divided four groups of the TBM cylinder (i.e., Cylinder right (A), lower (B), left (C), and top (D)); **b** installation of stroke sensor at the TBM

the forces that directly control the position of the TBM. Controlling these four thrust forces directly influences the deviations of the TBM. Herein, those four groups related to TBM operation parameters are set as the controllable variables for the proposed model to learn. Accordingly, the appropriate parameters measuring the TBM status are required to assist the model in data mining for TBM position control to facilitate tunneling. To include temporal characteristics of TBM operation, all variables considered are grouped in the form of time-series data to consider multiple historical time steps.

Data processing aims to handle the collected data to promote the usefulness and accessibility of the data. Data processing primarily includes data cleaning, data normalization, and data reconstruction, where data cleaning and data normalization are employed in this chapter to treat the collected TBM tunneling data. The cleaning of data aims to remove noise from the data to avoid interfering with the model performance. Specifically, it removes the invalid information containing zero values and null cells to prevent unnecessary effects on the model for TBM position control. As for the data normalization, this chapter achieves a balanced scale among input features by transforming units to avoid tedious data processing in the DRL algorithm.

3.2 *Environment Simulation with Spatial-Temporal Dynamics*

The configuration of the simulated environment with spatial-temporal dynamics is depicted in Fig. 3. The environment model is trained using actual environmental characteristics to capture spatial features. Additionally, by incorporating historical data through a time-series forecasting model, the temporal dynamics of the TBM

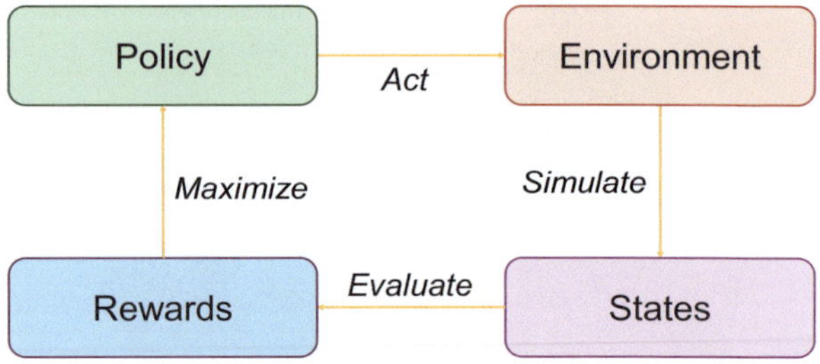

Fig. 3 Architecture of the simulated policy and environment networks

are also considered. In this chapter, the simulated environment created simulates the movement of the TBM, incorporating specified variables like thrust force and environmental pressure. The term "policy" denotes the set of rules dictating how the agent makes decisions based on the current state and receives rewards. Given the consideration of spatial-temporal dynamics in the environment concerning the TBM, the agent engages with the environment continuously, incorporating historical actions and states into each new step.

Owing to the complexity of the underground environment, dynamic changes over time arise and constantly affect the operation of the TBM, posing a challenge to constructing an accurate physical environment. However, the contentious historical data collected makes it possible to simulate the temporal dynamics of the TBM. Therefore, a time-series-based learning model is warranted. In view of the given dataset, a simulated environment is constructed based on an LSTM neural network to simulate the real-world dynamic environment. LSTM has proven to be a successful neural network applied to time series with the ability to process complex serial data for achieving the target task [12].

The key feature of LSTM lies in its memory cell, which is maintained in its state over a period of time where the weight of the self-loop changes by adding an input gate, forget gate, and output gate. In detail, the input gate determines the components of the incoming input and the previous cell state that should be updated and stored in the cell's memory. By taking the current and the previous hidden state as input, it computes the "input activation," indicating the new information to be incorporated into the cell's memory. The forget gate is responsible for determining the information from the previous cell state to be discarded. It takes the current and the previous hidden state as input and computes the "forget activation," representing the portion of the cell's memory to be erased. As for the output gate, it decides which parts of the updated cell state should be revealed as the output of the LSTM cell. Taking the current input and the previous hidden state as input, it calculates the "output activation," signifying the final output of the LSTM cell.

As can be seen from Fig. 4, the LSTM has three inputs, namely the cell state C_{t-1}, the hidden state h_{t-1}, the input vector X_t at time t, and two outputs, namely the cell state C_t, the hidden state h_t at time t. Herein, the involved three operations are as follows: (1) The cell state C_{t-1} information is passed, and the hidden layer state h_t at time t with the input X_t is modified appropriately for C_t and then passed to the next time; (2) C_{t-1} goes into the computation of the output h_t at time t; (3) The

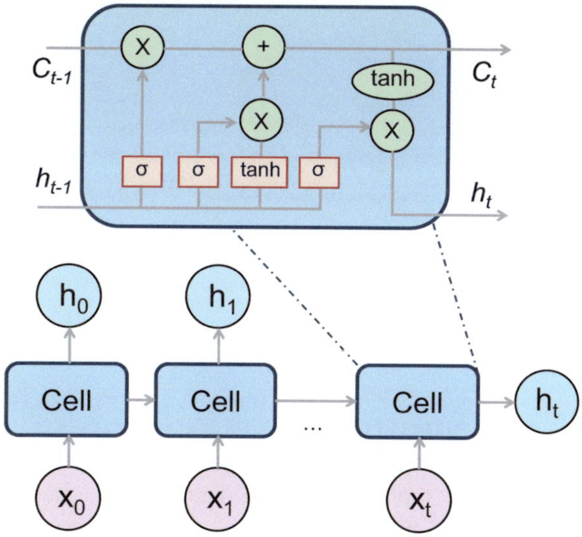

Fig. 4 Internal structure of LSTM cells

information of the hidden state h_{t-1} through the gate structure of the LSTM would modify the cell state and participate in the computation of the output. The LSTM relies on the cell state and the gate structure to process the input for the information transfer with long-term memory.

3.3 Intelligent Agent-Enabled Position Control

DRL is able to engage with the agent and offer feedback, thereby iteratively refining the agent's actions for enhanced development. To effectively control the position of the TBM, an agent-enabled position control model, i.e., the policy network, is developed to interact with the environment to provide strategies for adjusting the TBM's position. Derived from the Twin Delayed Deep Deterministic (TD3) algorithm [13], the model consists of an actor and a critic, which interact with the environment and feedback from the critic for continuous optimization to provide the forces of the four groups to achieve position control of the TBM. Critics are employed to evaluate the actors via Q-values in order to reduce the deviation between the real route and the simulated route.

In the DRL workflow, the deviations of the TBM are fed into the actor network as states, and then the output strategy is utilized as the actions to operate the TBM to generate four forces for the TBM position control. In this process, the actor interacts with the environment to generate rewards for the critics for evaluating the performance of the actor network. The actor network will be updated once after several steps of critic updating. This cycle is repeated to eventually produce the optimal strategy to aid in controlling the TBM's position for tunneling. The architecture of

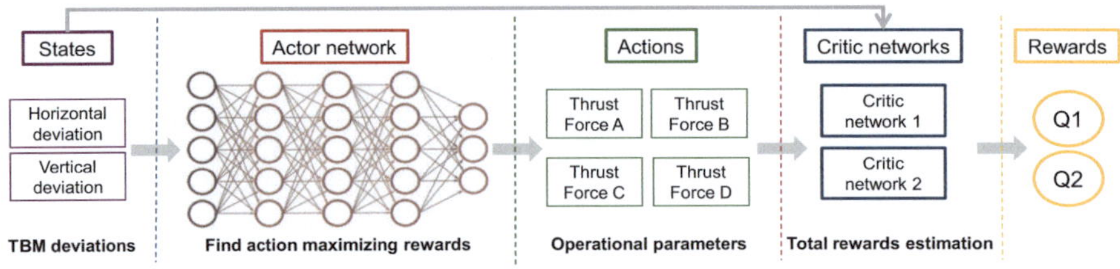

Fig. 5 Working flow of the DRL policy network

the policy network is presented in Fig. 5, which illustrates the basic function of each component and the relationship among them.

As a more advanced DRL method, TD3 evolves from Deep Deterministic Policy Gradient (DDPG) [14], which shows its superiority compared with traditional DRL methods by introducing techniques including the replay buffer and soft network updates. The replay buffer is a data structure that stores experiences encountered by the agent during interactions with the environment. Each experience consists of the current state, action, reward, next state, and whether the episode is finished. During training, the agent samples batches of experiences from the replay buffer to update its actor and critic networks. By using a replay buffer, the agent breaks the temporal correlation between consecutive experiences, reducing the possibility of learning from biased sequences and leading to more stable and efficient learning. The soft network update is used to stabilize the learning process and prevent the target networks from oscillating or diverging during training. Soft updates involve slowly blending the online network's parameters into the target network's parameters using a hyperparameter called the target network update rate.

As an advanced version of DDPG, TD3 keeps all the features above and introduces three new techniques including the double critic network, delayed updates, and target policy smoothing. There are two separate critic networks (Q-networks) to reduce the overestimation bias in TD3. The two critics independently estimate the action-value function, and their outputs are averaged during the training process. This averaging helps to mitigate the potential overestimation of Q-values and leads to more stable and accurate value function approximations. Delayed updates in TD3 refer to the infrequent updating of the actor networks. While the critics are updated more frequently (at each time step), the actor network is updated less often.

This decoupling allows the actor to explore the environment with a relatively unchanged policy for some time before updating it based on the most recent critic output. Delayed updates prevent the policy from diverging too quickly and help to stabilize the learning process. Lastly, the target policy smoothing is a regularization technique used to improve the stability of training in TD3. During the computation of target Q-values for the actor's loss function, noise is added to the target actions to make the target Q-values become smoother and less susceptible to sudden changes, which can lead to better convergence and more stable learning. A pseudocode explaining the logic of TD3 is presented in Algorithm 1.

Algorithm 1 TD3 algorithm with two separate critic networks.

Input: initialized policy network parameter θ, critic network parameter ϕ_1, ϕ_2, empty replay buffer D

Assign values $\theta_{targ} \leftarrow \theta, \theta_{targ,1} \leftarrow \phi_1, \theta_{targ,2} \leftarrow \phi_2$

repeat

  Observe state s and select action $a = clip\big(\mu_\theta(s) + \epsilon, a_{Low}, a_{High}\big)$, where $\epsilon \sim N$

  Interact with the environment using action a

  Obtain next state s', reward r, and done signal d to indicate termination

  Record (s, a, r, s', d) in replay buffer D

  If *done*, reset state and done signal

  if it's time to update**:**

    for j in range (number of updates) **do**

      Randomly select a batch of samples from the replay buffer, $B = \{(s, a, r, s', d)\}$

      Compute target actions

$$a'(s') = clip\big(\mu\theta_{targ}(s') + clip(\epsilon, -c, c), a_{Low}, a_{High}\big), \epsilon \sim N(0, \sigma)$$

      Compute targets

$$y(r, s', d) = r + \gamma(1 - d) \min_{i=1,2} Q_{targ,i}(s', a'(s'))$$

      Update Q-functions by gradient descent:

$$\nabla_{\phi_i} \frac{1}{|B|} \sum_{(s,a,r,s',d)\in \mathrm{B}} \big(Q_{\phi_i}(s, a) - y(r, s', d)\big)^2 \text{ for } I = 1, 2$$

      if j mod policy_delay $= 0$ **then**

        Update policy by gradient ascent:

$$\nabla_\theta \frac{1}{|B|} \sum_{s\in \mathrm{B}} Q_{\phi_i}(s, \mu\theta(s))$$

        Update target networks with

$$\phi_{targ,i} \leftarrow \rho\phi_{targ,i} + (1 - \rho)\phi_i \text{ for } I = 1, 2$$

$$\theta_{targ} \leftarrow \rho\theta_{targ} + (1 - \rho)\theta$$

      end if

    end if

  end if

until convergence

3.4 *Model Evaluation and Explanation*

In this DRL network, three losses are designed to assist the model in learning, namely, environmental loss, actor loss, and critic loss. For the environment model, it is considered a regression problem by analyzing the data. This chapter applies the mean squared error (MSE) loss as the environmental loss, which is expressed by Eq. (1).

$$MSE = \frac{1}{n}\sum_{i=1}^{n}(y_i - \hat{y}_i)^2 \quad (1)$$

where y_i represents the measured value, $\hat{y}_i$ stands for the predicted value, and n denotes the total number of samples.

For the agent-enabled position control model, the actor loss is actually a value-based loss function, which is computed by the critic network, which can be expressed by Eq. (2). The critic loss relies on the MSE function as shown in Eq. (1). The difference is that the critic loss is combined by two networks as shown in Eq. (3).

$$Actorloss = -\min(Critic_1(s, a), Critic_2(s, a)) \quad (2)$$

$$Criticloss = MSE(Critic_1(s, a), target) + MSE(Critic_2(s, a), target) \quad (3)$$

where s is the input state, a means the taken action, and the target indicates the planned target.

To evaluate the performance of the proposed model, several evaluation metrics are adopted, including the root mean squared error (RMSE), mean absolute error (MAE), and R^2. The RMSE is taken as the root of the MSE. The MAE is an instinct metric to show the deviation between the simulated route and the real route, which is shown in Eq. (4). R^2 can reflect the variance between the simulated route and the real route, which can be given in Eq. (5).

$$MAE = \frac{1}{n}\sum_{i=1}^{n} abs(y_i - \hat{y}_i) \quad (4)$$

$$R^2 = 1 - \frac{\sum_i (y_i - \hat{y}_i)^2}{\sum_i (y_i - \overline{y}_i)^2} \quad (5)$$

In order to find the importance of each variable in the trained environment model, this chapter applies the SHapley Additive exPlanations (SHAP) [15] technique. Based on this method, the importance of each feature will be computed with the aid of the additive feature attribution method, whose key is to compute the Shapley value, whose definition is as below:

$$\varphi_i = \sum_{S \subseteq N \setminus \{i\}} \frac{|S|!(K - 1 - |S|)!}{K!}\left[f_x(S \cup \{i\}) - f_x(S)\right] \quad (6)$$

where N is the set of all features, S contains non-zero entries, K is the number of features, and $f_x(S) = E[f(x)|x_S]$ is derived from the original prediction model outputs.

Additionally, to further examine the performance of the model, a comparison with human control is performed. For reflecting the performance between the man-controlled route and the autonomous-controlled route, a performance improvement ratio is computed, which is shown in Eq. (7).

$$Improvement\ Ratio = \frac{1}{n}\sum_{i}\left(1 - \frac{abs(Auto_i - y_i)}{abs(Man_i - y_i)}\right) \quad (7)$$

where $Auto_i$ represents the ith position by the proposed model, Man_i means the ith results by manual, and y_i stands for the ith planned position.

4 Experimental Studies

4.1 Data and Sources

A through-underground tunnel project belonging to the Thomson-East Coastline Tunnel Project in Singapore connects T308 Marine Terrace Station with T309 Siglap Station is selected for demonstration purposes in this chapter. It is a twin-lane tunnel with a length of 1.34 km, as shown in Fig. 6. This project is constructed using two Earth Pressure Balance (EPB) TBMs, whose cutterhead diameter is 6.67 m and length is 9.81 m. The EPB TBM is installed with a data collection system having hundreds of sensors monitoring, which is capable of collecting all real-time TBM parameters with intervals per 20 mm net stroke. This chapter focuses on data from 200 rings, i.e., ranging from 300 to 499 rings, to prepare a dataset for studies.

The available dataset prepared for this chapter contains approximately 6000 samples with about 200 operational parameters. Based on the feature selection, 11 features, and 2 targeted objectives are captured, which are described in detail in Table 1. These features are capable of capturing the interaction of the TBM machine with the environment and its changing factors during the TBM tunneling process with values that vary over time. To enhance the reasonability of features, some new features are generated based on the provided data. Instead of directly using the coordinate of the TBM, the change of the coordinate in each time step is computed to describe the route of the TBM, including the change in N coordinate (tail) (×1), change in E coordinate (tail) (×2), and change in Z coordinate (tail) (×3). Besides the position of the TBM, the penetration rate (×4) of the TBM is considered to control the speed of the excavation. The earth pressure (×5), which is computed as the average of the right and left sensors, is considered to describe the ground condition. All of the features mentioned above are used to describe the condition of the TBM.

As for the operational parameters to control the TBM attitudes, which are measured by the horizontal deviation of the TBM tail (y1) and the vertical deviation of the TBM tail (y2), the thrust forces from cylinder group A (×6), group B

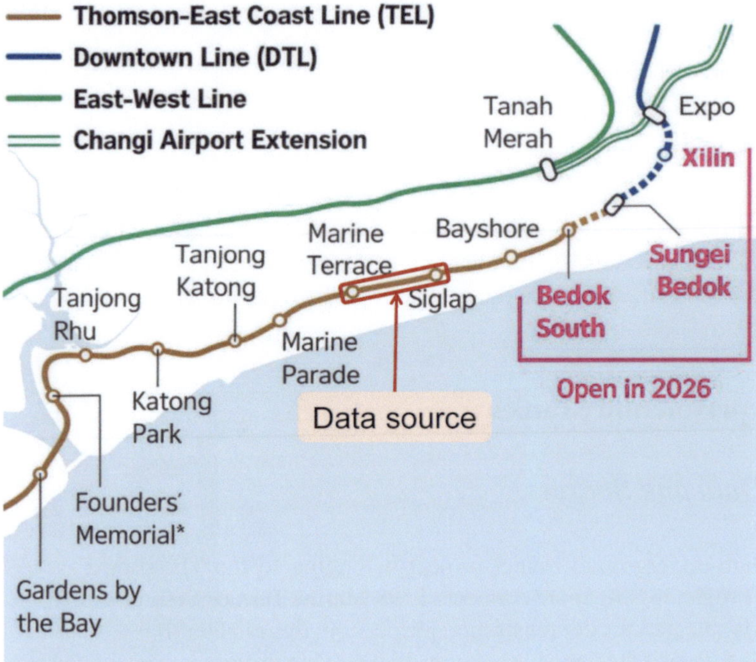

Fig. 6 Project profile of the Singapore's Thomson-east coastline [16]

(×7), group C (×8), and group D (×9) are considered. Therefore, in the DRL algorithm, there will be 4 actions coming from the thrust forces and 2 states measuring the deviations. To adjust the operational parameters in advance, the environment is trained as a time series prediction model with 2 historical steps. The data visualization of the action and state features is displayed in Fig. 7. It can be told that the thrust force from cylinders A and C does not differ too much, but that from group B is much larger than group D in order to balance the TBM self-weight. Additionally, the horizontal deviation is under less control as it varies much more compared with the vertical deviation.

4.2 Model Development

As training the environmental model, an Adam optimization algorithm with a learning rate of 0.005 is employed as an optimizer to continuously adjust the LSTM-based model to train the simulated environment. To enhance the learning capability of the model in the training process, this chapter designs a learning rate scheduling scheme that automatically changes the learning rate and monitors the validation loss to strengthen the training of the model. When the validation loss stops decreasing after

Table 1 Descriptions of the selected features

Categories	Parameter	Symbol	Unit	Description
Input	Change in N coordinate (tail)	× 1 (t − 1, t − 2)	mm	Change in north coordinate
	Change in E coordinate (tail)	× 2 (t − 1, t − 2)	mm	Change in east coordinate
	Change in Z coordinate (tail)	× 3 (t − 1, t − 2)	mm	Changer in depth
	Penetration rate	× 4 (t − 1, t − 2)	m/min	Penetration rate of the TBM
	Earth pressure	× 6 (t − 1, t − 2)	bar	The average earth pressure measured from the left and right sensors
	Thrust force from cylinder groups A, B, C, and D	× 7, × 8, × 9, × 10 (t − 1, t − 2)	*1000 kN	Push the machine forward to be determined by the DRL algorithm
	Horizontal deviation	× 11 (t − 1, t − 2)	mm	Horizontal deviation of the TBM tail
	Vertical deviation	× 12 (t − 1, t − 2)	mm	Vertical deviation of the TBM tail
Output	Horizontal deviation	y1 (t)	mm	Horizontal deviation of the TBM tail
	Vertical deviation	y2 (t)	mm	Vertical deviation of the TBM tail

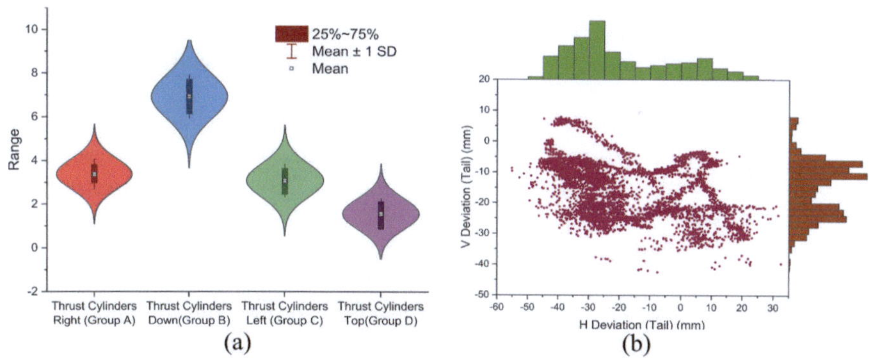

Fig. 7 Data visualization of the selected features: **a** operational parameters for deviation control, **b** horizontal and vertical deviations

10 epochs, the learning rate will be multiplied by 0.5 to promote the learning of the model. In addition, a grid search is used here to search appropriate hyperparameters to set the model for learning.

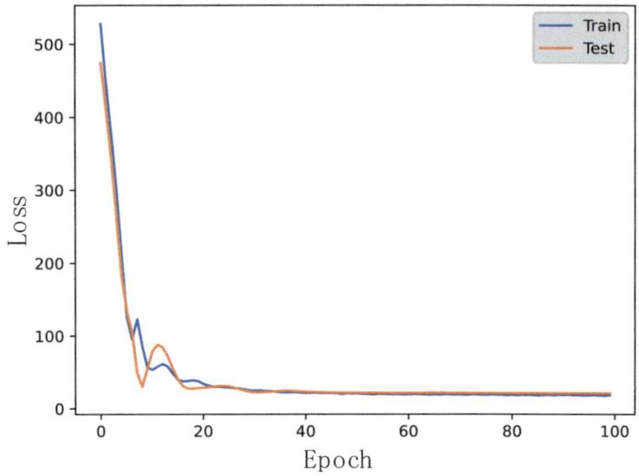

Fig. 8 Training and validation losses from the environment model

Based on the 6000 samples collected, the first 4500 of them are used for environment model training, with the following 1000 samples used for model testing. The last 500 samples are then utilized to examine the performance of the DRL algorithm. During the training process, the data in the training set undergoes a transformation that includes timesteps $t-1$ and $t-2$ to predict the TBM tail deviation at time step t. This data is consistently input into the created environment model with a batch size of 256. Concurrently, the training performance is assessed using loss and accuracy metrics for real-time tracking, extending up to 100 epochs. The results in Fig. 8 showcase the training progression, where the blue curve signifies the recorded training losses, and the orange curve represents the recorded validation losses. Evidently, the loss values initially decrease rapidly and then stabilize gradually, indicating the effective training of the developed model.

To validate the reasonability of using 2 historical time steps, the training performance using 1 and 3 historical steps is also tested. A higher number of historical time steps is not considered as it will take too long for the DRL algorithm to initialize. As can be seen from Table 2, using 1 historical step cannot provide adequate information to produce a precise model, whose R^2 is only 0.84 for horizontal deviation and 0.87 for vertical deviation. As more historical time steps are considered, the prediction accuracy increases, where 2 historical steps behave the best result with a R^2 of 0.86 for horizontal deviation and 0.92 for vertical deviation, respectively. The fitting results of prediction and original samples are plotted in Fig. 9. According to the SHAP analysis result plotted in Figs. 10 and 11, the horizontal deviation is more affected by the right and left cylinders (groups A and C) in the opposite manner. This proves the validation of the environment model as the behavior of the horizontal deviation fits well with the horizontal thrust forces. As in Fig. 1, a similar trend can also be observed that the top and bottom cylinders are influencing the vertical deviation in an opposite way.

Table 2 Performance of the developed environment model in the testing samples

Dataset	Historical steps	Horizontal deviation (mm)			Vertical deviation (mm)		
		RMSE	MAE	R^2	RMSE	MAE	R^2
Training	t − 1	3.47	2.22	0.96	0.85	0.57	0.99
	t − 1, t − 2	3.29	2.08	0.96	0.79	0.50	0.99
	t − 1, t − 2, t − 3	3.22	2.07	0.97	0.82	0.52	0.99
Testing	t − 1	6.66	4.58	0.84	1.39	1.08	0.87
	t − 1, t − 2	6.42	4.29	0.86	1.10	0.78	0.92
	t − 1, t − 2, t − 3	6.29	4.15	0.87	1.27	0.88	0.89

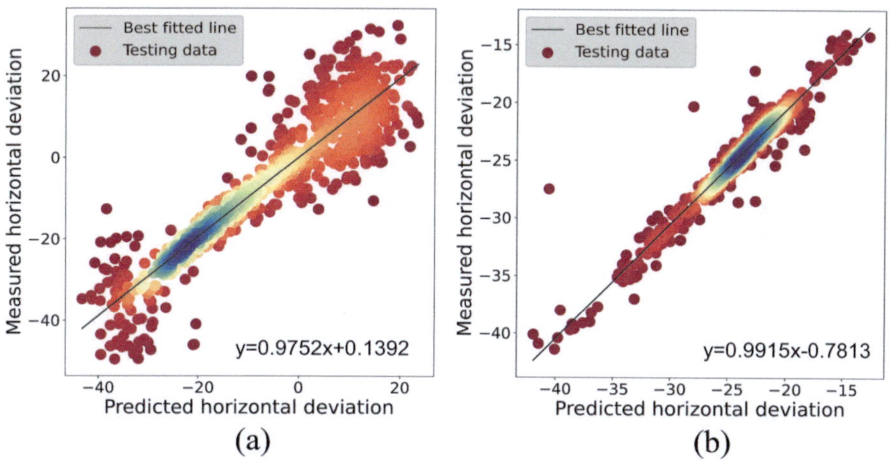

Fig. 9 Ground truth and prediction of the deviation in **a** horizontal and **b** vertical directions in the testing dataset

Further to examine the effectiveness and reasonability of using LSTM to forecast the TBM deviation, a comparative study is conducted by comparing LSTM with other state-of-the-art techniques, including a convolutional neural network (CNN), a hybrid CNN-LSTM network, an LSTM based transformer network, and a traditional ANN. All networks use similar structures (i.e. number of layers and neurons) and are all properly trained. The transformer network employs the LSTM network as its encoder and decoder to be more comparative. All models are trained using two historical steps. The R^2 value predicting the testing data using all models is summarized in Table 3. The LSTM is proven to have the highest accuracy in predicting the vertical deviation with comparative accuracy in predicting horizontal deviation. The results well indicate the reasonability of our methodology.

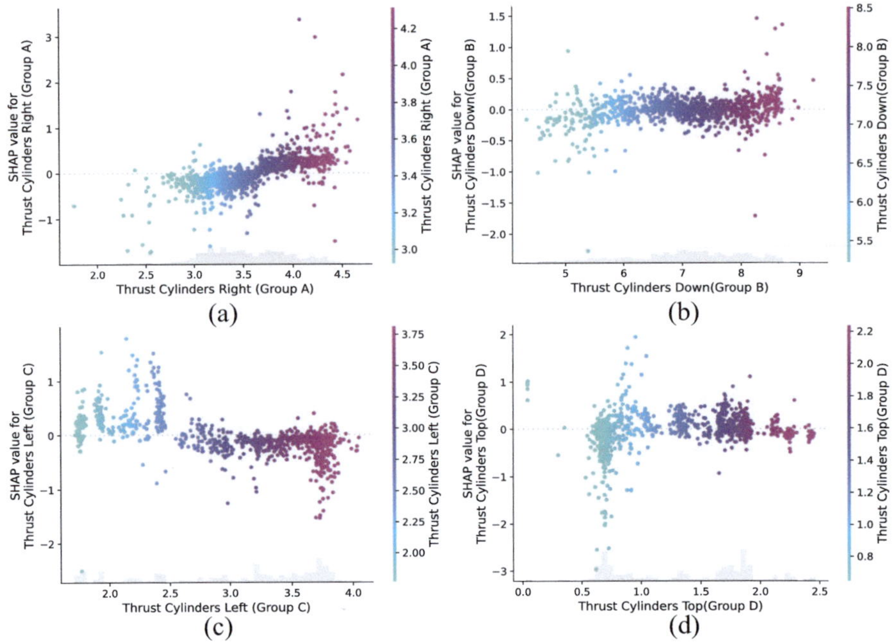

Fig. 10 SHAP value of **a** thrust cylinders group A, **b** thrust cylinders group B, **c** thrust cylinders group C, **d** thrust cylinders group D in modeling the horizontal deviation

4.3 Analysis of Results

In the implementation of DRL, the Adam optimization algorithm is applied as the optimizer to continuously adapt both the critic network and the actor network for training, where all networks are trained with a deep learning network with 3 hidden layers consisting of 128 neurons and activated by a sigmoid function. The discount value from the previous step is set to 0.99 as default. The policy noise is set to follow a normal distribution of $\epsilon \sim N(0, 0.1 * \sigma_x)$, where σ_x is the standard deviation (SD) of the horizontal and vertical deviations, respectively. Also, the parameters updating rate tau is set to 0.005 for a stable training process. The discount factor γ is set as 0.98, and the soft update factor ρ of the network's coefficients is 0.02. The lower and upper limit of the thrust cylinder forces follows the minimum and maximum values in the training samples to control the forces within a reasonable range.

To analyze and demonstrate the favorable performance of DRL, DRL is compared with manual operation in terms of horizontal and vertical deviations of the TBM tail, as well as the density and distribution of thrust forces. Table 4 lists the average horizontal and vertical deviations of the TBM tail, while Fig. 12 presents detailed visual representations of horizontal and vertical deviations of the TBM tail for both DRL and manual operation. Figure 13 illustrates the Bland-Altman plots comparing the horizontal and vertical deviations of the TBM tail between DRL and manual operation.

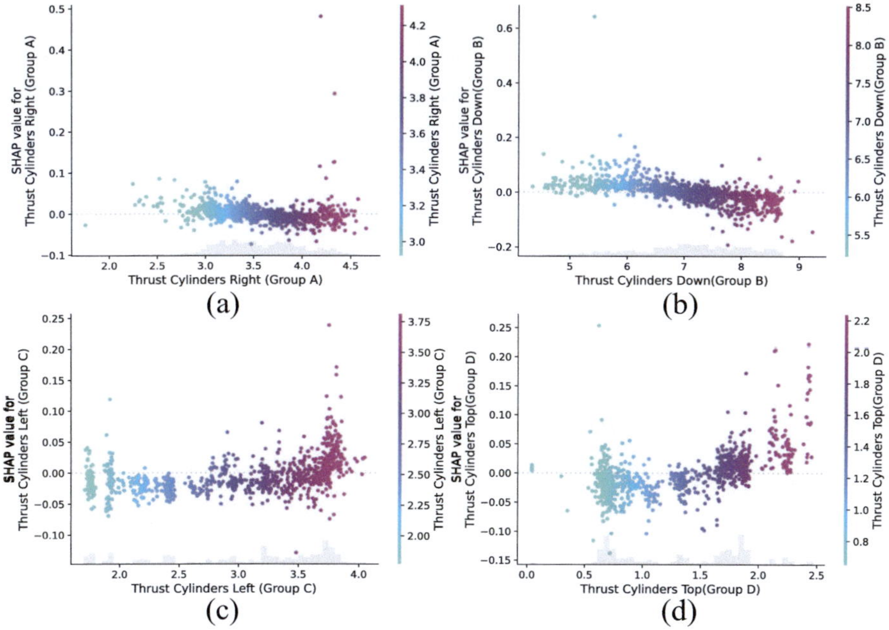

Fig. 11 SHAP value of **a** thrust cylinders group A, **b** thrust cylinders group B, **c** thrust cylinders group C, and **d** thrust cylinders group D in modeling the vertical deviation

Table 3 Comparative analysis of the prediction accuracy on testing data using LSTM and other models

Method	Horizontal deviation (mm)	Vertical deviation (mm)
LSTM	0.86	0.92
CNN	0.86	0.83
CNN-LSTM	0.88	0.80
Transformer	0.88	0.89
ANN	0.87	0.81

Figures 14 and 15 display the density of thrust forces from four thrust cylinder groups, comparing DRL and manual operation. Figure 16, along with Table 5, exhibits the distribution of thrust forces among four cylinder groups using DRL and manual operation, respectively. In view of these results, findings can be summarized as follows.

(1) The DRL algorithm significantly improves the deviation of the TBM tail compared with manual operation, where the vertical deviation is improved more than the horizontal deviation. As presented in Table 4, the average improvement of horizontal and vertical deviation is 35.1% and 66.3%, respectively, resulting

Table 4 Comparison of the proposed model and the manual method by average deviation

Method	H Deviation (mm)		V Deviation (mm)		Overall improvement
	Average[a]	Improvement from manual operation	Average*	Improvement from manual operation	
Manual	20.78	–	14.52	–	–
DRL	13.48	35.1%	4.89	66.3%	50.7%

[a] Average is computed based on the absolute value of the deviation in each step

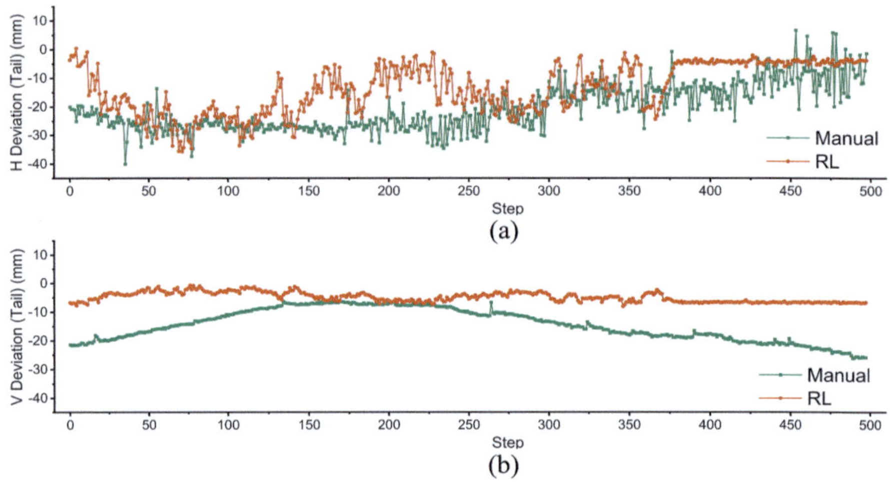

Fig. 12 Control performance for **a** horizontal and **b** vertical deviations of TBM tail using DRL and manual operation

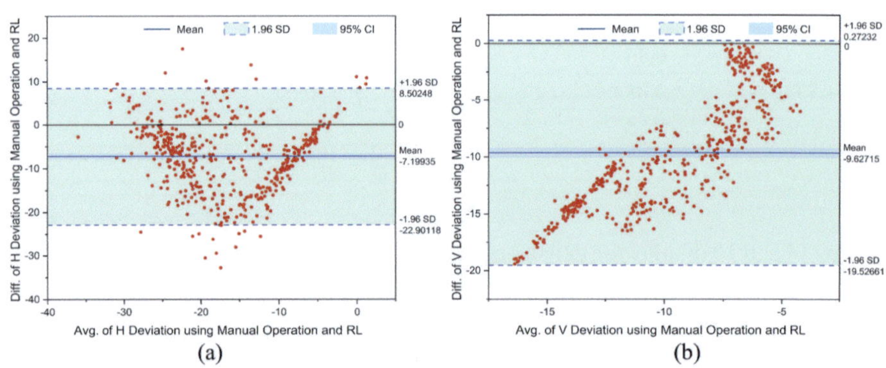

Fig. 13 Bland-Altman plot of manually operated result and DRL operated result reflecting **a** horizontal deviation and **b** vertical deviation of TBM tail

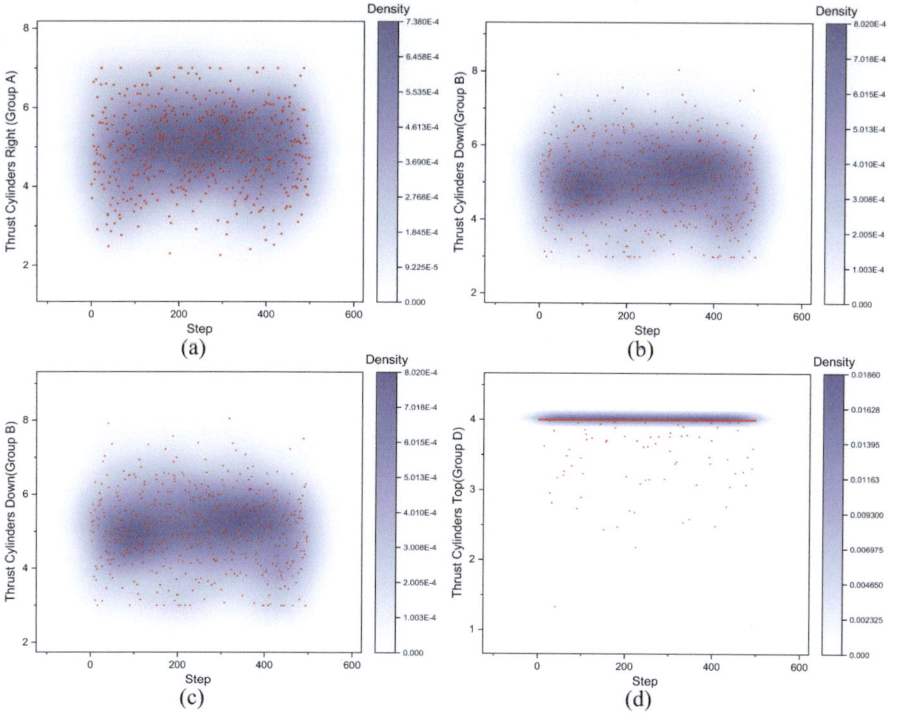

Fig. 14 Density plot of thrust forces in **a** thrust cylinders group A, **b** thrust cylinders group B, **c** thrust cylinders group C, and **d** thrust cylinders group D generated by DRL

in an overall improvement of 50.7%. Detailly focusing on the optimized deviations in each step, Fig. 12 indicates that the deviations resulting from DRL are more stable than manual operation, especially in the last 150 steps. Additionally, the vertical deviation is effectively constrained within 10 mm after automatic turning using DRL, which is a remarkable improvement compared with manual operation. This finding indicates that the DRL algorithm is significantly effective in controlling the TBM attitudes.

(2) The DRL algorithm maintains good consistency while automatically turning the TBM, which proves the reasonability and validity of the methodology. Figure 13 presents a Bland-Altman plot of manually operated results and DRL operated results reflecting the horizontal deviation and vertical deviation of the TBM tail. Firstly, the improvement in TBM deviations using DRL can also be proved as the mean of the difference between the manually operated result and the DRL generated result is far below 0 (− 0.720 mm for horizontal deviation and − 0.963 mm for vertical deviation). Moreover, it can be said that most of the samples stay within the range of a 95% range of consistency ($\pm 1.96SD$). In other words, the DRL makes remarkable improvements but effectively maintains the validity and reasonability of the TBM attitude.

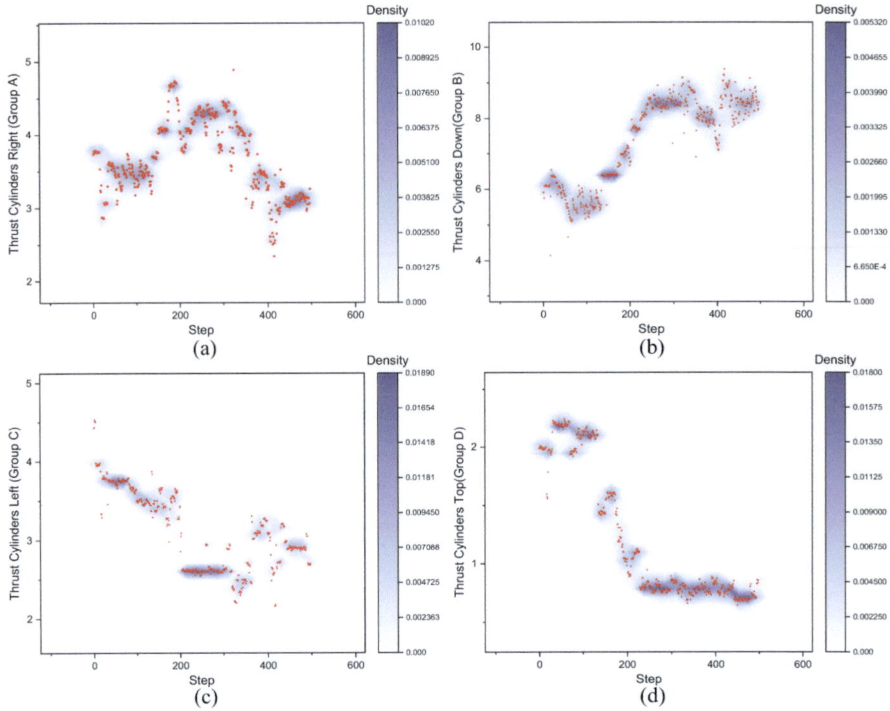

Fig. 15 Density plot of thrust forces in **a** thrust cylinders group A, **b** thrust cylinders group B, **c** thrust cylinders group C, **d** thrust cylinders group D from manual operation

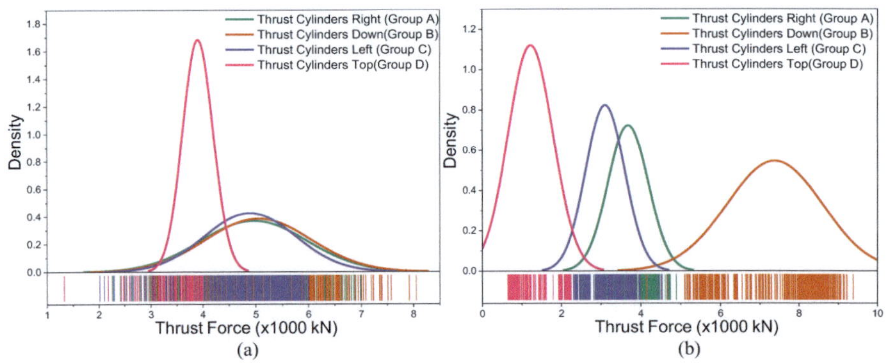

Fig. 16 Distribution of thrust forces from four cylinders from **a** DRL adjustment and **b** manual operation

Table 5 Comparison of the proposed model and the manual method on thrust forces

Parameters	Manual				DRL			
	Mean	SD	Min	Max	Mean	SD	Min	Max
Thrust cylinders right (group A)	3.68	0.51	2.34	4.89	4.96	1.00	2.26	7.00
Thrust cylinders down (group B)	7.38	1.21	4.14	9.38	5.05	0.99	3.00	8.05
Thrust cylinders left (group C)	3.10	0.49	2.16	4.53	4.89	0.86	2.01	6.00
Thrust cylinders top (group D)	1.22	0.57	0.65	2.24	3.90	0.29	1.33	4.00

(3) The DRL algorithm makes improvements from a more effective exploration of the space of action. Figures 14 and 15 present the density of actions taken in all steps using DRL and manual operation, respectively. The DRL is capable of taking actions from the whole design space while manual operation concentrates on a particular value densely. Further referring to Fig. 16 and Table 5, with the same upper and lower bounds set, the range it can be told that the range of actions taken by DRL is generally much larger than manual operation. The reason behind this finding implies a major shortcoming of manual operation compared with automatic control by algorithms. Practically, manual adjustments on TBM parameters are always passive, where actions are taken only after anomalies occur. That's the main reason that the manual actions do not vary a lot over time. Additionally, such a delay in manual operation can be risky as there may not be adequate time to adjust the TBM when severe deviations are noticed. However, these problems can all be solved by the DRL algorithm, as it is based on a time-series forecasting model. The algorithm can automatically explore the whole design space to find the optimal action in each step and make adjustments in advance.

5 Merits Analysis

The proposed DRL method based on the TD3 algorithm is proven to be effective in TBM attitude control, outperforming manual operation. Considering that TD3 is evolved from Deep Deterministic Policy Gradient (DDPG), this chapter tests the performance of the algorithm using DDPG to further prove the superiority of our proposed model. Compared with DDPG, TD3 employs three strategies including a double critic network, delayed updates, and a target policy smoothing strategy. Therefore, it is expected that our proposed methodology using TD3 has a better performance. Detailed findings are summarized as follows.

(1) Employing DDPG in the DRL algorithm improves the TBM deviation compared with manual operation but is significantly outperformed by TD3. As illustrated in Table 6, DRL using DDPG achieves similar performance on horizontal deviation with an improvement of 34.4%. However, the improvement in vertical deviation is merely 21.6%, resulting in an overall improvement of 28.0%. Detailly

referring to Fig. 17, it can be said that the two algorithms show very similar performance, which becomes quite stable in the last 150 steps. The main difference lies in the first 350 steps where TD3 remarkably outperforms DDPG on the vertical deviation. Additionally, results produced by DDPG show very high variance compared with TD3, especially in the first 250 steps. The high variance resulting from DDPG can also be observed in Figs. 18 and 19, which shows the distribution of horizontal and vertical deviations using different methods.

(2) Actions taken by DDPG are much more variant than those by TD3. As presented in Table 6, the SD of thrust forces in cylinder groups A, B, C, and D are 2.04 × 1000 kN, 1.98 × 1000 kN, 1.73 × 1000 kN, and 1.10 × 1000 kN, respectively, which are significantly larger than those resulted from TD3. Regarding the superiority of improvement made by TD3, it is reasonable to conclude that DDPG fails to precisely locate the optimal values of actions, which is due to its overestimation of rewards by using a single critic network and poor exploration of the action space with no noises introduced to the action. Practically, a

Table 6 Performance of the developed model and other methods by averaged deviation

Method	Horizontal deviation (mm)		Vertical deviation (mm)		Overall improvement
	Average[a]	Improvement from manual operation	Average[a]	Improvement from manual operation	
Manual	20.78	–	14.52	–	–
TD3	13.48	35.1%	4.89	66.3%	50.7%
DDPG	13.63	34.4%	11.39	21.6%	28.0%

[a] Average is computed based on the absolute value of the deviation in each step

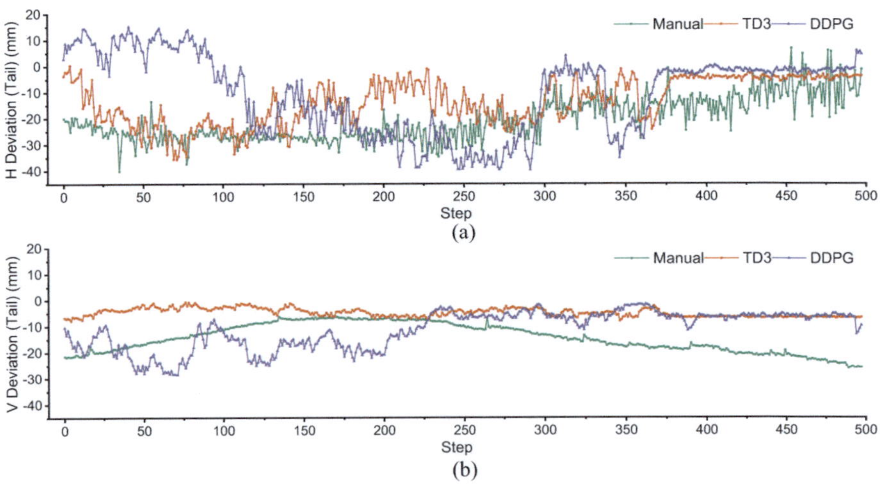

Fig. 17 The deviation of the TBM tail in **a** horizontal and **b** vertical directions by manual operation, TD3 and DDPG

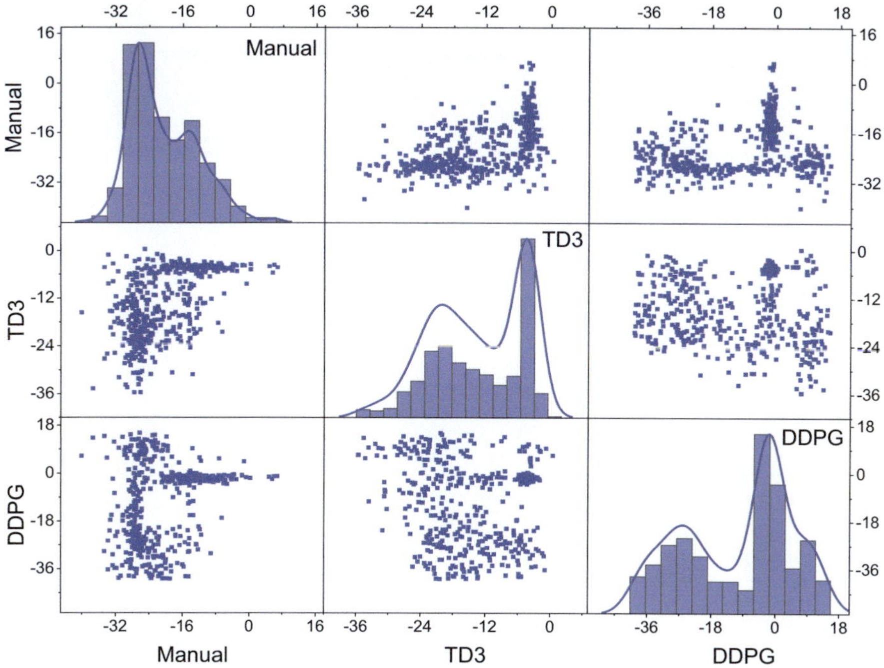

Fig. 18 Distributions for horizontal deviations of the TBM tail using manual operation, TD3 based DRL, and DDPG-based DRL

considerably high variance of operational parameters may cause instability in the underground space, which is unbeneficial in real projects (Table 7).

6 Chapter Summary

This chapter proposes a DRL method with consideration of the spatial-temporal dynamics of the TBM to perform position control in tunneling works. The method consists of feature selection and data processing, an environment model, an agent-enabled position control model, and coupled loss functions and evaluation metrics, With the aid of the LSTM technique, the environment model is designed to simulate the special-temporal dynamics of the TBM in the working environment to provide accurate forecasting results to assist the position control. By doing so, the agent is able to interact with the environment dynamically in both space and time to gradually obtain an accurate TBM position control strategy. The state-of-the-art DRL algorithm TD3 is employed as the dynamic solver for optimal operations on the TBM. In the training process, the loss function and reward model-coupled evaluation metrics are

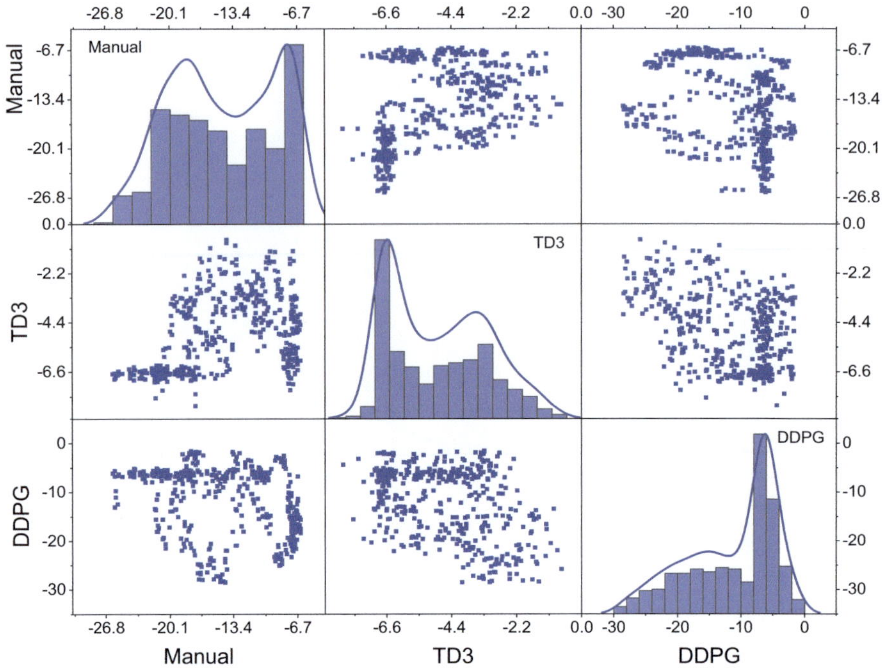

Fig. 19 Distributions for vertical deviations of the TBM tail using manual operation, TD3 based DRL, and DDPG-based DRL

Table 7 Summary of thrust forces using DRL based on TD3 and DDPG

Parameters	TD3 (×1000 kN)				DDPG (×1000 kN)			
	Mean	SD	Min	Max	Mean	SD	Min	Max
Thrust cylinders right (group A)	4.96	1.00	2.26	7.00	3.63	2.04	0.10	7.00
Thrust cylinders down (group B)	5.05	0.99	3.00	8.05	6.48	1.98	3.01	10.00
Thrust cylinders left (group C)	4.89	0.86	2.01	6.00	3.09	1.73	0.14	6.00
Thrust cylinders top (group D)	3.90	0.29	1.33	4.00	2.13	1.10	0.12	4.00

designed to assist the model in learning. The performance of the proposed method is validated in a real tunnel project collected from Singapore.

The contribution brought by this chapter can be portrayed as the incorporation of the spatial-temporal dynamics along with TBM excavation into DRL to achieve better position control for informed tunneling operations with higher efficiency and reliability. The developed approach explores novel and efficient control strategies that human-designed control systems can alert and adjust in advance. Practically, the methodology is capable of giving feasible solutions to achieve the position control of the TBM as well as potentially reducing the burden of on-site workers in tunnel construction. The custom-designed reward mechanisms, practical application, and demonstrated improvement in predictive accuracy collectively contribute to advancing the capabilities of intelligent systems in complex, real-world tunneling works.

References

1. Wang M, Yin X (2022) Construction and maintenance of urban underground infrastructure with digital technologies. Autom Constr 141:104464
2. Xu C, Liu X, Wang E, Wang S (2021) Prediction of tunnel boring machine operating parameters using various machine learning algorithms. Tunn Undergr Space Technol 109:103699
3. Lin P, Wu M, Xiao Z, Tiong RLK, Zhang L (2024) Physics-informed deep reinforcement learning for enhancement on tunnel boring machine's advance speed and stability. Autom Constr 158:105234
4. Tang F, Feng Z, Li Y, Yang C, Sun B (2023) A constrained multi-objective deep reinforcement learning approach for temperature field optimization of zinc oxide rotary volatile kiln. Adv Eng Inform 58:102197
5. Liu Z, Li L, Fang X, Qi W, Shen J, Zhou H, Zhang Y (2021) Hard-rock tunnel lithology prediction with TBM construction big data using a global-attention-mechanism-based LSTM network. Autom Constr 125:103647
6. Ren J, Wang Z, Pang Y, Yuan Y (2022) Genetic algorithm-assisted an improved AdaBoost double-layer for oil temperature prediction of TBM. Adv Eng Inform 52:101563
7. Zeng J, Roy B, Kumar D, Mohammed AS, Armaghani DJ, Zhou J, Mohamad ET (2022) Proposing several hybrid PSO-extreme learning machine techniques to predict TBM performance. Eng Comput 38:3811–3827
8. Liu W, Li A, Liu C (2022) Multi-objective optimization control for tunnel boring machine performance improvement under uncertainty. Autom Constr 139:104310
9. Qin C, Liu M, Zhang Z, Yu H, Jin Y, Sun H, Tao J, Liu C (2023) An adaptive operating parameters decision-making method for shield machine considering geological environment. Tunn Undergr Space Technol 141:105372
10. Zhang P, Li H, Ha QP, Yin Z-Y, Chen R-P (2020) Reinforcement learning based optimizer for improvement of predicting tunneling-induced ground responses. Adv Eng Inform 45:101097
11. Erharter GH, Hansen TF, Liu Z, Marcher T (2021) Reinforcement learning based process optimization and strategy development in conventional tunneling. Autom Constr 127:103701
12. Fu X, Pan Y, Zhang L (2024) A causal-temporal graphic convolutional network (CT-GCN) approach for TBM load prediction in tunnel excavation. Expert Syst Appl 238:121977
13. Dankwa S, Zheng W (2020) Twin-delayed DDPG: a deep reinforcement learning technique to model a continuous movement of an intelligent robot agent. In: Proceedings of the 3rd international conference on vision, image and signal processing, Article 66
14. Qiu C, Hu Y, Chen Y, Zeng B (2019) Deep deterministic policy gradient (DDPG)-based energy harvesting wireless communications. IEEE Internet Things J 6:8577–8588
15. Antonini AS, Tanzola J, Asiain L, Ferracutti GR, Castro SM, Bjerg EA, Ganuza ML (2024) Machine Learning model interpretability using SHAP values: application to igneous rock classification task. Appl Comput Geosci 23:100178
16. Wang HZX, Zhu M, Zhang L, Ju JW (2024) Delayed opening of TEL stage 5 irks some residents; others resigned to the longer wait, in. The Straits Times

Synchronous Excavation and Segment Assembly

1 Introduction

The rising demands for urban transportation have resulted in a heightened focus on the development of urban subway systems, oil and gas pipelines, as well as various other engineering tunnels. TBMs offer significant advantages, such as automation and reliable operations, which are widely used in various tunnel engineering projects [1]. However, traditional TBMs are structurally constrained and primarily operate in an intermittent manner, meaning that they halt after excavating one ring and await the assembly of segments before resuming the excavation process [2]. A defining characteristic of traditional TBMs is the alternating execution of the excavation and segment assembly processes, which consequently prolongs the construction duration and elevates the overall construction cost. Therefore, more attention has been paid to the exploitation of TBMs which can excavate and assemble segments simultaneously and are reported to enhance construction efficiency by 30 ~ 50%.

To achieve synchronous operations of excavation and segment assembly, the TBM should design a synchronous excavation and segment assembly (SESA) method. There are currently several popular methods for achieving SESA, including the front navigate (F-NAVI) method, the Lattice-type hydraulic cylinder method, the dual-hydraulic cylinder method, and others. Among these methods, the F-NAVI method has a limited scope in applications, while the Lattice-type hydraulic cylinder method requires too high control accuracy. The dual-hydraulic cylinder method is popular in many countries, including China and Japan, due to its simple structure, reliable operation, and wide range of applications. During the operation, the front excavation hydraulic cylinders and cutterhead perform the excavation work, while the rear assembly hydraulic cylinders and segment assembly machine perform synchronous segment assembly work.

Although the dual-hydraulic cylinder SESA method is operationally efficient and structurally simple, it also poses certain challenges during the operation process. Especially during different segment assemblies, the rear assembly hydraulic cylinders

L. Zhang et al., *Intelligent Construction in Tunnels*, Lecture Notes in Civil Engineering 679, https://doi.org/10.1007/978-981-96-8785-5_9

need to retract some cylinders to free up space for the segment [3]. This can result in the loss of balance and stress concentration in the TBM excavation process, especially near the retracted cylinders, which can lead to adverse effects on the TBM and segment. The stress concentration caused by cylinder retraction is not conducive to the normal operation of TBM, which is particularly prominent in SESA TBM [4]. To address these challenges, it is essential to develop a framework to achieve the status detection of SESA TBM and the redistribution of cylinder pressure.

As an advanced intelligent method, the multi-objective optimization method can find the optimal engineering decision in a data-driven manner. In the past decades, multi-objective optimization has been widely used in TBM active control due to its high efficiency, accuracy, and reliability. For example, Guo et al. [5] used the genetic algorithm to complete the active optimization design of tunnel alignment with the target of project budget and user comfort. Although multi-objective optimization has many applications in TBM active control, most of these studies focus on the excavation process, and there are few studies on the active control of segment assembly [6]. Besides, due to the continuity of the segment assembly process, the traditional offline optimization algorithm usually cannot obtain optimal optimization results in real time.

As a novel machine learning strategy, online learning can provide accurate and real time scheme guidance through continuous updating of samples [7]. The online multi-objective optimization method is a combination of multi-objective optimization and online learning strategy, which is widely used in the optimization of various continuous operation scenarios. The online multi-objective optimization can fully consider the optimization parameters in continuous scenarios and give the optimal solution in real time based on historical data. Cylinder pressure is also a kind of continuous operation s can be well combined with the online multi-objective optimization method. Therefore, to reduce the stress concentration of the hydraulic cylinder in the segment assembly process, an active control algorithm relying on online multi-objective optimization should be designed to redistribute the assembly hydraulic cylinder pressure.

This chapter introduces the main structure and working principle of the SESA TBM and presents a hydraulic cylinder pressure redistribution optimization framework. The Bayesian optimization extreme gradient boosting (BO-XGBoost) model is constructed to realize the detection of the cylinder status, and an online Archimedean optimization algorithm (AOA) algorithm is designed to realize the cylinder pressure redistribution. The novelty to be highlighted is the development of the active pressure control approach with the online multi-objective optimization algorithm incorporated, which is helpful in effectively improving the stress concentration phenomenon in the SESA TBM.

2 Hydraulic Cylinder Control

As a direct driving mechanism for TBM actuator components, the performance of hydraulic cylinders directly affects the efficiency and reliability of the TBM. Specifically, the cutter excavation and segment assembly are directly completed by the hydraulic cylinder. Thus, it is necessary to control the hydraulic cylinder accurately. In the research related to hydraulic cylinder control, there are two research directions, namely traditional methods and machine learning methods.

The traditional research methods for hydraulic cylinder control mainly include empirical, numerical, and experimental methods. The empirical method refers to the artificial judgment of hydraulic cylinder status parameters based on existing relevant expert experience, which is used to provide corresponding decision-making schemes. The numerical method refers to calculating the relevant status parameters of the hydraulic cylinder based on fluid mechanics principles and making execution decisions based on numerical fluctuations. The experimental method refers to reproducing engineering scenarios through experiments and guiding engineering practices based on the experiments. Among them, the expert experience method has a high degree of risk uncertainty since it is directly made by people, especially when different experts may express different preferences. The numerical method requires a large amount of computational cost and cannot be used on a large scale in practical engineering projects. Although the experimental method has the most reliable accuracy, it needs to consider various situations and design various experiments for complex engineering problems, which has poor engineering promotion. Thanks to the development of machine learning methods, data-driven research based on hydraulic cylinder related data has become a research hotspot [8]. By combining advanced machine learning methods and a control principle of hydraulic cylinders, the hydraulic cylinder status can be predicted and detected with high efficiency and precision.

In machine learning areas, predicting various hydraulic cylinder parameters using machine learning methods has become a hot topic. As the direct performance indicator of hydraulic cylinders, pressure can be accurately predicted through machine learning, which helps guide the control system and ensures the stable operation of the mechanism. The extension and retraction of the hydraulic cylinder are often closely related to the precise position of the actuator. Predicting the movement of the hydraulic cylinder can provide reliable action plans for the operation trajectory control of the entire machine. In addition to predicting hydraulic cylinder's pressure and movement, specific parameter predictions may be required for different engineering scenarios, such as the stroke difference or the response time of the hydraulic cylinder. By introducing appropriate machine learning methods and combining historical data with environmental parameters, hydraulic cylinder parameters can typically be accurately predicted, ensuring the efficient and reliable control of the hydraulic cylinder.

Although there have been many studies on predicting and detecting hydraulic cylinder status, there is still a research gap in the active control of multiple hydraulic cylinders, particularly in the TBM synchronous excavation and segment assembly process. In the current related research, there is a lack of analysis and detection of the cooperative operation of multiple cylinders, and the optimal control and decision-making of the cylinders cannot be realized. Specifically, due to the existence of stress concentration, how to accurately detect different cylinder statuses and give corresponding active control schemes is an essential issue. Therefore, this chapter efficiently combines the detection model with the multi-objective optimization algorithm, which achieves accurate identification and pressure control of the multi-cylinder system.

3 Synchronous Excavation and Assembly Framework

To improve the operational stability of the SESA TBM and reduce the stress concentration during the segment assembly process, this chapter designs an online cylinder pressure optimization framework. The framework consists of 3 main steps: (1) The construction scenario is described and the different hydraulic cylinder status is analyzed; (2) The BO-XGBoost is used to detect the status of the assembly hydraulic cylinder; (3) The online AOA is designed to optimize the cylinder pressure and obtain the best cylinder pressure distribution scheme. An overview of the proposed framework is shown in Fig. 1.

3.1 *Stress Concentration in SESA TBM*

The SESA TBM studied in this chapter is a kind of double-hydraulic cylinder machine. As shown in Fig. 2, to realize the synchronous excavation and segment assembly, two sets of hydraulic systems are designed. The excavation hydraulic system is located in the front shield, working with the cutterhead to complete the

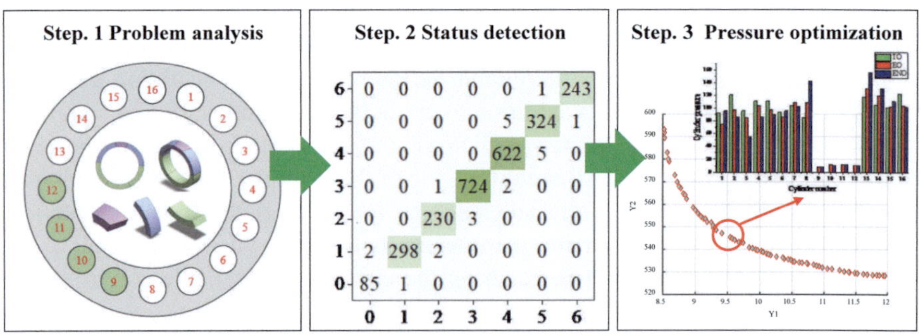

Fig. 1 The overview of the proposed framework

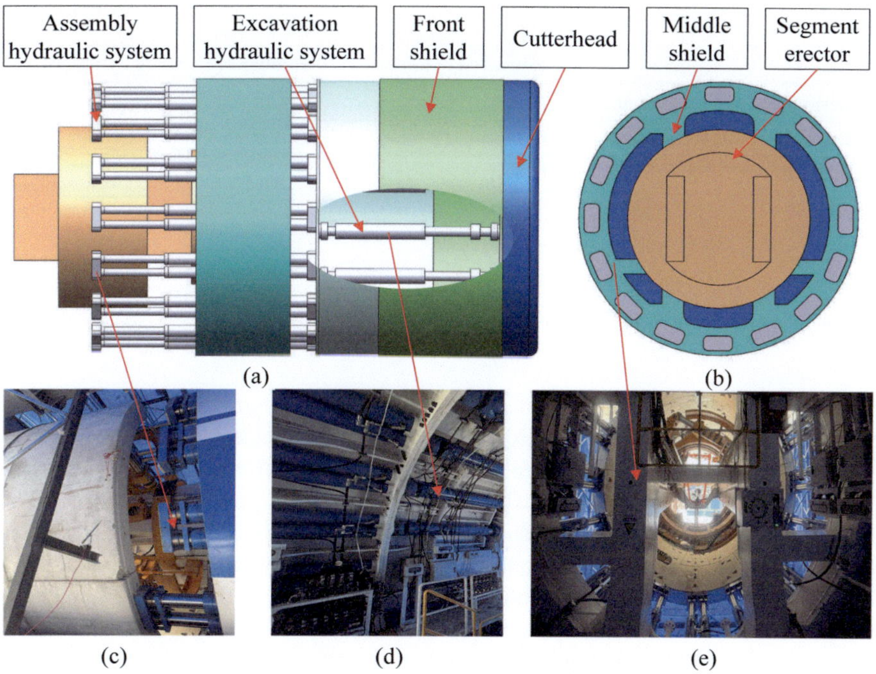

Fig. 2 The structure of SESA TBM: **a** main view; **b** side view; **c** assembly hydraulic system; **d** excavation hydraulic system; **e** middle shield

excavation of the soil and rock. The assembly hydraulic system is located in the middle shield, and the installation of the segments is completed by the assembly hydraulic system and the segment erector. To ensure the stability of the operation, the front shield and the middle shield are hinged, which can be locked or reset according to the actual scenario.

The front excavation module and the back segment assembly module of the SESA TBM are basically the same as the traditional TBM. The main difference is that the traditional TBM excavation module and the segment assembly module are directly connected. The SESA TBM adds a set of hydraulic cylinders between them, and the excavation and segment assembly can operate freely at the same time.

During the segment assembly, the hydraulic cylinder status is closely related to the style and position of the segment. There are mainly three types of segments, namely B, H, and K segments. One ring is composed of three B, two H, and one K segments. Due to the style and position of the segment, the actual assembly hydraulic cylinders can be divided into 7 statuses. That is to say, no segment assembly status, B-1 assembly status, B-2 assembly status, B-3 assembly status, H-1 assembly status, H-2 assembly status, and K assembly status.

When assembling different segments, the assembly hydraulic cylinders that need to be retracted are different. At the beginning of a ring, it is necessary to first select

Table 1 The construction scenarios and the corresponding cylinder status

Scenarios	Segment	Retracted cylinder	Number of working cylinders
Scenario 0	No segment	None	16
Scenario 1	B-1 segment	9, 10, 11, 12	12
Scenario 2	B-2 segment	6, 7, 8, 9	12
Scenario 3	B-3 segment	13, 14, 15, 16, 6	11
Scenario 4	H-1 segment	3, 4, 5, 6, 16	11
Scenario 5	H-2 segment	16, 1, 2, 3	12
Scenario 6	K segment	3	15

the corresponding K point position and assemble the B-1 segment at the place where it is symmetrical to the K point, and then assemble the adjacent segment. Assuming that the position of the No. 3 hydraulic cylinder is K point, the B-1 segment assembly needs to be retracted to the No. 9, 10, 11, and 12 hydraulic cylinders. The second step is to retract No. 6, 7, 8, and 9 hydraulic cylinders and assemble the B-2 segment. The third step is to retract No. 13, 14, 15, and 16 hydraulic cylinders and assemble the B-3 segment. At the same time, considering the assembly of the L-1 segment at the next moment, the invalid extension and retraction of the cylinders should be avoided. Hence, the No. 6 hydraulic cylinder is also in the retraction status in the third step. The fourth step is to retract No. 3, 4, 5, and 6 hydraulic cylinders and assemble the L-1 segment, which is similar to the third step, the No. 16 hydraulic cylinder is also in the retraction status in the fourth step. The fifth step is to retract No. 16, 1, 2, and 3 hydraulic cylinders and assemble the L-2 segment. The sixth step is to retract the No. 3 hydraulic cylinder and assemble the K segment. The construction scenarios and the corresponding cylinder status are shown in Table 1.

From the perspective of the number of hydraulic cylinders, there may be 11 (B-3, L-1), 12 (B-1, B-2, L-2), 15 (K), and 16 jacking-up hydraulic cylinders. For different scenarios, it is necessary to design a classification algorithm to detect the corresponding hydraulic cylinder status and optimize the cylinder pressure to reduce stress concentrations caused by partial cylinder retraction. Since hydraulic cylinders have different statuses, status detection and pressure optimization are both essential aspects, which will be addressed in the following two sections.

3.2 Hydraulic Cylinder Status Detection

The hydraulic cylinder status needs to be clarified before optimization. This section combines an advanced machine learning model to detect hydraulic cylinder status. The structure of the detection model and the related evaluation indicators are presented below in detail.

3.2.1 XGBoost Model with Bayesian Optimization

Based on Sect. 3.1, the detection of cylinder status can be regarded as a classification problem. To achieve pressure optimization, it is essential to determine the 7 operating statuses of the cylinder in advance according to the decision variables. As an extensible machine learning method based on the tree model, XGBoost has a good effect on classification problems with its advantages of effective tree pruning, regularization, and parallel computing. The core design of XGBoost is to minimize the regularization cost function shown below:

$$Obj^{(n)} = \sum_{i=1}^{m} F\left(v_i, \overline{v}_i^{(n)}\right) + \sum_{j=1}^{n} \emptyset(h_n) \tag{1}$$

where i is the i-th sample of the dataset. $F()$ represents the loss function. m is the total data in the j-th tree. n represents all trees established. v_i is the true value, which is the real hydraulic cylinder status in this chapter. $\overline{v}_i^{(n)}$ is the predicted value of the n-th prediction, which is the predicted hydraulic cylinder status in this chapter. $\varnothing()$ represents the model complexity, as shown in Eq. (2). h indicates the model structural item.

$$\emptyset(h_n) = \beta N + \frac{1}{2}\delta \sum_{k=1}^{N} \omega^2 \tag{2}$$

where β is the control parameter of leaf nodes. N indicates the number of leaf nodes of the tree. δ represents the normalized coefficient. ω is the weight of the tree model.

To further improve the model's reliability, this chapter introduces the Bayesian optimization (BO) algorithm to optimize the hyperparameters [9]. The process of BO-XGBoost is shown in Fig. 3. BO makes full use of the information from the previous sampling point by learning the objective function and finds the parameters that improve the result to the global optimum. XGBoost is optimized by BO to ensure the best accuracy of the hydraulic cylinder status detection model.

3.2.2 Detection Model Evaluation

In the hydraulic cylinder status detection model, the classification model is constructed by the training set, and the reliability of the model is verified by the test set. To effectively evaluate the reliability of the hydraulic cylinder status detection model, the corresponding model evaluation indicators should be introduced. At present, the common evaluation indicators are listed as follows:

$$\text{Accuracy}(i) = \frac{(TP_i + TN_i)}{(TP_i + TN_i + FP_i + FN_i)} \tag{3}$$

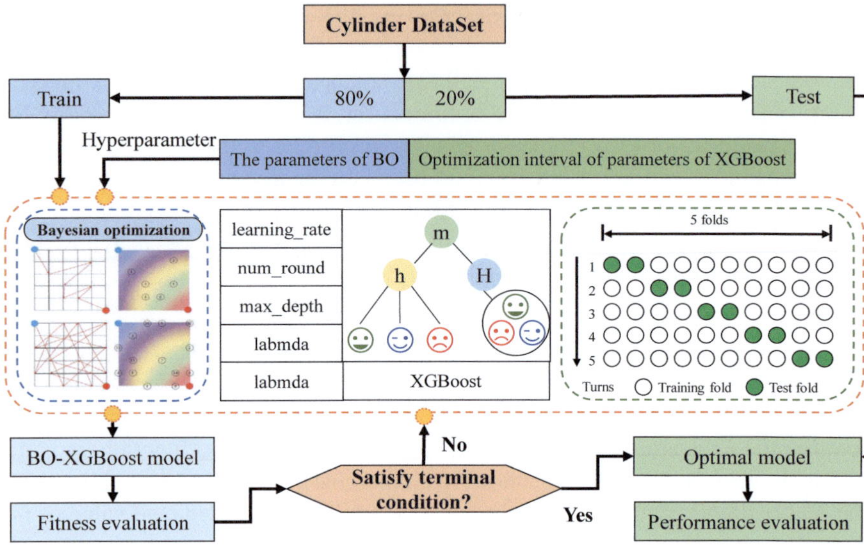

Fig. 3 The flowchart of BO-XGBoost

$$\text{Precision}(i) = \frac{TP_i}{TP_i + FP_i} \tag{4}$$

$$\text{Recall}(i) = \frac{TP_i}{TP_i + FN_i} \tag{5}$$

$$\text{F1} - \text{score}(i) = \frac{2 * \text{Precision}(i) * \text{Recall}(i)}{\text{Precision}(i) + \text{Recall}(i)} \tag{6}$$

where i indicates the hydraulic cylinder status scenario, which is 7 in this chapter and is consistent with Table 1. *TP* represents that the value is positive and the classification result also is positive. *TN* represents the value is negative and the classification result also is negative. *FP* represents the value is negative but the classification result is positive. *FN* represents that value that is positive but the classification result is negative. Since this chapter is a multi-classification problem, each indicator is in the form of the mean.

3.3 *Hydraulic Cylinder Pressure Optimization*

After completing the detection of the hydraulic cylinder status, this section carries out the pressure redistribution optimization. Specifically, the optimization model of pressure redistribution is clarified, and an online optimization method is designed for the model. The selection method of the Pareto optimal point and the corresponding algorithm evaluation indicators are followed.

3.3.1 Optimization Model Description

In order to ensure the optimal redistribution of hydraulic cylinder pressure, it is necessary to construct an optimization model of redistribution. The model is mainly composed of three parts: decision variables, objective functions, and constraints. The decision variable is the pressure of the redistribution hydraulic cylinders. According to the scenario analysis in Sect. 3.1, the decision variables of this chapter may be 11, 12, or 15, which need to be matched with the corresponding scenario. In the optimization model, the number of decision variables is directly determined by the detection model in Sect. 3.2. The decision variables are as follows:

$$X = [x_1, x_2 \ldots x_n], n = \begin{cases} 11 \text{ Scenarios } 1, 2, 5 \\ 12 \quad \text{Scenarios } 3, 4 \\ 15 \qquad \text{Scenario } 6 \end{cases} \tag{7}$$

After determining the decision variables, it is essential to design the objective functions to meet the effect of hydraulic cylinder pressure redistribution. Due to the existence of multiple hydraulic cylinders, a single hydraulic cylinder parameter (such as maximum pressure) cannot truly represent the whole cylinder pressure.

This chapter designs two objective functions to ensure that hydraulic cylinder pressure redistribution is effective and efficient. Firstly, the variance of the hydraulic cylinder pressure after redistribution is selected as the first objective function (Y1) in this chapter. It can ensure that each hydraulic cylinder bears similar pressure. Secondly, it should be ensured that the change of each hydraulic cylinder pressure before and after the redistribution is minimal. Taking Fig. 4 as an example, the hydraulic cylinder pressure is optimized from [100, 20] to [40, 80] or [80, 40], which both are allowed for Y1. But the optimization scheme 2 is obviously better than the optimization scheme 1, because it brings less burden to the control system, and the process is more efficient. Thus, the Euclidean distance before and after the pressure change of each cylinder is constructed as the second objective function (Y2). By optimizing Y1 and Y2, the efficient and reliable redistribution of hydraulic cylinder pressure can be ensured. The objective functions are described as follows:

$$\begin{cases} Y1 = Var(X) = \frac{1}{n}\sum_{i=1}^{n}(x_i - \overline{x})^2 \\ Y2 = \frac{1}{n}\sum_{i=1}^{n}(x_i - x_{oi})^2 \end{cases} \tag{8}$$

where x_i represents i-th variable in X. $\overline{x}$ is the mean of X. x_{oi} represents i-th variable in original dataset.

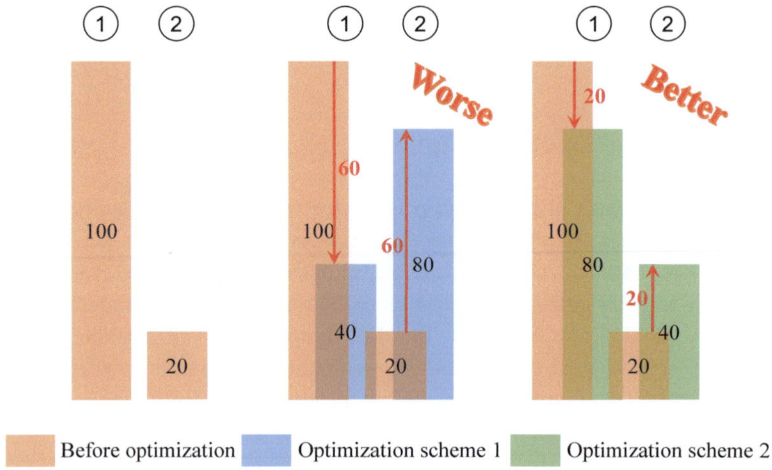

Fig. 4 The comparison of two pressure optimization schemes

The constraints are core to ensuring the feasibility of the redistribution scheme. By setting reasonable constraints, the optimization model can be guaranteed to be consistent with the real scene. In order to avoid instability and out-of-control, constraints should be added to ensure that the TBM before and after optimization has the same mechanical status. This is to say, force conservation and moment conservation should be achieved before and after cylinder pressure optimization. The constraints are described as follows:

$$\begin{cases} \sum_{i=1}^{n} x_i + F_{add}^a = \sum_{i=1}^{16} x_{oi} + F_{add}^b \\ \sum_{i=1}^{n} x_i * r * \sin(i * angle) + Mx_{add}^a = \sum_{i=1}^{16} x_{oi} * r * \sin(i * angle) + Mx_{add}^b \\ \sum_{i=1}^{n} x_i * r * \cos(i * angle) + My_{add}^a = \sum_{i=1}^{16} x_{oi} * r * \cos(i * angle) + My_{add}^b \end{cases} \tag{9}$$

where *angle* is the interval angle in the hydraulic cylinder, which is 22.5°(360/16) in this chapter. r is the moment radius. F_{add}^a is additional force after pressure optimization. F_{add}^b is additional force after pressure optimization.

Due to the pressure optimization is an instantaneous process, this chapter treats F_{add}^a and F_{add}^b as equal. Similarly, Mx_{add}^a and Mx_{add}^b, My_{add}^a and My_{add}^b both are regarded as equal. These parameters are the additional force and torque generated by the surrounding environment, and will not change with the cylinder pressure. The detailed mechanical status of SESA TBM is shown in Fig. 5, and related symbols are as follows:

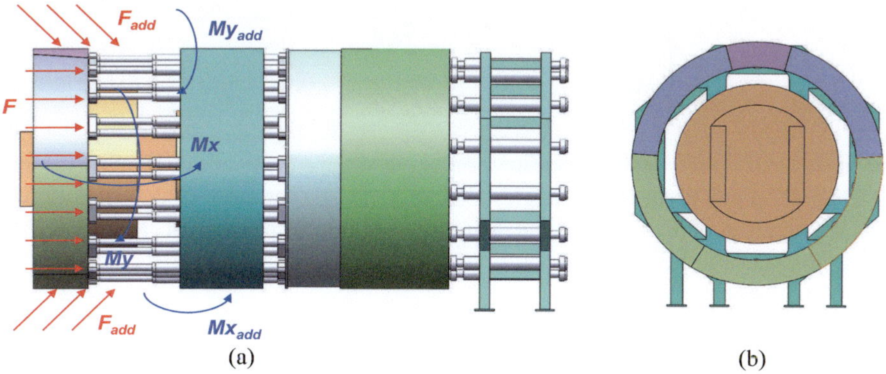

Fig. 5 The mechanical status of SESA TBM. **a** Main view; **b** side view

$$\begin{cases} F = \sum_{i=1}^{n} x_i \\ Mx = \sum_{i=1}^{n} x_i * r * \sin(i * angle) \\ My = \sum_{i=1}^{n} x_i * r * \cos(i * angle) \end{cases} \quad (10)$$

where F is the total force of the hydraulic cylinders. Mx is the total horizontal moment of hydraulic cylinders. My is the total vertical moment of hydraulic cylinders.

3.3.2 An Improved Online Updating AOA Method

In this chapter, the evolutionary algorithm can be used to optimize the pressure redistribution optimization model. As an evolutionary algorithm with fast convergence speed and strong adaptive ability, AOA has been widely used in optimization problems. The optimization idea of AOA mainly comes from the application of arithmetic operators. The process of AOA can be divided into two stages: exploration and exploitation. Exploration refers to the process of exploring the local optimal solution interval, and exploitation refers to the process of finding the optimal solution in each local optimum.

Before AOA begins to optimize, it is necessary to define the corresponding initial solution set and search strategy, and the search strategy is as shown in Eq. (11). The AOA first uses the D and M operators to obtain the highly distributed values or decisions of the region, as shown in Eq. (12). The purpose of this stage is to obtain dense regions where there may be an optimal solution set.

$$\begin{cases} MOA(C_{Iter}) = Min + C_{Iter} \times \left(\frac{Max-Min}{M_{Iter}}\right) \\ MOP(C_{Iter}) = 1 - \frac{C_{Iter}^{\frac{1}{\alpha}}}{M_{Iter}^{\frac{1}{\alpha}}} \end{cases} \tag{11}$$

where $MOA(C\_Iter)$ denotes the function value at the t-th iteration. $MOP(C_{Iter})$ is a coefficient, which denotes the function value at the t-th iteration. C_Iter denotes the current iteration, which is between 1 and the maximum number of iterations (M_Iter). Min and Max denote the minimum and maximum values of the accelerated function, respectively.

$$x_{i,j}(C_{Iter}+1) = \begin{cases} best(x_j) \div (MOP+\epsilon) \times \left((UB_j - LB_j) \times \mu + LB_j\right), & r2 < 0.5 \\ best(x_j) \times MOP \times \left((UB_j - LB_j) \times \mu + LB_j\right), & otherwise \end{cases} \tag{12}$$

where $x_{i,j}(C_{Iter}+1)$ denotes the j-th position of the i-th solution at the next iteration, and $best(x_j)$ is the j-th position in the best-obtained solution so far. ϵ is a small integer number, UB_j and LB_j denote to the upper bound value and lower bound value of the j-th position, respectively. μ is a control parameter to adjust the search process, which is fixed equal to 0.5 according to this chapter.

After the completion of the exploration, the exploitation of optimization objectives is carried out. Using the low dispersion of A and S operators, the search area is explored in depth in several dense areas, as shown in Eq. (13).

$$x_{i,j}(C_{Iter}+1) = \begin{cases} best(x_j) - MOP \times \left((UB_j - LB_j) \times \mu + LB_j\right), r3 < 0.5 \\ best(x_j) + MOP \times \left((UB_j - LB_j) \times \mu + LB_j\right), otherwise \end{cases} \tag{13}$$

Considering the real time scene of TBM, the traditional AOA method does not have the ability of real-time update. Therefore, to enhance the timeliness and reliability of the optimization results, this chapter adds an online optimization strategy based on the traditional AOA. Algorithm 1 illustrates the pseudo-code of the online AOA. As shown in Algorithm 1, the online AOA starts synchronously with the cylinder action. The cylinder pressure data in the first 10 s is collected in real time to optimize, and the cylinder pressure is updated using the optimization results in the first 10 s. The detailed optimization process is to rely on exploration and exploitation in AOA to find the current optimal solution. Until the current segment assembly is completed (S = 0), the online AOA ends the optimization and moves on to the following segment assembly scenario.

Algorithm 1 Pseudo-code of the online AOA.

01 Initialize the Arithmetic Optimization Algorithm parameters α, μ
02 Initialize the solutions' positions randomly. (Solutions: $i = 1,\ldots,N$.)
03 **while** cylinder action **do** # online strategy

(continued)

(continued)

```
while (C_Iter < M_Iter) do
    Calculate the Fitness Function (FF) using X in the first 10 s
    Find the best solution (Determined best so far)
    Update the MOA value using Eq. (11)
    Update the MOP value using Eq. (11)
    for (i = 1 to Solutions) do
        for (j = 1 to Positions) do
            Generate random values between [0, 1] (r1, r2, and r3)
            if r1 > MOA then
                Exploration phase
                if r2 > 0.5 then
                    (1) Apply the Division math operator (D "÷")
                    Update ith solutions' positions using first rule in Eq. (12)
                else
                    (2) Apply the Multiplication math operator (M "×")
                    Update ith solutions' positions using second rule in Eq. (12)
                end if
            else
                Exploration phase
                if r3 > 0.5 then
                    (1) Apply the Subtraction math operator (S "−")
                    Update ith solutions' positions using first rule in Eq. (13)
                else
                    (2) Apply the Addition math operator (A "+")
                    Update ith solutions' positions using second rule in Eq. (13)
                end if
            end if
        end for
    end for
    C_Iter = C_Iter + 1
end while
Return the current best solution X_c
if S = 0 then
    Update the final solution X_f value using solution X_c
    break up
else
    Update the X value using solution X_c
    continue while
```

(continued)

(continued)

42 **end if**
43 **end while**
44 Return the final solution X_f

3.3.3 Ideal Solution Determination

For the Pareto fronts obtained by multi-objective optimization, the Technique for Order Preference by Similarity to Ideal Solution (TOPSIS) is the most common method to evaluate individuals [10]. The general idea of the TOPSIS method is to compare the distance to the utopian point and dystopian point, and then the individuals are evaluated accordingly. Specifically, the target weight is given according to the importance of the target by selecting the utopian point Z_j^+ and the dystopian point Z_j^-. The Euclidean distance D_i^+ between the Pareto point x_{ij} and the utopian point, and the Euclidean distance D_i^- between Pareto point x_{ij} and the dystopian point are calculated, as shown in Eqs. (14) and (15), respectively. Finally, the fitness C is determined by Eq. (16), and the point with the best fitness is selected as the actual optimal solution.

$$D_i^+ = \sqrt{\sum_{j=1}^{m}\left(x_{ij} - Z_j^+\right)^2} \tag{14}$$

$$D_i^- = \sqrt{\sum_{j=1}^{m}\left(x_{ij} - Z_j^-\right)^2} \tag{15}$$

$$C = \frac{D_i^-}{D_i^+ + D_i^-} \tag{16}$$

Furthermore, in order to evaluate the advantages and disadvantages of the optimization algorithm, it is also necessary to introduce the corresponding algorithm evaluation indicators. Specifically, this chapter selected the three evaluation indicators, namely Time (T), Generational Distance (GD), Spacing ($Spacing$), which are calculated as follows:

$$T = Time(A) \tag{17}$$

$$GD\left(F^*, F\right) = \frac{\sum_{x \in F^*} \min dis(x, F)}{|F^*|} \tag{18}$$

$$Spacing = \frac{d_e + d_b + \sum_{i=1}^{N-1}\left|d_i - \overline{d}\right|}{d_e + d_b + (N-1)\overline{d}} \tag{19}$$

where F^* is the ideal solution. F is the whole solution. d_e denotes the distance between the extreme solution. d_b denotes the distance between the boundary solution. T represents the time of the algorithm, and its unit is s. GD represents the convergence of the algorithm. *Spacing* represents the spatial distribution of the solution set. Both GD and *Spacing* are dimensionless.

4 Experimental Studies

To demonstrate the actual performance and operation effect of SESA TBM and test the reliability of the detection method and optimization method, this chapter builds a relevant experimental bench. This section presents a comprehensive overview of the experimental setup, hydraulic cylinder detection, and hydraulic cylinder pressure optimization, as well as the analysis of the results.

4.1 Experimental Setup

In order to verify the effectiveness of the method, this chapter builds a 1:1 restoration SESA TBM experimental bench with a diameter of 6 m and a length of 10 m, as shown in Fig. 6. Since the experimental stage cannot truly simulate rock and soil cutting, the corresponding reaction frame is designed to replace the real cutting process. By designing different pressure forms of reaction hydraulic cylinder, different soil and rock forms can be simulated. The segment assembly process is consistent with the actual construction process.

During the experiment, with the assembly of segments, the whole shield will move forward accordingly. After the experiment is completed, the shield body will be reset. The specific TBM structure and experimental process are consistent with Sect. 3.1. The experimental data is collected by hundreds of sensors and transmitted directly to the control terminal. The algorithms and data processing for this experiment are performed by a DELL G15 laptop, which completes the communication and data transfer with the TBM control terminal via Modbus.

4.2 Detection Results Analysis

To accurately identify the operating status of different hydraulic cylinders, a series of dynamic parameters are selected as the decision variables, and the identification

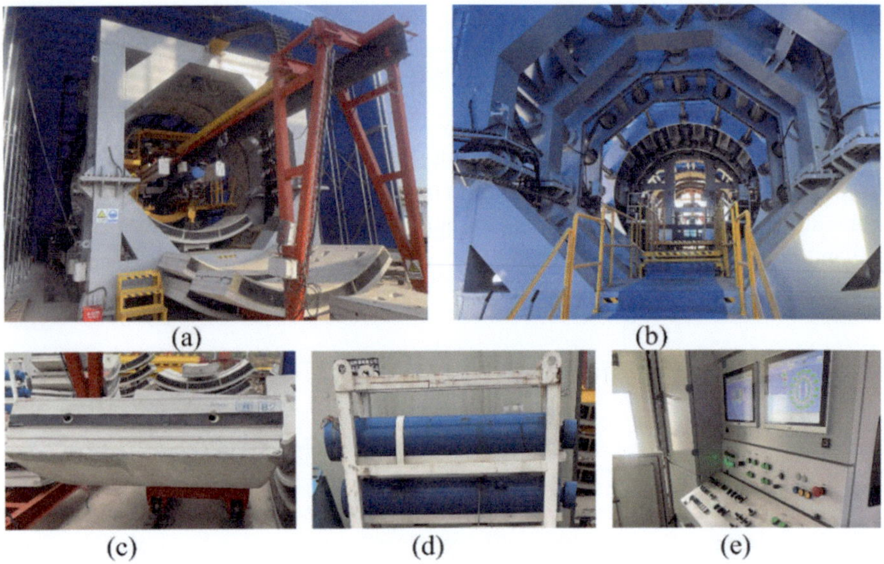

Fig. 6 The experimental bench. **a** Face view; **b** back view; **c** segment; **d** oil cylinder; **e** console

of hydraulic cylinder status is realized by the status detection algorithm. According to the analysis in Sect. 3.1, scenario numbers 0 ~ 6 are selected as the detection objective because the scenario number corresponds to the operating status of different hydraulic cylinders. The decision variables selected in this chapter are shown in Table 2. The first 70% data is used for model training, and the last 30% data is used for model testing. To verify the reliability of the selected decision variables, the correlation analysis is further carried out. From Fig. 7, it can be found that the correlation between all the decision variables and the objective is lower than 0.7, which indicates that there is no strong correlation between the decision variables and the objective. Also, the correlations between different decision variables are all less than 0.8, which indicates that there are no redundant variables.

The detection model used in this chapter is BO-XGBoost, and the hyperparameters of the model are all found by BO autonomously, as shown in Table 3. To verify the effectiveness of the hyperparameter, this chapter compares it with other hyperparameter schemes. By comparing different combinations of hyperparameters, the rationality of the hyperparameters selected in this chapter is verified. The results of this chapter are shown in Figs. 8, 9 and Table 4.

The proposed detection model provides a reliable result of TBM cylinder status. As shown in Fig. 8 and Table 4, for the detection model of the hydraulic cylinder status, the Accuracy, Precision, Recall, and F1-score indicators are 0.9893, 0.9897, 0.9885, and 0.9891, respectively, indicating that this model has a good ability to distinguish different samples. The training set and test set both have good detection results, indicating that the model does not have over-fitting or under-fitting problems. As shown in Fig. 9, the area under curve (AUC) values of each detection target are all

Table 2 Description of the studied variables in the optimization model

Symbol	Description	Location	Unit
MP1	Main excavation hydraulic cylinder pressure	Bottom	bar
MP2	Main excavation hydraulic cylinder pressure	Left-Bottom	bar
MP3	Main excavation hydraulic cylinder pressure	Left-Top	bar
MP4	Main excavation hydraulic cylinder pressure	Right-Bottom	bar
MP5	Main excavation hydraulic cylinder pressure	Right-Top	bar
MP6	Main excavation hydraulic cylinder pressure	Top	bar
MF	Main excavation driving force		TF
RP1	Reaction hydraulic cylinder pressure	Bottom	bar
RP2	Reaction hydraulic cylinder pressure	Left	bar
RP3	Reaction hydraulic cylinder pressure	Right	bar
RP4	Reaction hydraulic cylinder pressure	Top	bar
A	Assembly angle		°
No	Scenario number		

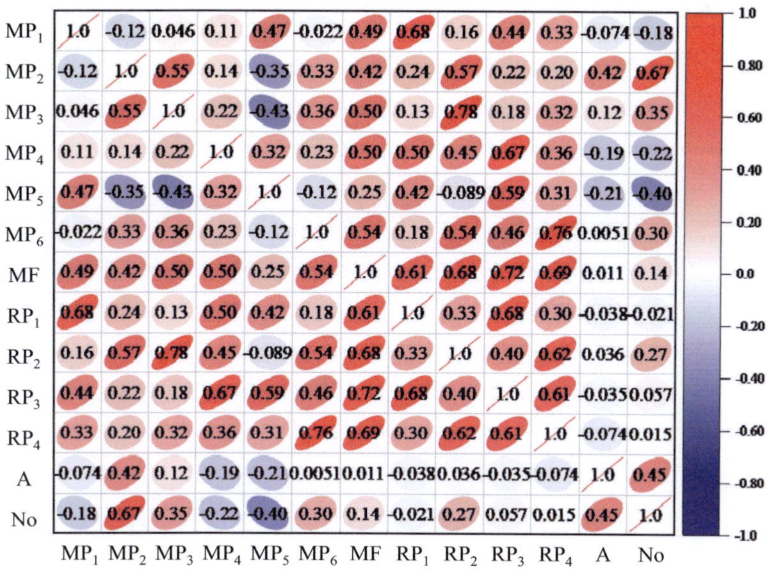

Fig. 7 The correlation map between all the variables and the objective

greater than 0.9995, which indicates that the number of samples of different detection targets has no effect on the reliability of the model. Besides, the model converges after 23 generations of training, indicating that it gives detection results in a short time.

Table 3 The hyperparameters of BO-XGBoost

Hyperparameters	Settings
Max depth	12
Eta	0.23
Min child weight	1
Subsample	0.8
Num class	7
Epoch	1000
Num parallel tree	8

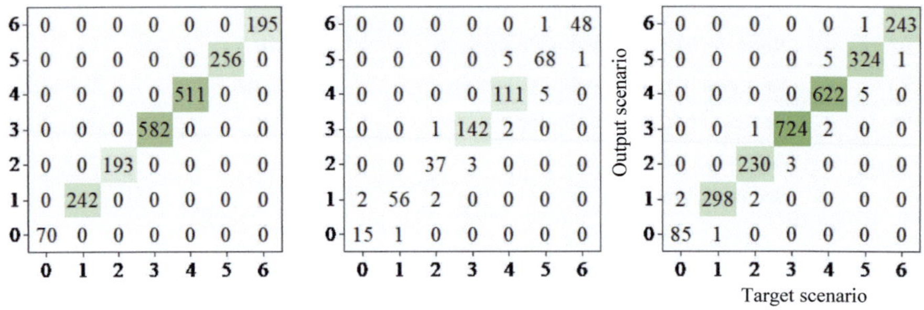

(a) (b) (c)

Fig. 8 The detection results of XGBoost. **a** Training; **b** test; **c** whole samples

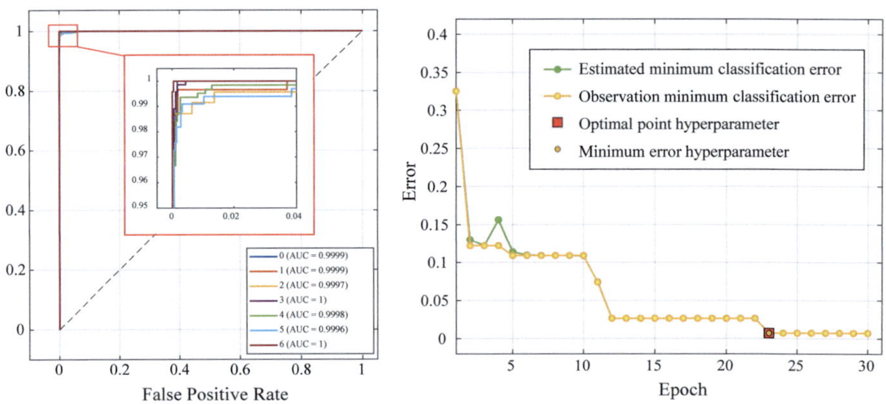

(a) (b)

Fig. 9 The evaluating indicator of XGBoost. **a** AUC; **b** astringency

Table 4 Detection results of different scenarios

Scenarios	Accuracy	Precision	Recall	F1-score
Scenario 0	0.9827	0.9884	0.9770	0.9827
Scenario 1	0.9917	0.9868	0.9967	0.9917
Scenario 2	0.9871	0.9871	0.9871	0.9871
Scenario 3	0.9959	0.9959	0.9959	0.9959
Scenario 4	0.9904	0.9920	0.9857	0.9888
Scenario 5	0.9818	0.9818	0.9818	0.9818
Scenario 6	0.9959	0.9959	0.9959	0.9959
Average	0.9893	0.9897	0.9885	0.9891

Thanks to BO's hyperparameter optimization of XGBoost, it gives full play to the maximum potential of the integrated tree model and ensures the highest accuracy efficiently. In summary, BO-XGBoost has higher efficiency and lower computational cost, which perfectly fits the engineering scenario and subsequent online optimization requirements.

4.3 Optimization Results Analysis

After the accurate identification of the hydraulic cylinder status is achieved, the hydraulic cylinder pressure needs to be redistributed and optimized. Since the hydraulic cylinder withdrawal is instantaneous, this chapter uses an online optimization method. The method continuously collects the hydraulic cylinder pressure data every 10 s at any time, calculates the pressure optimization scheme and stores it in the computer. Until the segment is assembled, the hydraulic cylinder retracts, and the result is directly transmitted to the SESA TBM controller, which greatly avoids the error caused by the calculation time of the offline algorithm. The online algorithm can often better match the actual scenario by its continuous input variables and constraints. The offline algorithm only focuses on the initial status and the final status, which does not pay attention to the changes of variables and targets in the optimization process. The offline algorithm will have a large deviation from the actual situation in the continuous operation scenario.

The optimization algorithm used in this chapter is the online AOA algorithm, which has certain advantages in optimization problems with high-dimensional variables. The ideal point is obtained directly by the TOPSIS method for the Pareto solution. Since there are 7 scenarios of SESA TBM operation, among which no optimization is required when no segment is assembled, 6 optimization schemes will be generated correspondingly. To illustrate the reliability of the optimization method, this chapter conducts a control experiment, i.e., an experimental process without adding the pressure redistribution optimization algorithm, and the specific results are compared as shown in Fig. 10. The optimization results of offline AOA and online AOA are shown in Tables 5, 6 and 7. The following conclusions can be obtained:

(1) The introduction of the pressure redistribution optimization algorithm can significantly improve the stress concentration phenomenon during the segment assembly, making each cylinder uniformly bear the sudden increase in pressure caused by the withdrawal of the partial assembly hydraulic cylinder. As shown in Fig. 10 and Tables 5 and 6, for the 6 optimization scenarios, the variance of the hydraulic cylinder pressure (Y1) is improved by 44.06%, 46.87%, 31.53%, 43.76%, 31.14%, and 12.56%, respectively. The change of hydraulic cylinder pressure (Y2) is reduced by 46.73%, 15.48%, 17.40%, 52.40%, 23.71%, and 18.45%, respectively. The stress concentration of the hydraulic cylinder is improved by 16.03%, 14.29%, 8.84%, 31.10%, 15.47%, and 11.11%, respectively. In the whole segment assembly cycle, the maximum pressure is reduced from 254 to 175 bar, which reduces the bearing burden of the hydraulic cylinder and the segment. The relevant results show the reliability of the optimization method, and the error between the theoretical optimization results and the experimental results is mainly caused by the hardware response time and the sensor accuracy.
(2) The optimization method will have different optimization effects for different segment assembly scenarios. In the scenario where four hydraulic cylinders are missing (Scenarios 1, 2, and 5), this method can reduce the maximum pressure by about 15%. In the scenario of only one hydraulic cylinder missing, this method can only reduce the maximum pressure by about 10%. In the scenario where five cylinders are missing, the improvement effect is related to the position of the hydraulic cylinders. For example, in Scenario 3, the hydraulic cylinders are symmetrically missing, and the improvement effect is less effective. The hydraulic cylinders in Scenario 4 are missing, and the improvement is obvious, and the maximum pressure is reduced by about 30%. Besides, in the experimental scheme without the online optimization algorithm, the pressure inheritance phenomenon occurs, resulting in some extreme pressure as the segment assembly proceeds. The online optimization algorithm significantly improved this phenomenon, so that there is essentially no pressure inheritance during the assembly of the previous segment with the next segment.
(3) Compared with the offline AOA, the online AOA has a higher optimization effect and more reliable optimization results. It can be found in Table 5 that the online AOA can effectively control the maximum pressure of the hydraulic cylinder through multiple online optimizations. Compared with the unoptimized experimental scheme, online AOA and offline AOA can be improved by 31.10% and 19.69%, respectively. In addition, since each input of the online AOA is real scenario data, multiple iterations can avoid calculation errors. The errors between the theoretical results and the experimental results of the online AOA and offline AOA methods are 17.98% and 25.87%, respectively, where online

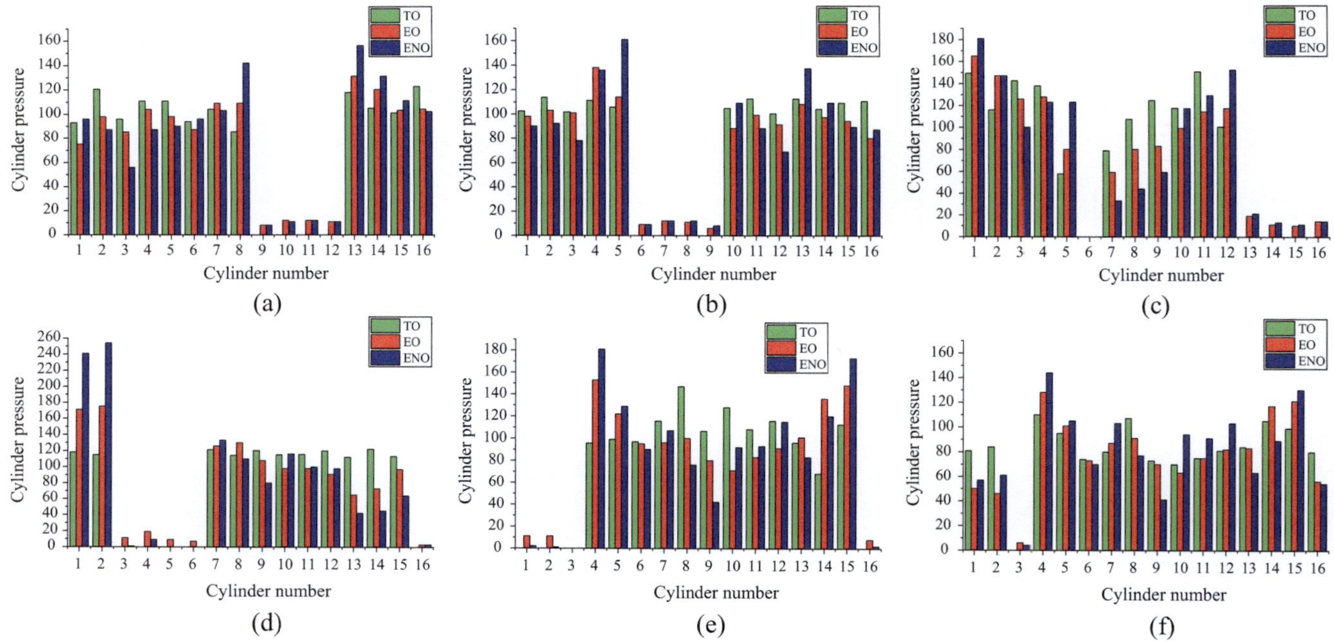

Fig. 10 The comparison of optimization results. **a** Scenario 1; **b** Scenario 2; **c** Scenario 3; **d** Scenario 4; **e** Scenario 5; **f** Scenario 6. (TO is theoretical results with online AOA optimization, EO is experimental results with online AOA optimization, and ENO is experimental results without online AOA optimization)

Table 5 The comparison of optimization results

Scenario	Y1				Y2			
	TO	EO	ENO	Improvement (%)	TO	EO	ENO	Improvement (%)
Scenario 1	9.86	15.21	27.19	44.06	544.38	705.5	1324.5	46.73
Scenario 2	7.56	14.08	26.50	46.87	688.47	956.67	1131.83	15.48
Scenario 3	28.01	30.56	44.63	31.53	1524.68	1135.23	1374.31	17.40
Scenario 4	34.24	40.40	71.83	43.76	1318.55	1179.08	2477.00	52.40
Scenario 5	18.60	25.92	37.64	31.14	940.14	2767.92	3628.00	23.71
Scenario 6	12.76	24.43	27.94	12.56	534.60	687.00	842.44	18.45
Average	18.51	25.10	39.29	36.11	925.14	1238.57	1796.35	31.05

Note TO is theoretical results with online AOA optimization, EO is experimental results with online AOA optimization, and ENO is experimental results without online AOA optimization. "Improvement" represents the comparison between EO and ENO

Table 6 The comparison of stress concentration

Scenario	Max pressure (bar)			
	TO	EO	ENO	Improvement (%)
Scenario 1	122	131	156	16.03
Scenario 2	111	138	161	14.29
Scenario 3	150	165	181	8.84
Scenario 4	122	175	254	31.10
Scenario 5	147	153	181	15.47
Scenario 6	110	128	144	11.11
Average	127	148	180	17.36

Table 7 The comparison between offline AOA and online AOA

Method	Max pressure (bar)	Error between theory and experiment (%)
Without optimization	254	
Online AOA	175 (31.10%)	17.98
Offline AOA	204 (19.69%)	25.87
Improvement	36.69%	30.50

Note "()" represents the comparison between the optimization method and the without optimization method. "Improvement" represents the comparison between online AOA and offline AOA

AOA improves the error by 30.50%. In terms of optimization effect and reliability, the online optimization method is superior to the offline optimization method and can better match the actual scenario.

5 Merits Analysis

To verify the reliability of the hydraulic cylinder detection model, this chapter compares several types of current machine learning algorithms, such as AdaBoost, artificial neural network (ANN), gaussian process (GP), support vector machine (SVM), decision tree (DT), and radial basis function neural network (RBF). All of them use the same training and test datasets, and the hyperparameters are sought by BO. The specific detection results and accuracy are shown in Table 8. It is clear to see that BO-XGBoost achieves the highest accuracy, precision, recall, and F1-score, as highlighted in bold in Table 8, demonstrating its superior predictive performance.

After completing the comparison of different detection methods, the differences among different evolutionary algorithms are further compared. Four algorithms, namely NSGA-II, HREA, NSGA-III, and MOEA, are selected to compare with the AOA algorithm selected in this chapter. All the comparison algorithms come from an open-source package on GitHub called plat-EMO. Each algorithm adopts the form of online optimization. This section selects the optimization scenarios of Scenario 1, that is, the scenarios where No. 9, 10, 11, and 12 hydraulic cylinders are missing. The relevant optimization results are shown in Table 9. All the results are discussed below.

Table 8 The comparison of different methods

Index	Accuracy	Precision	Recall	F1-score
BO-XGBoost	**0.9893**	**0.9897**	**0.9885**	**0.9891**
AdaBoost	0.9626	0.9637	0.9597	0.9617
ANN	0.9398	0.9421	0.9264	0.9342
GP	0.9723	0.9731	0.9703	0.9717
SVM	0.9775	0.979	0.9741	0.9766
DT	0.8656	0.8683	0.8331	0.8503
RBF	0.5644	0.5855	0.5327	0.5578

Table 9 The comparison of different evolutionary algorithms

Indicators	AOA	HREA	NSGA-II	NSGA-III	MOEA
T (s)	8.12	10.56	12.14	13.54	25.42
GD	5.43	6.24	5.69	6.04	42,800
$Spacing$	4.28	5.52	9.53	8.34	45,804

(1) Compared with the traditional machine learning methods, the BO-XGBoost method has better detection results. As shown in Table 8, the F1-score of the BO-XGBoost, AdaBoost, ANN, GP, SVM, DT, and RBF are 0.9891, 0.9617, 0.9342, 0.9717, 0.9766, 0.8503, and 0.5578, respectively. Compared with the other six methods, the detection results of BO-XGBoost are improved by 2.85%, 5.88%, 1.79%, 1.28%, 16.32%, and 77.32%, respectively. The remaining indicators of Accuracy, Precision, and Recall also show similar performance, and the results show that BO-XGBoost is superior to other machine learning algorithms in all aspects of this chapter. In the star detection problem of reference [11], XGBoost also embodies the advantages of high precision and generalization. The conclusions of this chapter further illustrate the advantages of the tree model and ensemble learning methods in detection problems.
(2) Compared with the state-of-the-art evolutionary algorithms, AOA has better algorithm convergence and can quickly find the optimal solution set. As shown in Table 9, the MOEA is in the form of failure in this problem, and the minimum GD of the other four algorithms is 5.43, 6.24, 5.69, and 6.04, respectively, indicating that each algorithm has a good convergence. The AOA algorithm can achieve the best GD in about 1000 iterations. Compared with HREA (2000 iterations), NSGA-II (5000 iterations), and NSGA-III (4000 iterations), AOA has a faster and better convergence effect. Meanwhile, the calculation time of AOA is the fastest, which is 8.12 s. It shows that the AOA has the lowest computational complexity and can better reduce the project cost.
(3) Compared with the state-of-the-art evolutionary algorithms, AOA has a better spatial distribution, where the solution set can cover a wider range of sample intervals. As shown in Table 9, AOA and HREA have the best solution set distribution, and their minimum spacing values are 4.28 and 5.52, respectively. Similar to the convergence degree of the algorithm, the AOA algorithm can also complete the best spatial distribution within 1000 iterations, while HREA, NSGA-II, and NSGA-III algorithms require 4000, 5000, and 3000 iterations, respectively. The results show that the AOA algorithm selected in this chapter can obtain a better spatial solution set distribution and better meet the needs of engineering scenarios.

6 Chapter Summary

Considered as a new type of SESA TBM equipment, this chapter proposes a hydraulic cylinder pressure redistribution framework for hydraulic cylinder status detection and hydraulic cylinder pressure optimization, which can better reduce the stress concentration during segment assembly. The redistribution framework integrates BO-XGBoost and online AOA methods to achieve real-time optimization of hydraulic cylinder pressure. The online learning module is incorporated to ensure the efficiency and practicability of optimization.

To verify the feasibility of the redistribution framework, the SESA TBM experimental bench is built, and the research method is verified by field experiments. Through the proposed online optimization method, the stress concentration of the hydraulic cylinder can be effectively reduced. The innovations of this method include: (1) designing a kind of TBM that can realize synchronous excavation and segment assembly and building the simulation and experimental model of the TBM to analyze the operation process, (2) constructing an online optimization framework that integrates BO-XGBoost and online AOA methods to improve the stress concentration phenomenon of hydraulic cylinder pressure. The developed approach breaks through the defects of the traditional TBM segment assembly and active control methods in engineering applications, providing a theoretical reference for realizing synchronous excavation and segment assembly technology.

References

1. Wang KY, Wu XG, Zhang LM, Song XQ (2023) Data-driven multi-step robust prediction of TBM attitude using a hybrid deep learning approach. Adv Eng Inf 55:101854
2. Pan Y, Fu XL, Zhang LM (2022) Data-driven multi-output prediction for TBM performance during tunnel excavation: An attention-based graph convolutional network approach. Automat Constr 141:104386
3. Xu J, Ding L, Luo H, Chen EJ, Wei L (2019) Near real-time circular tunnel shield segment assembly quality inspection using point cloud data: A case study. Tunn Undergr Space Technol 91:102998
4. Li Y, Sun Q, Zhang L, Wang Z, Feng W (2023) Online multi-objective optimization for tunnel boring machine segment assembly considering stress concentration. Autom Constr 156:105127
5. Guo K, Zhang L (2022) Adaptive multi-objective optimization for emergency evacuation at metro stations. Reliab Eng Syst Saf 219:108210
6. Huang T, Wang XL, Liu HT, Yang YH (2016) Force analysis of an open TBM gripping-thrusting-regripping mechanism. Mech Mach Theory 98:101–113
7. Hu X, Chen Y, Ren L, Xu Z (2023) Investor preference analysis: an online optimization approach with missing information. Inf Sci 633:27–40
8. Pang JY, Wang SB, Tang ZF, Qin YM, Tao XF, You XH, Zhu JK (2022) A new 5G radio evolution towards 5G-advanced. Sci China Inf Sci 65:45
9. Guo Y, Wang H, Guo Y, Zhong M, Li Q, Gao C (2022) System operational reliability evaluation based on dynamic Bayesian network and XGBoost. Reliab Eng Syst Saf 225:108622
10. Zhao DW, Bezgans Y, Vdonin N, Du WH (2021) The use of TOPSIS-based-desirability function approach to optimize the balances among mechanical performances, energy consumption, and production efficiency of the arc welding process. Int J Adv Manuf Tech 112:3545–3559
11. Chao L, Wen-hui Z, Ji-ming L (2019) Study of star/galaxy classification based on the XGBoost algorithm. Chin Astron Astrophy 43:539–548

TBM Fault Detection

1 Introduction

Statistical analyses indicate that shield cutter head failures in TBM are prevalent in nearly every tunnel project, constituting approximately fifty percent of all accident categories in tunnel engineering practices in China [1]. In the Kranji tunnel project in Singapore, the combination of mixed face conditions and high abrasivity of granite led to a significant reduction in cutter lifespan, with each cutter excavating less than 100 m^3 of material. Similarly, during the excavation of a pipe tunnel beneath the Weser River in Germany, the abrasiveness of the soil exerted considerable stress on TBM components, particularly the cutter heads, resulting in accelerated eccentric wear during the drilling process. Furthermore, the Oporto Metro project in Portugal experienced severe wear damage to cutter heads, which triggered three surface collapses and tragically resulted in the death of a resident. These incidents collectively caused nearly nine months of project delays. Consequently, it is imperative to investigate the failure mechanism of shield cutter heads and identify the primary driving factors to develop effective risk mitigation strategies [2].

Fault tree analysis is widely recognized as the most prevalent quantitative methodology for investigating accident mechanisms. Fault trees have achieved broad acceptance in the dependability and safety assessments of complex and critical systems due to their straightforward manipulation and the availability of robust software tools that facilitate both qualitative and quantitative analyses [3]. Traditional static fault trees (SFTs) illustrate the combinations of component failures that can precipitate a system-wide failure through the use of Boolean logic gates, such as AND, OR, and voting gates. They are typically resolved using binary decision diagram techniques. However, the traditional SFTs are limited in their ability to model the dynamic behaviors associated with system failure mechanisms. These limitations arise because such failure mechanisms often involve sequence-dependent events, the utilization of spare components, dynamic redundancy management, and the prioritization of failure events [4].

L. Zhang et al., *Intelligent Construction in Tunnels*, Lecture Notes in Civil Engineering 679, https://doi.org/10.1007/978-981-96-8785-5_10

To address the above challenges, dynamic fault trees (DFTs) were introduced as an extension of the traditional fault tree approach, incorporating sequential dependencies [5]. The DFT employs the linear modularization algorithm to decompose the entire fault tree into multiple independent subtrees. This makes the DFT a suitable choice for modeling time-dependent risk scenarios, as it provides an intuitive representation that leverages dynamic gates to construct complex time-dependent logic while maintaining the clarity of the fault structure. However, the sequential dependencies and dynamic interactions among system components present significant challenges for the DFT analysis using conventional SFT algorithms. Typically, the DFT is transformed into a corresponding Markov chain model for analysis [6]. Nevertheless, this approach suffers from the state space explosion problem, where the number of states in the Markov chain increases exponentially with the number of components in the DFT. This issue is a common obstacle in Markov process modeling, leading to exceedingly complex computations and limiting the practical applicability of the DFT for large-scale systems [7].

In response to the aforementioned problem, temporal Bayesian networks have recently emerged as a robust method for modeling dynamic systems. Discrete-time Bayesian networks (DTBNs) have been proposed as a viable alternative to DFTs, effectively circumventing the limitations associated with Markov chains, such as state-space explosion and the error-prone nature of converting the DFT to Markov chains. This chapter explores the integration of DFT and DTBN to simulate and analyze diverse system component behaviors and interactions. In this chapter, a hybrid simulation framework combining the DFT and DTBN methods is developed to enable both predictive and diagnostic analysis of shield cutter head failures in tunnel excavation projects. To validate the applicability and effectiveness of the proposed approach, a case study involving a tunnel project in China is conducted.

2 TBM Failure Mechanism

In soft ground conditions, TBM is primarily classified into two types: earth pressure balance (EPB) TBM and slurry TBM. The EPB TBM is favored over slurry TBMs due to their relatively lower cost and reduced spatial requirements, making them highly prevalent in a wide range of applications, particularly in the construction of urban rail transport systems [8]. Figure 1 illustrates the internal structure of an EPB TBM. Specifically, the EPB method is a mechanized tunneling technique that maintains balanced pressure at the TBM face without the necessity of using slurry. Excavated material is transported from the base of the excavation chamber to a belt conveyor via a screw conveyor. The synchronization between the screw conveyor and the TBM's advance rate ensures precise control of the soil paste's support pressure, thereby enhancing the stability and efficiency of the tunneling process.

Failures of shield cutter heads are frequent in tunneling operations due to factors such as faulted zones, blocky rock masses, mixed face conditions, and high groundwater pressure. Figure 2 presents photographic evidence of shield cutter head failures

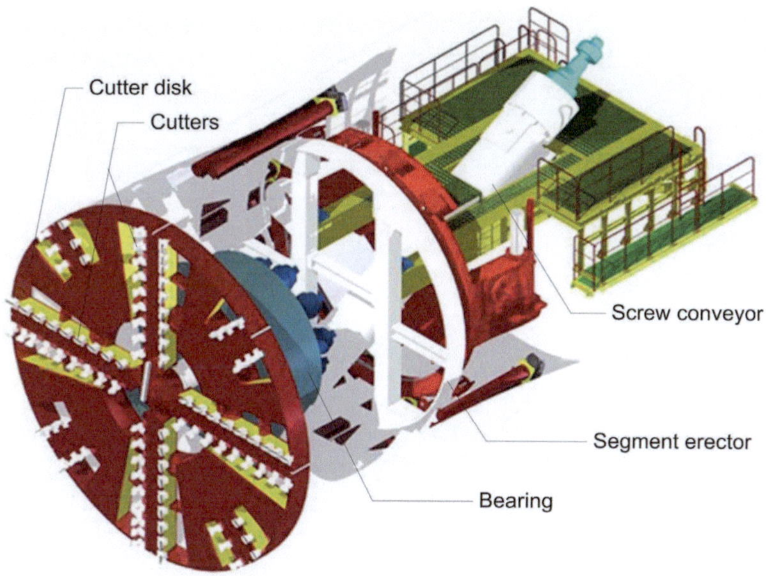

Fig. 1 Internal structure of the EPB TBM

observed at construction sites. The performance failures of shield cutter heads can be broadly categorized into the following three types:

(1) ***Cutter disk failure* (CDF)**. This is a prevalent form of failure, typically manifesting as *cutter disk wear* (CDW, see Fig. 2a) and *mud cake formation* (MCF, see Fig. 2b). Such failures can result in significant losses of time and resources. The occurrence of CDF is influenced by both internal and external variables. Internal factors include *fault in cutter disk* (X1) and *fault in the dreg outlet* (X4). External factors are associated with geological conditions, such as *mixed face conditions* (X2) and *unexpected clayey soils* (X3).
(2) ***Cutter tool failure* (CTF)**. The shield cutter, a hard rock tool, functions by exerting extruding, shearing, and tensile forces to crush soils or rocks through cutter rolling. Cutter failures primarily occur as *cutter eccentric wear* (CEW, see Fig. 2c) and *cutter crack* (CC, see Fig. 2d). Key contributing factors to CTF include *faults in the working cutter* (X5), *faults in the spare cutter* (X6), *faults in the cutter tool itself* (X7), and *faults in the thrust loading system* (X8).
(3) ***Bearing and other system failures* (BSF)**. The bearing system is crucial for ensuring the reliability of TBM performance. However, bearings and other drilling components are subject to severe wear during prolonged tunneling operations. Failures typically present as *bearing wear* (BW, see Fig. 2e) and *screw wear* (SW, see Fig. 2f). Common causes of BSF include *faults in the thrust loading system* (X8), *faults in the sealing materials* (X9), and *fault in the screw mechanism* (X10).

Fig. 2 Failures of shield cutter heads: **a** cutter disk wear; **b** mud cake formation; **c** cutter eccentric wear; **d** cutter crack; **e** bearing wear; and **f** screw wear

3 TBM Fault Detection Framework

The DFT is capable of modeling time-dependent risk factors, providing a foundational framework to address the dynamic characteristic underlying failure occurrence. The DTBN can be utilized to mitigate the computational complexity associated with DFT calculations. Consequently, a hybrid approach that combines the DFT and DTBN has been developed to facilitate predictive and diagnostic analysis within a complex system. The overall workflow of the proposed approach comprises four main steps.

3.1 TBM Failure Mechanism

A failure mechanism refers to a physical defect or condition that leads to system failure. Exploring the failure mechanism entails collecting information about the causes and consequences of failures within system components, identifying risk or hazard factors that may impact project performance, and thoroughly documenting these attributes. The identification of the specific failure mechanism typically marks the conclusion of the failure analysis process. After pinpointing the mechanism, efforts can focus on uncovering the root causes of the failure and implementing measures to prevent recurrence.

With the rapid advancements in tunneling construction worldwide, a substantial body of diverse knowledge has emerged. Those resources offer valuable prior knowledge for the exploration of failure mechanisms. The exploration process in tunnel construction consists of three key procedures.

(1) Acquire prior resources by reviewing existing construction standards and technical manuals, and tacit knowledge obtained through consultations with domain experts.
(2) Determine relevant variables that depict system conditions, flow processes, and performance states based on the acquired knowledge resources.
(3) Establish relationships and interdependencies among these variables, which process varying attributes, using probable failure-consequence scenarios of the targeted event as a framework.

3.2 DFT Construction

The SFT is structured by experts in accordance with the HAZard and Operability Analysis (HAZOP) report. However, it has limitations in addressing temporal dependencies within a dynamic system. To overcome this, the DFT methodology was developed as a reliability assessment tool that models sequential dependencies and interactions among system components while maintaining the robust representation of the traditional SFT. Within a DFT framework, four fundamental dynamic gates are utilized: the priority AND (PAND), the sequence enforcing (SEQ), the standby or spare (SPARE), and the functional dependency (FDEP). The gates facilitate the modeling of temporal dependencies and interactions among system components. By integrating static fault gates (such as AND/OR gates) with dynamic fault gates (including PAND, SEQ, SPARE, FDEP, and other gates), comprehensive failure logic diagrams are constructed. These diagrams delve into component-level failures and operational modes of the system, offering a detailed analysis of the system's reliability and dynamic behavior.

To explicitly incorporate time and account for sequential dependencies among components, the timeline $[0, +\infty]$ is segmented into $n + 1$ intervals. Each interval corresponds to a particular state reflecting the progression of failure over time [8], as illustrated in Fig. 3. This study assumes that an exponential density function characterizes a basic event within the DFT model. Based on the terminology established by [9], if a basic random event A is stated to be in a state i $(1 \leq i \leq n)$, denoted as $A = i$, it indicates that A has failed during the ith interval. The prior probability of A being in state i $(1 \leq i \leq n)$, represented as $P(A = i)$, can be calculated using Eq. (1). Similarly, if A is described as being in state $n + 1$, it signifies that A has successfully survived the mission time T. The prior probability of A being in state $n + 1$, denoted as $P(A = n + 1)$, can be determined via Eq. (2).

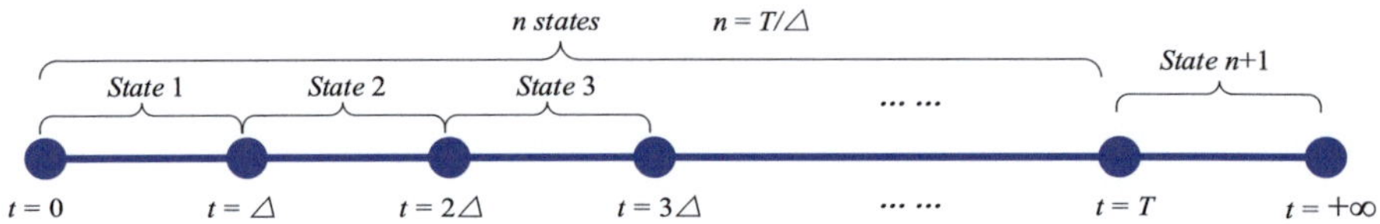

Fig. 3 Divisions of time intervals along with the failure evolution

$$P(A = i) = P((i-1)\Delta < t_A \le i\Delta) = \int_{(i-1)\Delta}^{i\Delta} f_A(t)\mathrm{d}t$$
$$= e^{-\lambda_a i\Delta}\left(e^{\lambda_a\Delta} - 1\right), \quad 1 \le i \le n \tag{1}$$

$$P(A = n+1) = P(t_A > T) = \int_T^{\infty} f_A(t)\mathrm{d}t = e^{-\lambda_a T} \tag{2}$$

where t_A denotes the failure time of the basic random event A, while λ_a represents the failure rate of A. The term f_A corresponds to the density distribution function. Additionally, $\Delta = T/n$ indicates the interval length, and n is the time granularity.

3.3 DTBN Formation

Successfully mapping the dynamic gates of the DFT into a Bayesian network has traditionally posed challenges, particularly in avoiding the creation of large multi-dimensional probability tables. To address this, a new general formalism has been developed to model probability distribution functions of dynamic gates, including PAND, SEQ, SPARE, and FDEP gates. Figure 4 illustrates the rules for mapping these dynamic gates of DFT into the DTBN, with detailed descriptions of the probability distribution functions provided below.

(1) **PAND gate transition**

A PAND gate is commonly utilized to model interactions between dependent components within complex systems. It reaches a failure state only when all its input components fail in a specified sequence, from left to right in graphical notation. Figure 4a illustrates a PAND gate along with its equivalent structure in the DTBN. In the context of transitions, a critical task in simulating the PAND gate within the DTBN is deriving an analytical expression for the conditional probability functions associated with the node C_{PAND}, defined by $P_{ijk}^{C_{PAND}}$, as described by Eq. (3).

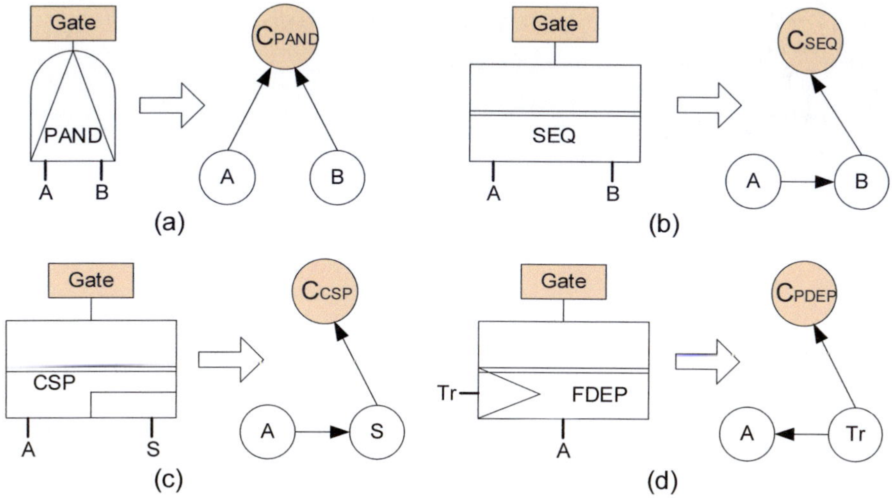

Fig. 4 Mapping dynamic gates of DFT into DTBN: **a** PAND gate; **b** SEQ gate; **c** CSP gate; **d** PDEP gate

$$
P_{ijk}^{C_{PAND}} = \begin{cases} P(C_{PAND} = [(k-1)\Delta, k\Delta) | A = [(i-1)\Delta, i\Delta), B = [(j-1)\Delta, j\Delta)) \\ \qquad = \begin{cases} 1, & 0 < i \le j = k \le n \\ 0, & 0 < j < i \le n \end{cases} \\ P(C_{PAND} = [T, \infty) | A = [(i-1)\Delta, i\Delta), B = [(j-1)\Delta, j\Delta)) \\ \qquad = \begin{cases} 0, & 0 < i \le j = k \le n \\ 1, & 0 < j < i \le n \end{cases} \\ P(C_{PAND} = [T, \infty) | A = [(i-1)\Delta, i\Delta), B = [T, \infty)) = 1 \\ P(C_{PAND} = [T, \infty) | A = [T, \infty), B = [(j-1)\Delta, j\Delta)) = 1 \\ P(C_{PAND} = [T, \infty) | A = [T, \infty), B = [T, \infty)) = 1 \\ \qquad 0, \quad otherwise \end{cases} \tag{3}
$$

where $P_{ijk}^{C_{PAND}}$ denotes the probability of the node C_{PAND} failing in the kth interval, given that A has failed in the ith interval and B has failed in the jth interval. $C_{PAND} = [(k-1)\Delta,\ k\Delta)$ indicates that the node C_{PAND} fails in the kth interval, $1 \le k \le n$, and $C_{PAND} = [T,\ \infty)$ indicates that the node C_{PAND} fails in the $(n+1)$th interval. For simplicity, $n = 2$ is used (i.e., three time intervals) as an example to show the meaning of $P_{ijk}^{C_{PAND}}$ clearly. Table 1 provides the conditional probability table (CPT) for the node C_{PAND} given both A and B are inputs in a PAND gate using Eq. (3).

Table 1 CPT of the node C_{PAND} given A and B are input in a PAND gate

A (i)	B (j)	C_{PAND} (k)		
		[0, Δ)	[Δ, T)	[T, ∞)
[0, Δ)	[0, Δ)	1	0	0
[0, Δ)	[Δ, T)	0	1	0
[0, Δ)	[T, ∞)	0	0	1
[Δ, T)	[0, Δ)	0	0	1
[Δ, T)	[Δ, T)	0	1	0
[Δ, T)	[T, ∞)	0	0	1
[T, ∞)	[0, Δ)	0	0	1
[T, ∞)	[Δ, T)	0	0	1
[T, ∞)	[T, ∞)	0	0	1

(2) **SEQ gate transition**

A SEQ gate requires its inputs to fail in a predetermined order. When incorporated into a DFT, the SEQ gate operates only if the failures occur in this specific order and does not activate if they occur otherwise. Figure 4b depicts a SEQ gate and its corresponding structure within the DTBN. Equation (4) defines the conditional probability function for node B falling in the jth interval, given that node A has failed in the ith interval, denoted by P_{ij}^{B}. Equation (5) defines the conditional probability function for the node C_{SEQ} falling in the kth interval, given that node B has failed in the jth interval, denoted by $P_{jk}^{C_{SEQ}}$.

$$P_{ij}^{B} = \begin{cases} P(B = [(j-1)\Delta, j\Delta)|A = [(i-1)\Delta, i\Delta)) \\ \quad = \begin{cases} 0, & 0 < j < i \le n \\ 1 - \frac{\lambda_a (e^{(\lambda_a - \lambda_b)\Delta} - 1)}{(\lambda_a - \lambda_b)(e^{\lambda_a \Delta} - 1)}, & 0 < j = i \le n \\ \frac{\lambda_a e^{-(\lambda_a - \lambda_b)(j-i)\Delta}(e^{(\lambda_a - \lambda_b)\Delta} - 1)(e^{\lambda_a \Delta} - 1)}{(\lambda_a - \lambda_b)(e^{\lambda_a \Delta} - 1)}, & 0 < i < j \le n \end{cases} \\ P(B = [T, \infty)|A = [(i-1)\Delta, i\Delta)) = 1 - \sum_{0<j\le n} P_{ij}^{B} \\ P(B = [T, \infty)|A = [T, \infty)) = 1 \\ \quad 0, \quad otherwise \end{cases} \tag{4}$$

$$P_{jk}^{C_{SEQ}} = \begin{cases} P(C_{SEQ} = [(k-1)\Delta,\ k\Delta)|B = [(j-1)\Delta,\ j\Delta)) = 1 \\ P(C_{SEQ} = [T,\ \infty)|B = [T,\ \infty)) = 1 \\ \quad 0, \quad otherwise \end{cases} \tag{5}$$

where $C_{SEQ} = [(k-1)\Delta,\ k\Delta)$ signifies that the node C_{SEQ} fails within the kth interval, $1 \le k \le n$; $C_{SEQ} = [T,\ \infty)$ indicates that the node C_{SEQ} fails in the $(n+1)$th interval; λ_a and λ_b represent the failure rates of A and B, respectively.

(3) **CSP gate transition**

Spare gates are employed to model scenarios where one or more principal components can be replaced by one or more backup components (spares) that provide identical functionality. In the context of a CSP gate, the failure rate of a dormant spare is zero, meaning that the spare cannot fail or degrade before or simultaneously with the primary component. Figure 4c illustrates a CSP gate along with its equivalent structure in the DTBN. Equation (6) describes the conditional probability function for the node S falling in the jth interval, given that A has failed in the ith interval, denoted by P_{ij}^{S}. Similarly, Eq. (7) describes the conditional probability function for the node C_{CSP} falling in the kth interval, given that S has failed in the jth interval, denoted by $P_{jk}^{C_{CSP}}$.

$$P_{ij}^{S} = \begin{cases} P(B = [(j-1)\Delta, j\Delta) | A = [(i-1)\Delta, i\Delta)) \\ \quad = \begin{cases} e^{-\lambda_a (j-i)\Delta}(e^{\lambda_a \Delta} - 1), & 0 < i < j \le n \\ 0, & otherwise \end{cases} \\ P(B = [T, \infty) | A = [(i-1)\Delta, i\Delta)) = 1 - \sum_{0<j\le n} P_{ij}^{S} \\ P(B = [T, \infty) | A = [T, \infty)) = 1 \\ \quad 0, \quad otherwise \end{cases} \tag{6}$$

$$P_{jk}^{C_{CSP}} = \begin{cases} P(C_{CSP} = [(k-1)\Delta, k\Delta) | B = [(j-1)\Delta, j\Delta)) = 1 \\ P(C_{CSP} = [T, \infty) | B = [T, \infty)) = 1 \\ \quad 0, \quad otherwise \end{cases} \tag{7}$$

where $C_{CSP} = [(k-1)\Delta, k\Delta)$ indicates that the node C_{CSP} fails in the kth interval, $1 \le k \le n$; $C_{CSP} = [T, \infty)$ signifies that the node C_{CSP} fails in the $(n+1)$th interval; and λ_a denotes the failure rate of A.

(4) **PDEP gate transition**

An FDEP gate is employed to model scenarios where a triggering event (which could be a basic event or the output of another gate within a fault tree) causes one or more dependent events to become inoperable or inaccessible. Figure 4d illustrates a CSP gate along with its corresponding structure in the DTBN. Equation (8) provides the conditional probability function for the occurrence of the trigger event Tr in the jth interval, given that A has failed in the ith interval, represented by P_{ij}^{A}. Equation (9) describes the conditional probability function for the node C_{FDEP} in the kth interval, given that Tr has failed in the jth interval, denoted by $P_{jk}^{C_{FDEP}}$.

$$P_{ij}^{A} = \begin{cases} P(A = [(i-1)\Delta, i\Delta) | Tr = [(j-1)\Delta, j\Delta)) \\ \quad = \begin{cases} e^{-\lambda_{Tr} i\Delta}(e^{\lambda_{Tr}\Delta} - 1) & 0 < i < j \le n \\ 1 - \sum\limits_{i<j} P_{ij}^{A} & 0 < i = j \le n \\ 0 & otherwise \end{cases} \\ P(A = [(i-1)\Delta, i\Delta) | Tr = [T, \infty)) = e^{-\lambda_{Tr} i\Delta}(e^{\lambda_{Tr}\Delta} - 1) \\ \quad 0, \quad otherwise \end{cases} \tag{8}$$

$$P_{jk}^{C_{FDEP}} = \begin{cases} P(C_{FDEP} = [(k-1)\Delta,\ k\Delta) | Tr = [(j-1)\Delta,\ j\Delta)) = 1 \\ P(C_{FDEP} = [T,\ \infty) | Tr = [T,\ \infty)) = 1 \\ \quad 0, \quad otherwise \end{cases} \tag{9}$$

where $C_{FDEP} = [(k-1)\Delta,\ k\Delta)$ indicates that the node C_{FDEP} fails in the kth interval, $1 \le k \le n$; $C_{FDEP} = [T,\ \infty)$ denotes that the node C_{FDEP} fails in the $(n + 1)$th interval; λ_{Tr} represents the failure rate of the trigger event *Tr*.

3.4 Fault Diagnosis and Detection

The DTBN has gained recognition as a robust method for modeling dynamic systems by addressing both sequential and functional dependencies among components. Unlike traditional Markov chains, which often suffer from state-space explosion and error-prone conversion from the DFT, the DTBN provides a more efficient and accurate approach. According to Khakzad, et al. [10] a key application of the DTBN is updating probabilities based on actual event observations. This capability enables the computation of probability distributions for potential risks and identification of the most likely causes of accidents. This study conducts an in-depth predictive and diagnostic analysis using DTBNs.

(1) **Predictive analysis**

Predictive analysis is employed to predict the unreliability of a failure represented by *R*, which is determined by a combination of basic events (X_1, X_2... X_n). When *R* fails within a time interval [0, *T*), it suggests that the failure might occur in one of the following sub-intervals: $[0, \Delta)$, $[\Delta, 2\Delta)$, ..., $[(n-1)\Delta, n\Delta)$, where n equals T/Δ. Thus, the unreliability of *R* within the time interval [0, *T*), denoted as $P(R = [0, T))$, is the sum of the probabilities of *R* failing in each specific sub-interval. Equation (10) can be used to calculate $P(R = [0, T))$, serving as a measure to assess the unreliability of *R* and assist decision-makers in taking appropriate actions before an accident occurs.

$$\begin{aligned} P(R = [0,\ T)) &= \sum_{1 \le i \le n} P(R = [(i-1)\Delta, i\Delta)) \\ &= \sum_{1 \le i \le n} P(R = [(i-1)\Delta,\ i\Delta) | X_1 = x_1, X_2 = x_2, \ldots, X_J = x_J) \end{aligned}$$

$$\times P\left(X_1 = x_1, \ldots, X_j = x_j, \ldots, X_J = x_J\right),$$
$$x_j \in \{[0, \Delta), [\Delta, 2\Delta), \ldots, [(n-1)\Delta, T), [T, \infty)\} \tag{10}$$

where $P(R = [(i - 1)\Delta, i\Delta))$ denotes the probability that R fails in the ith time interval $[(i - 1)\Delta, i\Delta)$. The term $P(R = [(i - 1)\Delta, i\Delta)| X_1 = x_1, X_2 = x_2, \ldots, X_J = x_J)$ represents the conditional probability distribution of R given specific values for events X_1 through X_J. Meanwhile, $P(X_1 = x_1, \ldots, X_j = x_j, \ldots, X_N = x_N)$ signifies the joint probability distribution of the basic events X. The variable x_j belongs to the set $\{[0, \Delta), [\Delta, 2\Delta), \ldots\ldots, [(n - 1)\Delta, T), [T, \infty)\}$, which defines the $n + 1$ possible states of the basic event X_j $(1 \leq j \leq J)$ along the timeline.

Moreover, considered a central measure of the distribution when it fits the data adequately, the mean time to failure (MTTF) serves as a valuable indicator for estimating the expected operational lifespan of a system. As the interval length Δ approaches an infinitesimal value, $P(R = [0, T))$ can illustrate the cumulative probability of a failure occurring within a specified time period. The MTTF of a TBM system is particularly significant for engineers in construction practice, as it provides insights into the system's expected performance duration underground. The MTTF for the failure of R can be computed using Eq. (11).

$$MTTF = \int_0^\infty P(R = [0, T))\mathrm{d}t \tag{11}$$

(2) **Diagnostic analysis**

Diagnostic analysis aims to determine the posterior probability distribution of basic events following a failure. This posterior probability distribution allows suspected causes to be quickly identified, thereby aiding engineers in performing real-time fault diagnosis. The DTBN leverages Bayes' theorem to update the probabilities of events based on new observations, resulting in the posterior probability. The capability of diagnostic analysis within the Bayesian framework is unparalleled and cannot be replicated in either the SFT or DFT frameworks. When a failure is observed in the ith interval $[(i - 1)\Delta, i\Delta)$, the posterior probability distribution of a basic event X_j $(1 \leq j \leq J)$, represented by $P\left(X_j = x_j | P(R = [(i-1)\Delta, i\Delta))\right)$, can be calculated using Eq. (12).

$$P(X_j = x_j | P(R = [(i-1)\Delta, i\Delta))) = \frac{P(X_j = x_j) \times P(R = [(i-1)\Delta, i\Delta) | X_j = x_j)}{P(R = [(i-1)\Delta, i\Delta))} \tag{12}$$

where $P(R = [(i - 1)\Delta, i\Delta))$ denotes the probability that R fails within the ith time interval $[(i - 1)\Delta, i\Delta)$. The variable x_j corresponds to the $n + 1$ possible states, $[0, \Delta)$, $[\Delta, 2\Delta)$, …, $[(n - 1)\Delta, T)$, $[T, \infty)$, of the jth event X_j $(1 \leq j \leq J)$ along the timeline.

4 Experimental Studies

4.1 Data and Resources

A tunnel case from the Wuhan metro system (WMS) in China is employed to validate the applicability of the developed approach. WMS is a combination of elevated and underground urban metro systems located in Wuhan, China. Metro Line 7 (ML7) is a new addition with 19 stations, and the Dongfangmacheng to Changfeng Section (DCS) serves as the case study to assess the reliability of the shield cutter head performance. The DCS is a double-track tunnel, measuring 1714 m on the right track and 1717 m on the left track. Two EPB TBMs, each with a diameter of approximately 6.2 m, are employed to excavate the soil between the Dongfangmacheng Station and the Changfeng Station at an average depth of 20 m below ground level. Figure 5 provides a photograph of an EPB TBM operating during the construction of DCS.

Since 2007, researchers at Huazhong University of Science and Technology have been developing safety control systems for metro construction and operation tasks across several Chinese metro systems, including those in Shenyang, Zhengzhou, Shenzhen, and Wuhan. They have implemented web-based early warning systems tailored to the safety control needs of each project, accumulating a diverse range of knowledge resources in the process [11, 12]. In alignment with the mechanistic analysis of shield cutter head failures, potential hazards, faults, and their causal relationships are initially identified. The Delphi technique is employed to achieve expert consensus.

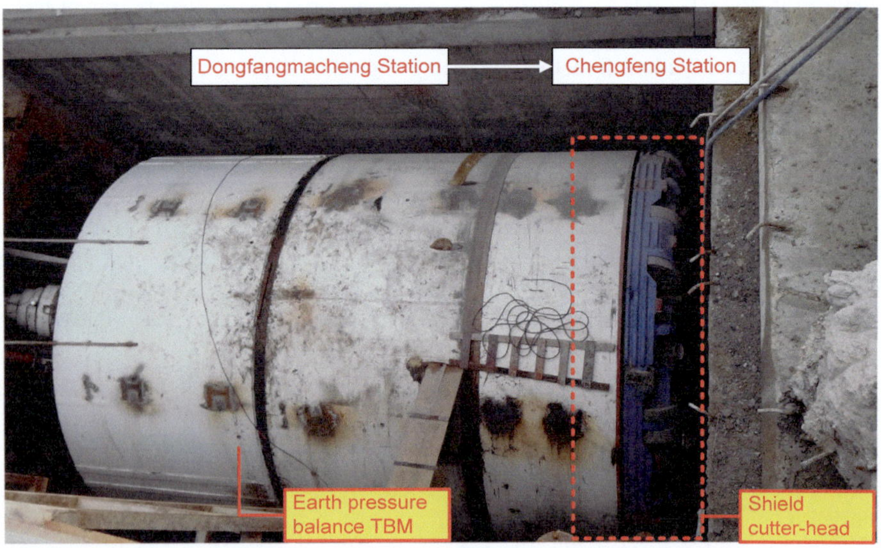

Fig. 5 Photograph of an EPB TBM operating during the construction of DCS

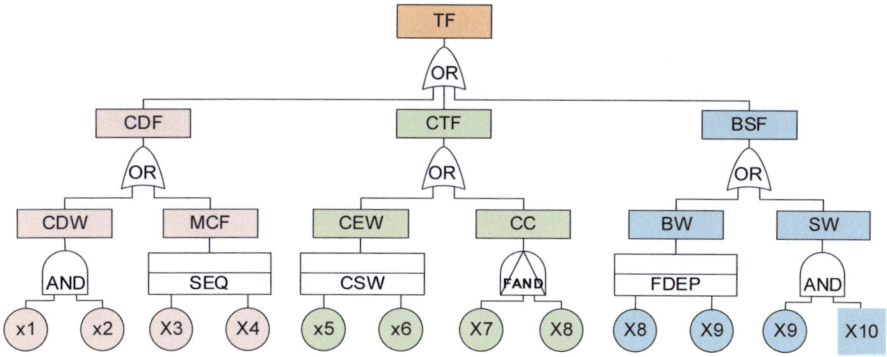

Fig. 6 A DFT model for the shield cutter head failure in tunnel construction

Table 2 Descriptions of basic events in the established DFT model

Event	Description	Event	Description
X1	Fault in cutter disk	X6	Fault in the spare cutter
X2	Mixed conditions	X7	Fault in the cutter tool
X3	Undiscovered clayey soils	X8	Fault in the thrust loading system
X4	Fault in the dreg outlet	X9	Fault in the sealing materials
X5	Fault in the working cutter	X10	Fault in the screw

The process begins with interviewing a risk engineer to create the fault tree structure for shield cutter head failures. The structure is then reviewed by a senior risk coordinator, whose feedback informs further revisions. The revised structure is sent back to the risk engineer for additional review. Incorporating prior expert knowledge, a fault logic diagram is developed to construct a failure-consequence scenario, progressing from top to bottom events using logical AND/OR and dynamic gates. Figure 6 illustrates the final DFT model, highlighting the shield cutter head failure in tunnel construction, with ten basic events contributing to the top event failure, as detailed in Table 2.

4.2 Model Development

To address the challenges of state-space explosion and error-prone conversion procedures associated with traditional DFT calculations, the DFT model (see Fig. 6) for shield cutter head failures in tunnel construction is transformed into a DTBN model. Figure 7 illustrates this equivalent DTBN model, derived following the transition rules for mapping a DFT into a DTBN. It is important to note that the failure rate of specific system components may vary depending on the type of TBM and actual geological conditions. Failure rates for system components such as the cutter disk

(X1), cutter tool (X7), and screw (X10) should be provided by construction suppliers, achieved by continuously operating one unit until failure occurs. Regarding geological conditions, such as mixed conditions (X2) and undiscovered clayey soils (X3), failure rates are typically determined in consultation with field domain experts. Failure rates for all basic events, expressed in terms of a tunnel ring, are presented in Table 3. In this context, a tunnel section ring is used as a unit for interval length to align with the daily measurement of operational parameters in practice.

Once the equivalent DTBN model for shield cutter head failures is established, various what-if experiments can be conducted to investigate causal-effect interactions under different scenarios. Predictive and diagnostic analysis techniques within Bayesian inference are employed to demonstrate both feedforward and feedback control. To reduce computational complexity and achieve significant time savings, a MATLAB program has been developed for this purpose. All risk-based calculations and analysis are performed using the developed MATLAB program.

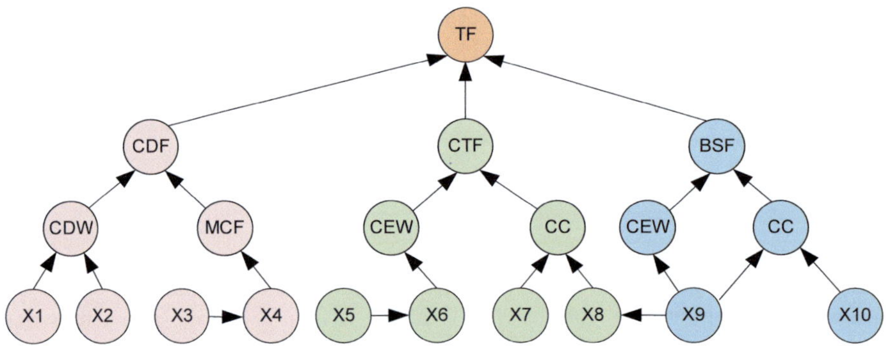

Fig. 7 Equivalent DTBN model of DFT model for shield cutter head failures

Table 3 Failure rates of basic events in the established DTBN model

Event	Failure rate λ (per ring)	Event	Failure rate λ (per ring)
X1	0.0001	X6	0.0005
X2	0.0002	X7	0.0005
X3	0.0003	X8	0.0004
X4	0.0003	X9	0.0002
X5	0.0002	X10	0.0001

Note It is assumed that the failure rate of each event performs an exponential density function

4.3 Feedforward Control and Analysis

Feedforward control seeks to predict the probability distribution of a failure before it occurs using the predictive analysis technique within Bayesian inference. By propagating given information, such as the failure rates of root nodes, the outcomes of other nodes in the network can be updated over time. This approach enables a quantitative estimation of potential failure timing, thereby allowing the proposal and timely implementation of control measures to mitigate risks.

In this case, the failure rates of the ten root nodes (as listed in Table 3) are input into the DTBN model for shield cutter head failures (see Fig. 7). This allows for the calculation of the unreliability of the failure nodes over time using Eq. (10). To achieve more precise results, the interval length Δ is set to be very short, that is 200 rings in this instance. Figure 8 displays the unreliability results of the failure nodes over time within the network, including TF, CDF, CTF, and BSF. Generally, these unreliability curves exhibit significantly different growth rates among the failure nodes (refer to the horizontal axis in Fig. 8), suggesting considerable variation in the actual failure points of those nodes. As previously mentioned, the MTTF serves as a valuable metric for estimating the operational lifespan of a system, and the MTTF results for these four failure nodes are presented in Table 4. It is evident that the first node to fail is TF (MTTF = 2633 rings), followed by CTF (MTTF = 4687 rings) and CDF (MTTF = 7883 rings), with BSF as the last node to fail (MTTF = 10,238 rings). A detailed analysis of the results is provided below.

Regular maintenance and repair are crucial for enhancing the safety, reliability, and cost-effectiveness of manufacturing equipment. However, determining the optimal timing for these activities involves a tradeoff since waiting for a failure to occur is not advisable. Excessively frequent maintenance can lead to diminished productivity and increased costs. In some cases, repairs are not performed at the most opportune time, resulting in TBM accidents when the level of risk escalates beyond manageable limits. Given the limited working space underground, decisions regarding the timing of minor or routine repairs must be made with great caution.

To address the uncertainty surrounding optimal repair timing, a questionnaire is distributed to fourteen experts involved in the construction of the DCS. This group comprises nine domain experts, each with a minimum of five years of experience, and five professors specializing in this field. The experts are asked to respond to two questions: (i) At what level of unreliability is a minor repair necessary (0, 0.05, 0.10, 0.15, 0.20, 0.30, 0.40 or 0.50)? and (ii) At what level of unreliability is a normal repair necessary (0, 0.05, 0.10, 0.15, 0.20, 0.30, 0.40 or 0.50)? The findings revealed that at least 80% of participating experts concur that a minor repair is required when unreliability exceeds 0.10, and a normal repair is warranted when the unreliability exceeds 0.20. Consequently, two indicators, T_{10} and T_{20}, are proposed to represent the time for performing minor and normal repairs, respectively. The linear interpolation method is utilized to determine the specific time points for T_{10} (when unreliability reaches 0.10) and T_{20} (when unreliability reaches 0.20), as illustrated in Fig. 8. The results for the T_{10} and T_{20} time points are presented in Table 5.

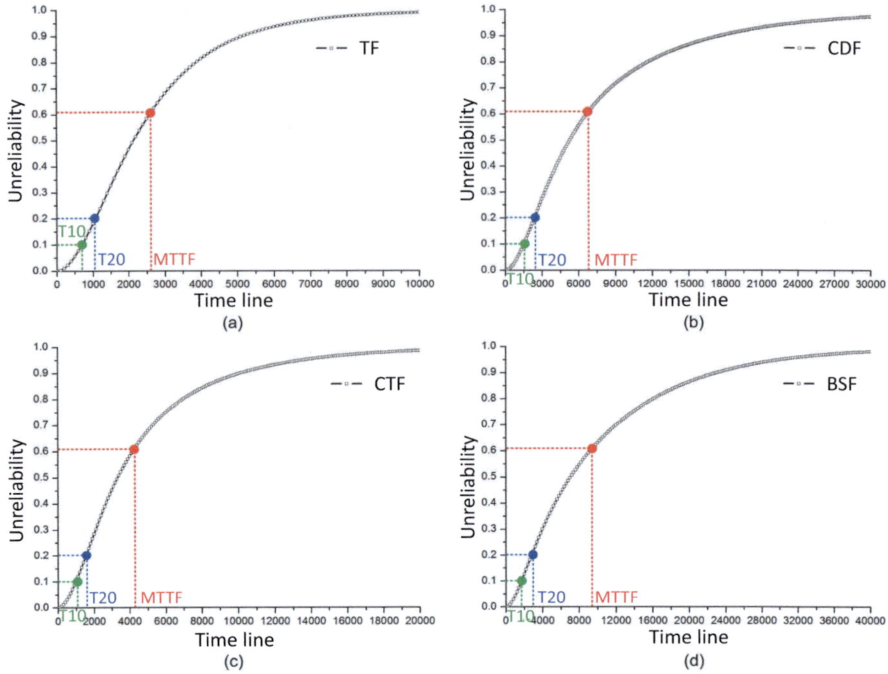

Fig. 8 System unreliability evaluation over time: **a** TF; **b** CDF; **c** CTF; **d** BSF

Table 4 Time to failure of the system and its components (unit: ring)

Failures	T_{10}	T_{20}	MTTF
TF	694	1064	2633
CDF	1506	2380	7883
CTF	981	1541	4687
BSF	1700	2815	10,238

Table 5 MTTF of root nodes in different scenarios (unit: ring)

Basic events	Prior	Posterior			
		Scenario 1	Scenario 2	Scenario 3	Scenario 4
X5	5000	3962	3418	3878	7410
X6	2000	1629	1487	1792	2618
X7	2000	503	1228	1820	2771
X8	2500	1395	1854	2806	2654

As shown in Table 4, the T_{10} and T_{20} values for the top node TF (they are 694 and 1604 rings, respectively) are the lowest among those four nodes, indicating that TF is more susceptible to failure compared to its three components. This is attributed to the fact that those components cause TF's failure through an OR gate mechanism. To decrease the failure probability of TF, greater emphasis should be placed on the reliability of its components, particularly for the CTF, as its T_{10} and T_{20} values (they are 981 and 1541 rings, respectively) are significantly lower than those of the other two components. It is prudent to schedule minor and normal repairs for the shield cutter tools when the TBM completes 981 and 1541 rings, respectively. Specifically, the CTF is likely to be the earliest component to fail among the three, should repairs not be conducted timely. Consequently, the CTF can serve as a focal point for identifying the most critical root causes in the subsequent feedback analysis.

4.4 Feedback Control and Analysis

Feedback control is intended to identify the most probable root cause(s) of a target event failure by utilizing diagnostic analysis within Bayesian inference. The propagation of the observed information from the target event facilitates the updating of the posterior probability distribution of root causes over time within the network. Greater attention should be directed toward those adverse causes with high posterior probabilities, should a failure be observed. In this context, the failure of CTF serves as the focal point for an in-depth exploration of the most significant root causes.

As depicted in the unreliability curve of CTF over time in Fig. 8c, three indicators, T_{10}, $T_{20,}$ and MTTF, can segment the CTF's performance into four distinct phases on the timeline: (i) *Safe* (δ_S) with an interval $[0, T_{10})$, (ii) *Slightly Dangerous* (δ_{SD}) with an interval $[T_{10}, T_{20})$, (iii) *Dangerous* (δ_D) with an interval $[T_{20}, \text{MTTF})$, and (iv) *Extremely Dangerous* (δ_{ED}) with an interval $[\text{MTTF}, \infty)$. Consequently, the failure node of CTF can be categorized into four separate states. Upon observation of a specific state of CTF, Eq. (12) is applied to calculate the posterior probability distribution of its four root causes, namely X5, X6, X7, and X8.

Figure 9 presents the prior and posterior probability distribution across different scenarios involving the four root nodes. In Scenario 1, the δ_S state of CTF is observed; in Scenario 2, the δ_{SD} state is observed; in Scenario 3, the δ_D state is observed; and in Scenario 4, the δ_{ED} state is observed. As evident in Fig. 9, there are significant differences in the posterior probability distributions among the various root nodes, indicating that the impact of CTF's failure on each root node varies considerably. Furthermore, for a specific root node, such as X5, the posterior probability distributions show considerable changes compared to the prior probability distribution, indicating a high sensitivity of the root node to observation. The failure model depicted in Fig. 8 is very responsive to these changes.

To clearly assess the impact of specific observations on root nodes, the MTTF metric transfers complex probability distributions into a precise and practical indicator for quantitative comparison. Equation (11) is employed to determine the MTTF

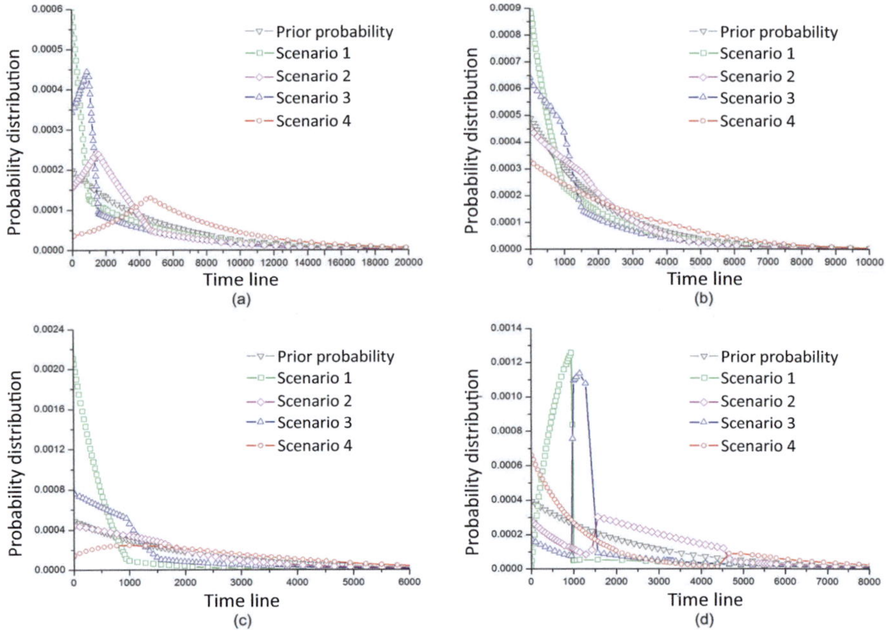

Fig. 9 Prior and posterior probability distribution in different scenarios of four root nodes: **a** X5; **b** X6; **c** X7; and **d** X8

time points for the four root causes, as detailed in Table 5. It is evident that in Scenarios 1 and 2, the root node X7 is most likely to be the most unfavorable cause, as it has the lowest MTTF values (503 rings in Scenario 1 and 1228 rings in Scenario 2) among the four root nodes. This suggests that X7 would be the earliest node to report a fault during TBM operation if the performance of CTF is less than slightly dangerous.

In Scenarios 3 and 4, the root node X6 is likely to be the most unfavorable cause, as it has the lowest MTTF values (1792 rings in Scenario 3 and 2618 rings in Scenario 4) among the four root nodes. This indicates that X6 would be the first node to report a fault during TBM operation if the performance of CTF exceeds slightly dangerous levels. In general, the most unfavorable cause must be reassessed as the observed performance of the target event changes. Therefore, the primary focus to ensure satisfactory TBM operation should be continuously updated throughout the tunneling excavation process.

5 Merits Analysis

Model validation is crucial for ensuring the plausibility of applications. Comparing simulation results with historical data of actual TBM failures would enhance the credibility of the simulation performance of the proposed approach. However, the historical data on actual TBM failures during the construction of the Wuhan metro system are scarce. This scarcity is partly due to maintenance and repairs being performed preemptively to avoid excessive costs before actual failures occur. Additionally, the historical data from other cities may not be applicable due to differences in TBM types and geological conditions. To address this challenge, domain experts can leverage their experience to predict when TBM failures are likely to occur, offering a potential avenue for validation.

Three indicators, T_{10}, $T_{20,}$ and MTTF, have been proposed to condense the complex probability distributions into specific and useful numbers for quantitative comparison. Domain experts are particularly confident in determining the values of T_{10} (the time to perform minor repairs) and T_{20} (the time to perform normal repairs) based on the actual tunneling construction practice. Six domain experts involved in the construction of the Wuhan metro system in China, including the project manager, the technical consultant, two site safety managers, and two workers, are invited to evaluate the effectiveness of the model outcomes. They assess the time required for minor and normal repairs on specific TBM components, including CDF, CTF, and BSF. Table 6 presents a comparison between the simulation results and expert judgments. The findings show that the deviation rates between the simulation results and expert judgments are all within a predefined threshold of 0.05. This threshold aligns with the statistical conventions where $p < 0.05$ is considered a significant difference. Thus, the simulation results are not significantly different from the actual construction practice, validating the rationality and reliability of applying the model to assess TBM performance on construction sites.

Table 6 Comparison between simulation results and expert judgments (unit: ring)

Item	CDF		CTF		BSF	
	T10	T20	T10	T20	T10	T20
Expert 1#	1529	2472	1000	1619	1761	2924
Expert 2#	1504	2452	950	1563	1743	3072
Expert 3#	1337	2260	1139	1502	2012	2953
Expert 4#	1724	2547	975	1630	1532	2897
Expert 5#	1552	2650	1008	1697	2012	3004
Expert 6#	1799	2557	1041	1567	1634	2674
Average	1574	2490	1019	1596	1782	2921
Simulation	1506	2380	981	1541	1700	2815
Deviation rate (%)	4.52	4.60	3.84	3.60	4.84	3.75

Note Deviation rate = (Aaerage – simulation)/simulation * 100

The principal advantage of the developed approach lies in its ability to conduct both predictive analysis for estimating TBM performance and diagnostic analysis when poor performance or failure is observed. Several models have been established to describe TBM performance, broadly categorized into: (i) cutter load models; (ii) statistical simulation models; and (iii) situ observation-based models. These developed models offer valuable insights into the prognosis and diagnosis of TBM performance. Consequently, they provide decision support for both feedforward and feedback control, ensuring satisfactory TBM operation over time.

6 Chapter Summary

A systematic approach for assessing the risk of shield cutter head failures in tunnel construction has been developed, comprising four major steps, namely failure mechanism exploration, DFT construction, DTBN formation, and risk analysis. A tunnel case in China is selected as a case study to validate the applicability of this approach. The following conclusions can be drawn below.

(1) DTBN provides a viable alternative for solving computation DFT without the need for Markov chains, thereby overcoming common challenges associated with Markov chains, such as state-space explosion and error-prone conversion processes. Based on the failure mechanism analysis, three distinct categories of shield cutter head failures have been identified, along with a list of ten influential fault variables that significantly affect cause-effect relationships. A causal network model with 20 nodes has been constructed to simulate shield cutter head failures during TBM operation in tunnel construction.
(2) Three indicators, T_{10}, $T_{20,}$ and MTTF, have been introduced to distill the complex probability distributions into specific and useful numbers for explicitly measuring system unreliability. These indicators enhance the understanding of when to perform minor or normal repairs on the TBM or its components. They enable the categorization of TBM or component performance into four distinct ranges, providing a unified comparative framework for validating the simulation performance of the proposed approach.
(3) The developed approach is capable of conducting both predictive analysis for estimating TBM performance and diagnostic analysis in the event of poor performance. It offers decision support strategies for both feedforward and feedback control on a quantitative basis, ensuring consistent and satisfactory TBM operation over time.

References

1. Jin D, Yuan D, Li X, Su W (2021) Probabilistic analysis of the disc cutter failure during TBM tunneling in hard rock. Tunn Undergr Space Technol 109:103744
2. Zhang L, Wu X, Skibniewski Miroslaw J (2016) Simulation-based analysis of tunnel boring machine performance in tunneling excavation. J Comput Civil Eng 30:04015073
3. Chen M, Xiao NC, Zuo MJ, Ding Y (2021) An efficient algorithm for finding modules in fault trees. IEEE Trans Reliab 70:862–874
4. Tripathi M, Singh LK, Singh S, Singh P (2022) A comparative study on reliability analysis methods for safety critical systems using petri-nets and dynamic flowgraph methodology: a case study of nuclear power plant. IEEE Trans Reliab 71:564–578
5. Wang C, Wang L, Chen H, Yang Y, Li Y (2021) Fault diagnosis of train network control management system based on dynamic fault tree and bayesian network. IEEE Access 9:2618–2632
6. Khan S, Katoen JP, Bouissou M (2024) A compositional semantics of boolean-logic driven Markov processes. IEEE Trans Depend Secure Comput 21:701–716
7. Xu Z, Guo D, Wang J, Li X, Ge D (2021) A numerical simulation method for a repairable dynamic fault tree. Eksploatacja i Niezawodność 23
8. Samadi H, Hassanpour J, Rostami J (2023) Prediction of earth pressure balance for EPB-TBM using machine learning algorithms. Int J Geo Eng 14:21
9. Boudali H, Dugan JB (2005) A discrete-time Bayesian network reliability modeling and analysis framework. Reliab Eng Syst Saf 87:337–349
10. Khakzad N, Khan F, Amyotte P (2013) Risk-based design of process systems using discrete-time Bayesian networks. Reliab Eng Syst Saf 109:5–17
11. Ding LY, Zhou C (2013) Development of web-based system for safety risk early warning in urban metro construction. Autom Constr 34:45–55
12. Zhang L, Wu X, Ding L, Skibniewski MJ, Yan Y (2013) Decision support analysis for safety control in complex project environments based on Bayesian Networks. Expert Syst Appl 40:4273–4282

Intelligent Proactive Decisions

Tunnel Line Alignment

1 Introduction

As construction projects, particularly large-scale tunnel projects, grow more complex. The intricate nature of tunnel projects demands a systematic approach capable of comprehensively analyzing and optimizing multiple objectives, such as minimizing cost, navigating challenging environments, and managing complex surrounding conditions [1]. Factors such as cost, time, quality, safety, and environmental impact often involve trade-offs. Moreover, the dynamic project conditions and diverse stakeholder perspectives highlight the need for a flexible decision-making framework to address potentially conflicting objectives [2]. Given the significance of multi-objective optimization (MOO) in construction project management, researchers have proposed various approaches. These models aimed to optimize project schedules, minimize costs, and balance resource allocation. Empirical studies have demonstrated the effectiveness of MOO models in providing trade-off solutions for complex construction projects [3].

Algorithms have been developed to solve MOO, among which multi-objective evolutionary algorithms (MOEAs) are the most prevalent. These methods include non-dominated sorting genetic algorithm (NSGA), Pareto archived evolution strategy (PAES), reference points-based evolutionary algorithm (RPEA), mixed integer linear programming (MMILP), strength-Pareto evolutionary algorithm (SPEA), preference-inspired coevolutionary algorithm (PICEA), and particle swarm optimization (PSO) [1]. These methods have been effectively applied to real-world problems such as planning, classification, economic load dispatch, scheduling, and cyber-physical systems [4].

Although MOO methods are recognized as effective tools for solving problems with conflicting objectives and supporting decision-making, their application in construction projects remains limited. One major obstacle is that MOO often yields numerous optimal solutions, complicating decision-making. For example, Koo et al. [5] identified up to 250 optimal solutions for a construction time–cost trade-off

L. Zhang et al., *Intelligent Construction in Tunnels*, Lecture Notes in Civil Engineering 679, https://doi.org/10.1007/978-981-96-8785-5_11

problem through GA optimization. This abundance of optimal solutions can overwhelm decision-makers in the construction field, making it difficult to choose the best option. Therefore, additional rules are needed to select specific solutions under given conditions, completing the optimization process to enhance MOO's applicability in construction [6].

This chapter proposes a GA-based approach for optimizing conflicting objectives in complex construction projects, offering a trade-off solution to support decision-making. It addresses two main questions: (1) how to develop a multi-objective optimization-based decision-making approach for construction projects, and (2) how the approach can provide insights for stakeholders when conditions and information change. A tunnel alignment project is used as a case study to demonstrate the effectiveness of the proposed approach, in which the decision-maker could identify objectives and decision variables, construct relationships between them, and execute the optimization process. The approach yields a set of optimal trade-off solutions, selects one closest to the ideal as a reference for decision-making, and accordingly presents the final optimal decision. The case study demonstrates the approach's ability to handle conflicting objectives, offering flexibility by providing a range of solutions and enabling further analysis across multiple dimensions.

2 Multi-objective Optimization (MOO)

Metro lines are essential to urban development, and optimizing their design has been a research focus for years. For example, Zhang and Chiaradia [7] analyzed metro network structures and found that a two-overlapping-ellipse design was more efficient than a simple ring, improving network efficiency and reliability. Nian et al. [8] applied a Tabu search algorithm to identify optimal corridors for new lines, aiming to enhance public transportation resilience. Researchers also explored the impacts of new metro lines on existing city infrastructure. For instance, Gu et al. [9] examined the effects on bike-sharing systems, and Fu and Gu [10] studied the influence on passenger flow and travel times in transit networks. While these studies highlighted the impact of new metro lines on urban development, few have focused on the alignment design between existing stations. This gap is significant, as designing line alignments involves complex factors due to the mature communities around stations.

In 1975, Professor Holland of the University of Michigan introduced the genetic algorithm (GA), a random search method inspired by biological evolution, which is now widely applied to optimize conflicting objectives in construction projects [11]. GA is widely used for decision problems with conflicting objectives due to its strong global search capabilities and low algorithmic complexity. It mimics natural selection, where the fittest individuals are chosen for reproduction to generate the next generation. The process starts with selecting the fittest individuals from an initial population, whose offspring inherit their traits, leading to improved fitness over iterations. The final generation, known as the Pareto front, represents the set of non-dominated solutions where no objective can be enhanced without compromising another.

As illustrated in Fig. 1, f1(x) and f2(x) are two fitness functions of objectives in the optimization problem. A and B are two random and optimal non-dominated solutions on the Pareto front. The solution A has a larger value in the fitness function $f_2(x)$ but it is inevitably dominated by the solution B in the fitness function $f_1(x)$. Without any preference, all solutions on the Pareto front are considered equally optimal. Rather than searching for a single solution, GA performs the search on a set of solutions, making it a powerful tool for complicated MOO problems [12]. RazaviAlavi and AbouRizk [13] proposed a GA-based optimization model to seek the optimum solutions between the construction site layout, construction operation, and cost. However, even searching for a set of solutions gives GA the advantage of solving the conflict between objectives, a relatively wide range of solutions (often more than two) would also bring difficulties for decision-making. This is because all solutions on the Pareto front, as presented in Fig. 1, are equally optimal. After gaining the optimal solutions from the Pareto front, methods are required to narrow down the scope of solutions. Despite the recognition of the importance of MOO in construction projects, most studies currently focus on two or three primary factors, such as scheduling, cost, and quality. However, other critical objectives, like system efficiency and user comfort, significantly influence the successful completion of construction projects. For example, in tunnel alignment projects, headway enhances transportation efficiency [14], and comfort design impacts ridership and investment recovery [15]. The construction of new metro lines affects various urban aspects, requiring consideration of factors such as city development strategies, traffic flow, and safety concerns related to existing buildings [10]. Thus, addressing multiple objectives in metro line construction with more objectives is essential. MOO in construction should encompass both generic factors (e.g., time and cost) and specific factors (e.g., system efficiency and comfort design). Establishing methods for optimal metro line alignment design requires a flexible optimization approach that accommodates multiple dimensions, which could thereby enrich the understanding of MOO challenges in the construction field.

3 MOO-Enabled Tunnel Line Alignment

This chapter aims to propose a GA-based approach to identify an optimal set of trade-off solutions for tunnel line alignment, ultimately providing a selected optimal solution to support decision-making. As illustrated in Fig. 2, the approach consists of three main steps: (1) problem analysis, (2) relation construction, and (3) optimization. Successful completion of construction projects hinges on numerous interrelated factors, necessitating a well-organized analysis to inform effective planning, design, and execution. The first step involves conducting a detailed analysis to identify decision variables, value scopes, objectives, constraints, and their interrelations. This

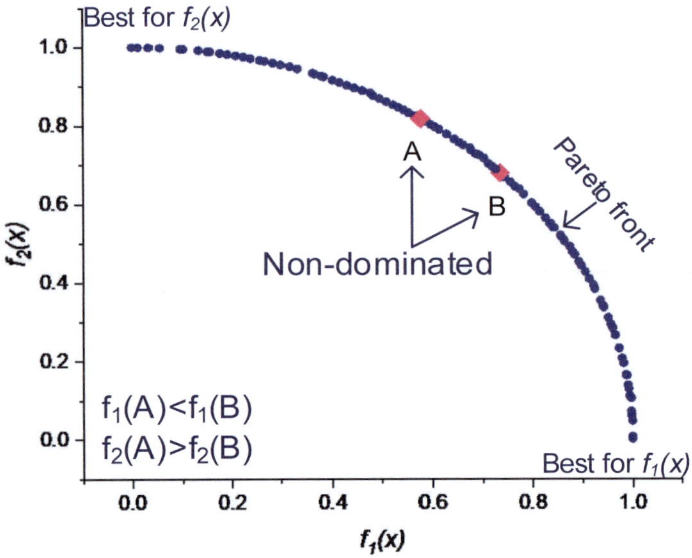

Fig. 1 Example of Pareto front solutions in GA

analysis enables the construction of appropriate relation functions between decision variables and objectives, expressed through fitness functions for optimization. Finally, the GA process is employed to perform the optimization.

3.1 Problem Analysis

A tunnel line alignment project between stations A and B is studied, as presented in Fig. 3. An arc is used to connect the two straight lines leading to stations A and B. The radius of the arc could be different, a small radius like r_1 or a large radius like r_2. Different arc radii result in varying total tunnel lengths, consequently affecting construction costs. Additionally, the choice of arc radius influences the project's impact on the surrounding environment, as each design may intersect distinct urban areas with different environmental conditions. For instance, certain locations may impose depth requirements for tunnel construction due to their surroundings, further influencing project costs. Moreover, the design of the tunnel line affects its efficiency, impacting passenger experience and traffic flow. Therefore, the alignment project involves more than merely selecting arc radii, and it necessitates careful consideration of multiple interrelated factors that influence one another.

To successfully complete a construction project, it is essential to first identify the objectives that must be achieved. Since conflicting relationships may arise among these objectives, MOO problems in construction generally can be summarized in the following functional forms.

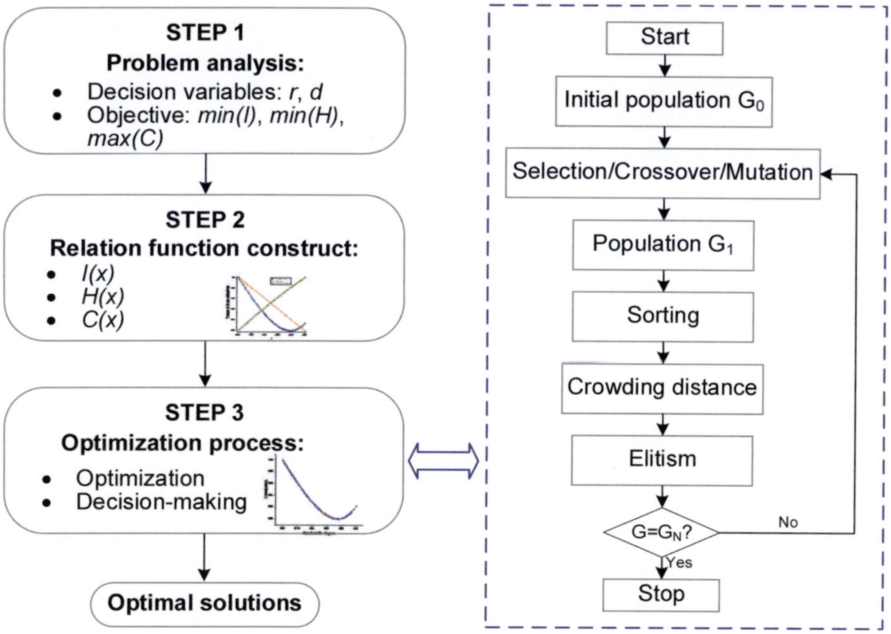

Fig. 2 Framework of the proposed approach for the design improvement of the tunnel constructions

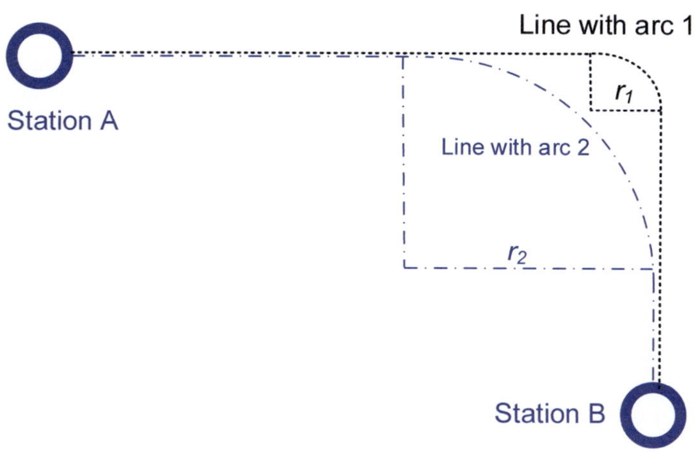

Fig. 3 Model for the tunnel line alignment project

$$\begin{cases} \min(F_i(x_1, x_1, \ldots, x_n)) \\ \max(F_i(p_1, p_2, \ldots, p_m)) \end{cases} \tag{1}$$

$$\begin{cases} g(x_1, x_1, \ldots, x_n) \leq 0 \\ G(p_1, p_2, \ldots, p_m) \geq 0 \end{cases} \tag{2}$$

where $F_i(x_i)$ and $F_i(p_i)$ $(i = 1, \ldots, n)$ are the objective functions of decision variables x_i and p_i, the minimum and maximum value of the functions are the targets that need to be fulfilled, $g(x_i)$ and $G(p_i)$ are the possible constraints that have to be satisfied with the project conditions.

In the tunnel line alignment project, investment is a key objective, as the tunnel section's length varies with different arc designs, directly affecting construction costs. This is particularly critical in densely populated cities, where high costs can significantly impact feasibility and create financial burdens for project owners, leading to increased economic risk. Effective allocation of limited funds is often prioritized in the construction market. Additionally, efficiency is vital, as it is influenced by the tunnel line's length and directly impacts the sustainability of the transportation system. In mature metro networks, factors such as train availability and average speed are relatively fixed, making tunnel length crucial for optimizing efficiency. Efficient public transportation attracts ridership, reduces personal vehicle use, alleviates traffic congestion, and decreases energy consumption. The headway is a critical measure of this efficiency: shorter headways can lead to higher ridership, which in turn enhances the project's investment recoupment. Thus, both investment and headway are essential objectives for optimizing the tunnel line alignment project.

The comfort degree is set as the third objective. The comfort magnitude is not a separate objective, as it is a key factor for the efficiency of the transportation and the recoup of the investment for the project. In addition to the time duration of the headway, the comfort design is studied as one of the major concerns considered by the passengers, and it is directly related to the satisfaction of the passengers, which would in turn influence the ridership of the tunnel line. Therefore, keeping the train in a comfortable scope has become one of the most important targets in train management. Because of this, the comfort magnitude has been set as an objective of the transportation system in many studies. Thus, the comfort design of the tunnel section is considered a vital objective in this project.

3.2 *Objective Function Construct*

As discussed, three objectives are identified, including the investment, headway, and comfort. With the analysis and identification of the decision variables, they should be expressed as Eqs. (3)–(5). In this chapter for the tunnel alignment project, the radius of the arc is inevitably a decision variable. Besides the radius, other factors could be considered according to the project condition. With further analysis, n decision variables could be identified, and how they are influencing the objectives can be expressed in the functions.

$$f_I(i_1 \ldots i_n) \tag{3}$$

$$f_H(h_1 \ldots h_n) \tag{4}$$

$$f_C(c_1 \ldots c_n) \tag{5}$$

where $f_I(i_n)$ is the function of how the investment is related to its possible n decision variables, $f_H(h_i)$ is the function of the headway and its decision variables, and $f_C(c_i)$ is the function of the comfort and its decision variables, i_n, h_n and c_n represent the n decision variables for the objectives.

Since a larger investment would be a heavy financial burden and a bigger headway would indicate a low efficiency of transportation, the optimization aims would be the less the better for both of these two objectives. Firstly, the comfort degree should be as high as possible. A higher comfort magnitude is able to attract a larger ridership, which is not only important for the recoup of the investment, but also very important to fully make use of the tunnel line. Thus, the optimization aim for the three objectives can be expressed as in Eq. (6): to minimize the investment and headway, and maximize the comfort of the tunnel section. The expression of these objectives with possible decision variables should be analyzed in detail according to the project condition.

$$\begin{cases} \min(I) \\ \min(H) \\ \max(C) \end{cases} \tag{6}$$

where I is the investment of the tunnel section, H is the headway, and C is the magnitude of the comfort.

3.3 Optimization for Alignment Decision

(1) Optimization process

In this chapter, NSGA-II, an algorithm widely used in the construction field due to its effectiveness, is adopted. NSGA-II adopts the fast non-dominated sorting algorithm, which greatly reduces the computational complexity and maintains the diversity of the population. It can prevent the loss of the optimal individual and improve the speed and robustness of the algorithm. In the optimization process, an initial population is generated randomly, and three operations, including selection, crossover, and mutation, are then utilized to produce a new generation in search of the solution (offspring). The search continues until the optimal solution is reached.

As presented in Fig. 4, an initial population, P_1, is randomly produced with N individuals. The fitness function is used to determine how fit an individual is, i.e., the ability of an individual to compete with other individuals. It gives a fitness value to each solution. The probability that an individual will be selected for reproduction is based on its fitness value. Next is the selection phase, where the objective is to select the fittest individuals and let them pass their genes to the next generation. Two pairs of individuals are selected based on their fitness values. Individuals with a higher fitness should have more chances to be selected for reproduction. Then, a crossover point is chosen randomly within the genes to create new offspring. When certain new offspring are formed, some of their genes can be subjected to a mutation with a low random probability. This mutation process is performed to maintain the diversity of the population. With the operations of selection, crossover, and mutation, Q_1 can be created.

The population, R_1, is combined with P_1 and Q_1. R_1 is sorted by the non-dominated rule. Next, solutions within the scope of F_1 are the best individuals and will be included in the next generation. If the size of F_1 is larger than P_1, the population P_2 is selected from F_1 with N individuals according to the crowding distance sorting rule. If the size of F_1 is smaller than P_1, the population P_2 is selected from F_1 to F_2 and F_3, until the population number equals N. This process will be repeated until the algorithm terminates when the stopping criterion is met, i.e. the process is repeated with a certain pre-determined number of times, or the population has converged (does not produce offspring which are significantly different from the previous generation). In the final generation P_n, the best set of solutions is the Pareto front (indicating the optimal solutions to the target problem).

(2) Decision making

The optimization algorithm navigates the search space to identify the best solutions that optimize all objectives. To refine the solution set to a feasible and optimal selection, a knee point is chosen as the final optimal solution. This knee point represents the minimum distance from the ideal point across all objectives. All the optimal trade-off solutions are ranked according to the distance to the ideal point. This method can help rank and select a solution from a finite set of optimal solutions that are in the presence of conflicting criteria in different objectives, becoming especially helpful

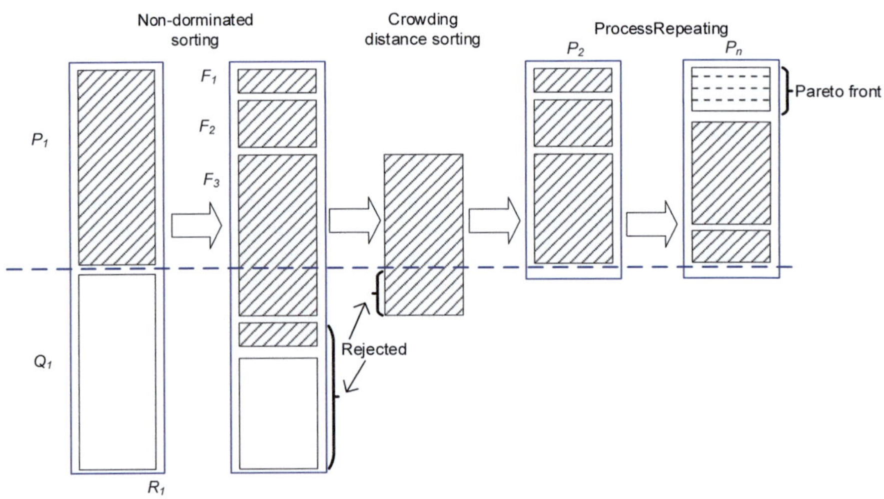

Fig. 4 Procedures of using NSGA-II

in a situation where the decision maker does not have specific priorities. The rule can be expressed in Eq. (7). This principle is based on the distance from every solution to the ideal one. Thus, the solution that is nearest to the ideal one is regarded as the ideal optimum solution.

$$D_i = \sqrt{(m_{1i} - i_{10})^2 + (m_{2i} - i_{20})^2 + \cdots + (m_{ni} - i_{n0})^2} \tag{7}$$

where D_i is the distance of the ith solution on the Pareto front, i is from 1 to n, n is the number of solutions on the Pareto front, m_{ni} is the value of the nth objective for the particular solution, and i_{n0} is the ideal value of that particular objective.

The values of different objectives often have varying units, necessitating standardization before optimization. This process ensures comparability among objectives and facilitates the effective handling of multiple factors in practical applications. All values of objectives can be standardized using Eqs. (8), and (9) can be used to get the original values of the objectives after obtaining the selected optimal solution.

$$S_i = \frac{Z_i - Z_{imin}}{Z_{imax} - Z_{imin}} \tag{8}$$

$$Z_i = (Z_{imax} - Z_{imin}) \times S_i + Z_{imin} \tag{9}$$

where S_i is the standardized value for ith value, Z_i is the original value of ith objective, Z_{imax} is the maximum value, and Z_{imin} is the minimum value of the ith objective.

In the tunnel line alignment project, three objectives are determined: minimizing investment and headway while maximizing comfort magnitude. With the standardization of the three objectives, the position of the ideal point can be set as (0,0,1), in which the first direction is the investment, the second one is the headway, and the last one is the comfort. Next, the rankings of each solution to the ideal point can be gained through Eq. (7), i.e., the distance of each solution to the ideal point. Then, the optimal solution from the Pareto front set of solutions, which has the minimum distance, can be selected. This selected solution can be regarded as the optimal solution recommended to the project owner for decision making.

4 Experimental Studies

To test the applicability of the proposed approach, a tunnel line alignment case in Wuhan city, China, is used to perform the optimization. The decision variables of the objectives and the math functions representative of their relations are identified first, and the optimization process is conducted by using the NSGA-II algorithm.

4.1 Data and Resources

The tunnel section is designed to connect the Xu Jia Peng station and Hubei University station in Wuhan, China. Xu Jia Peng station is the first station of Line 7 when the line crosses the Yangtze River, and it acts as the transfer station with metro lines 5 and 8. The Metro Line 7 starts from Xu Jia Peng, goes along Qingyuan Road, turns to Youyi Ave, and ends up at the Hubei University station. It could cross different buildings, such as residential areas of Yuhua Garden, Qingyuanju, Tuanjie Mingyuan, the City Village at Sanjiao Road, and commercial buildings.

As illustrated in Fig. 5, the major challenge confronting the project owner in this tunnel project is to select an optimal tunnel line with the arc turning from Qingyuan Road to Youyi Ave. Figure 5a is one tunnel line alignment plan with a radius of 320 m, and (b) is another plan with a radius of 400 m. It can be seen that the construction environment, under which the different tunnel lines would cross over, could be very different. For example, the line would cross several residential buildings, such as NO. 2 of Qingyuanju and No. A4 of the City village. Through careful investigation, the arc radius could range from 200 to 450 m, and this tunnel line would have to be constructed under a depth of at least 20 m to minimize the influence on those existing buildings. Thus, the first question coming with the tunnel line alignment project is what the advantages and disadvantages of tunnel lines with different radii are, and how should the design be optimized.

4.2 Constraints and Optimization

(1) Relation function determination

In this tunnel line alignment case, the arc connecting the tunnel line from Qingyuan Road to Youyi Ave is the key point. As presented in Fig. 6, there are two fixed sections L_1 and L_2, and they can be measured through field investigation. Then, the two straight-line sections l1 and l2 can be gained through the calculations of subtracting the radius, *r*, from L_1 and L_2, respectively. The length of the tunnel line then can be determined, and the value of *r* is a decisive factor. Besides, the depth, *d*, of the tunnel line may be varied because of different *r*. As presented in Fig. 5, the tunnel line may cross over different buildings when *r* varies. Thus, to guarantee the safety of the surroundings and the tunnel construction, *d* should be different when *r* is different. Therefore, these two factors, *r* and *d*, are currently determined as decision variables for this tunnel line alignment project.

With the two decision variables determined, the construct of the relation function is to find out how the three objectives are related to and influenced by *r* and *d*. Equation (10) is used to gain the total length of the tunnel line. To simplify the calculation, the height (depth) of station A and B is assumed the same (d_1). Then, the total length of the tunnel line can be estimated as the sum of l_1, l_2, the arc and length difference (i.e., *d* presented in Fig. 6) caused by the depth of the tunnel line

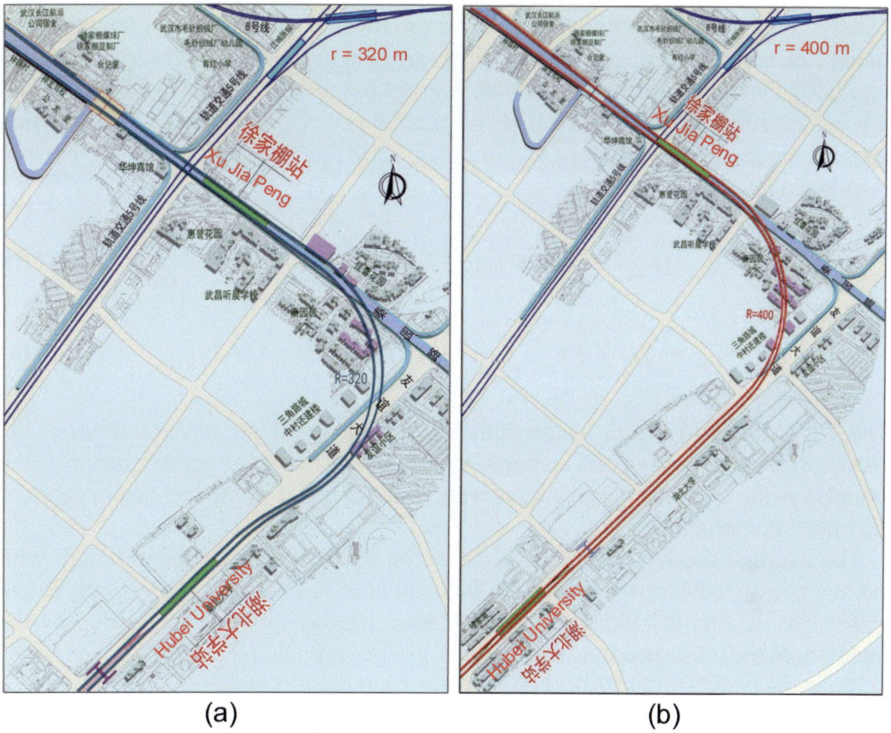

Fig. 5 Two alternatives for the tunnel line alignment between Xu Jia Peng and Hubei University stations: **a** the radius of the arc is 320 m; **b** the radius of the arc is 400 m

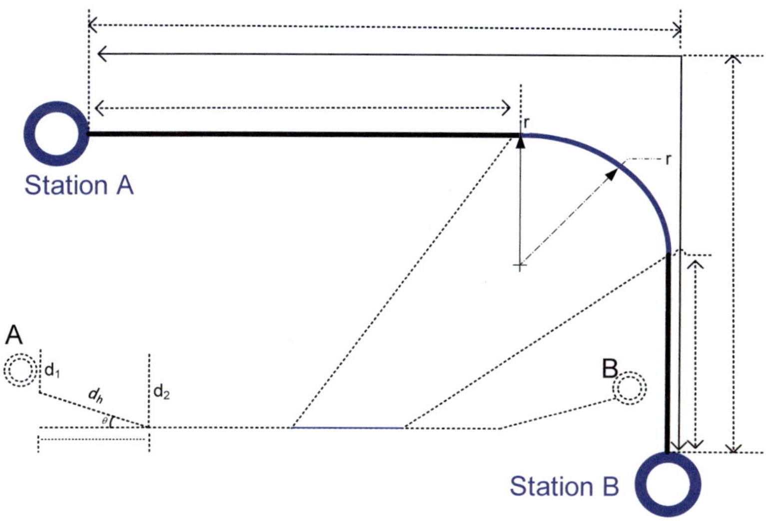

Fig. 6 Model for the analysis of the tunnel line alignment project between stations

from the two sides of station A and B. For the investment, I, the total length of the tunnel line can be calculated with different arc schemes, and the construction price per kilometer can be estimated from the average tunnel price in the center of Wuhan city, where more than 9 tunnel lines have already been constructed. The relation between the length and depth of the tunnel line and the investment can be expressed as in Eqs. (10) and (11).

$$L = (L_1 - r) + (L_2 - r) + 0.5 \times \pi \times r + 2 \times \left(\frac{d_2 - d_1}{\sin\theta} - \frac{d_2 - d_1}{\tan\theta}\right) \tag{10}$$

$$I_{sum} = \left(L - 2 \times \frac{d_2 - d_1}{\sin\theta}\right) + (d_2 - d_1) \times k + L \times p \tag{11}$$

where L is the total length of the tunnel line, d_2 is the depth at the bottom of the tunnel, d_1 is the depth of the stations A and B, p is the price per kilometer of the tunnel, k is the coefficient of the increasing cost per meter of the increasing depth of the tunnel construction.

The average headway, H, can be estimated by dividing the tunnel length with the average speed $\overline{v}$ of the train, and the available fleet size f_s. The length of the tunnel line is obtained above in the calculation for the investment. To simplify the calculation, the train speed and the fleet size can be estimated from the average values of the subway system in Wuhan city, where more than 9 tunnel lines have been in operation. The relation between them can be expressed as in Eq. (12).

$$H = (L/\overline{v})/f_s \tag{12}$$

where H is the headway, $\overline{v}$ is the average speed of the train, L is the total length of the tunnel section, and f_s is the fleet size.

In this tunnel line alignment project, the length is a critical factor, as the average train speed is fixed. A longer tunnel may require the train to reach a higher acceleration rate, and a too high acceleration rate could lead to discomfort. Thus, the comfort in this chapter is studied mainly from the acceleration rate perspective. In addition, since the tunnel length is decided by the r and d of the tunnel line from Eq. (10), the comfort degree is directly related to the tunnel length, then indirectly influenced by r and d. According to the study of Hou et al. [16], the relation between acceleration rate and the comfort magnitude can be expressed as in Eq. (13), and based on the study of Yu and Zhao [17], the change of the train speed with time can be estimated in Eq. (14).

$$\begin{cases} C(x; a, b, c) = \frac{1}{1+\left|\frac{x-c}{a}\right|^{2b}} \\ S = f(a_0, t) \\ a_0 = f(x, t) \end{cases} \tag{13}$$

$$v = q/3 \times t^3 + w/2 \times t^2 \tag{14}$$

where a, b, and c are parameters deciding the shape of the model for the comfort calculation, x is the jerk (the rate of the change of acceleration concerning time), v is the train speed used to calculate the jerk value when the length of the tunnel line is fixed, q and w are the coefficients for the train speed calculation, S is the displacement distance, t is the time required for the displacement, and a_0 is the acceleration. When calculating the comfort degree from Eq. (13), the a_0 can be gained from the derivation of the train speed in Eq. (14).

(2) Optimization process

Constraints should be required to guarantee the feasibility of the tunnel. One is that the difference between the depth of the tunnel and the station cannot be more than 15 m, otherwise, it would lead to an excessive slope degree. Besides, the coefficients, a, b, c, k, and q, for the jerk calculation are retrieved from the simulation experiment of the studies of Yu and Zhao [17] and Hou et al. [16]. Then, the initial population of the trade-off solutions can be produced by putting the decision variables into the GA algorithm. As presented in Fig. 7, the initial solutions scatter widely within the scope, and the simulated relation between the two decision variables and the three objectives are presented in Fig. 8. Next, a GA-based optimization is conducted on the parameters of the tunnel line alignment project from Table 1, where Table 2 illustrates the values of the variables selected in NSGA-II. Herein, decision variables, r and d, are changing within their possible scope, and functions between the decision variables and the three objectives [i.e., Eqs. (11)–(13)] are used as the fitness functions, the termination rule is set as 1000 generations. Since no priorities are required and to make the objectives comparable, the values of the three objectives are normalized by Eq. (18).

$$d_1 - d_2 \leq 15\,\text{m} \tag{15}$$

$$200\,\text{m} \leq r \leq 450\,\text{m} \tag{16}$$

$$0 \leq \theta \leq 3\% \tag{17}$$

$$x_n = \frac{x - x_{min}}{x_{max} - x_{min}} \tag{18}$$

where x_n is the normalized value of a specific objective value, x is the original value, x_{max} and x_{min} are the maximum value and minimum value from the list of the objective values, respectively.

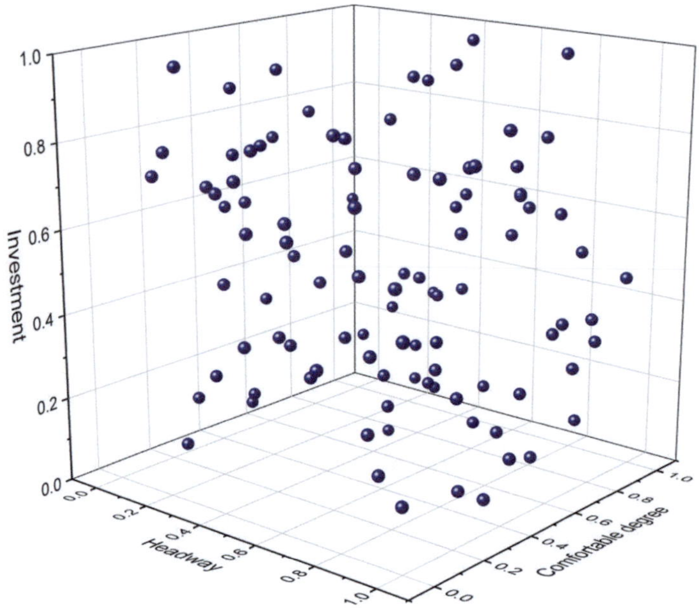

Fig. 7 Initial population of the solutions for the tunnel line project

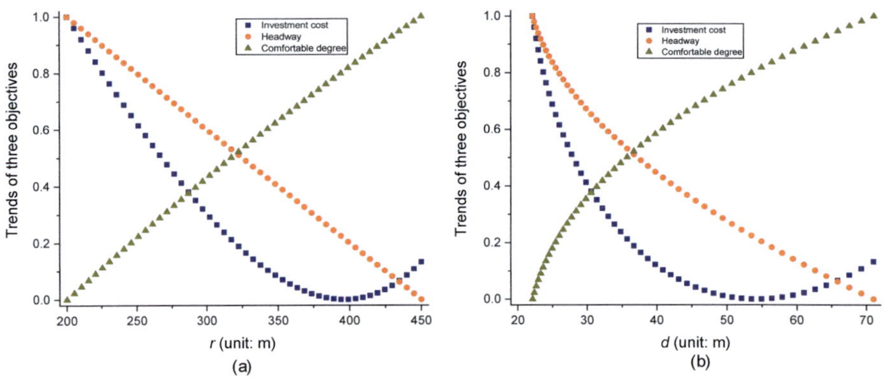

Fig. 8 Relations of two decision variables with the three objectives in tunnel line alignment: **a** r; **b** d

4.3 Tunnel Line Alignment Results

Through the optimization, the solution space is reduced to a feasible and optimal set. The Pareto front is built, and a selected solution is determined as the optimal reference solution for the tunnel line alignment project. Figure 9 presents the scatter graphs of the inputs for the Pareto front solutions, Fig. 10 demonstrates the Pareto

Table 1 Basic parameters of the tunnel line alignment project

Parameters	Description	Values
r	Radius of tunnel lines	200–450 m
d	Depth of tunnel lines	20–70 m
L_1	Straight line section from station A	740 m
L_2	Straight line section from station B	1300 m
p	Construction price for the tunnel line per kilometer	300,000 CNY
k	Acceleration coefficient 1	0.3
q	Acceleration coefficient 2	0.0015
$\overline{v}$	Average speed of the trains	40 km/h
f_s	Fleet size number	30
a	Coefficient a in the comfort model	1
b	Coefficient b in the comfort model	0.5
c	Coefficient c in the comfort model	0

Table 2 Values of the variables selected in NSGA-II

Parameters	Description	Values
n	Number of decision variables	2
M	Number of objective functions	3
Cn	Constraints	Equations (15–17)
In	Initial population size	100
N	Number of generations	1000
Cr	Cross over rate	0.8
Mr	Mutation rate	0.2

front and the selected optimal solution, and Fig. 11 shows the optimization process. The objectives and decision variables of the finally selected optimal solution are presented in Table 3. Those results are analyzed in detail below.

(1) The proposed approach effectively identifies one specific optimal solution from the Pareto front as the suggested design plan for the tunnel line alignment. As presented in Fig. 11, the selected solution maintains a stable minimum distance of approximately 0.3 from the ideal solution throughout the optimization process, indicating the reliability of the search. An optimal solution with a headway of 5.56 min, a comfort magnitude of 0.8646, and an investment of 559.81 million CNY, is not only non-dominated by any other solutions, but also closest to the ideal solution. This optimal solution makes a strong recommendation for the project owner, especially when no particular priority is set.

(2) The three objectives exhibit a high degree of conflict. The results show that comfort magnitude tends to have a negative correlation with both investment

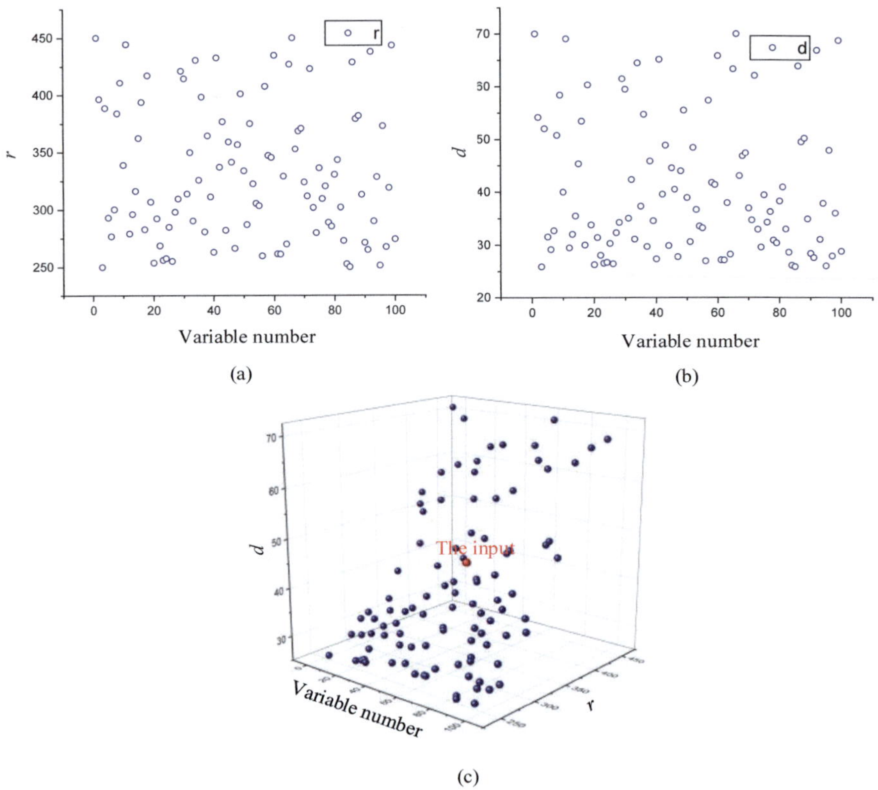

Fig. 9 Inputs for the Pareto front solutions and the input for the selected solution: **a** scatter plot of r; **b** scatter of d; **c** inputs for the Pareto front and the input for the selected solution

and headway, while investment and headway are positively correlated. When the comfort magnitude reaches the highest value of 0.8938, the corresponding investment and headway are 568.25 million CNY and 5.80 min, respectively. When the investment reaches the minimum value of 558.81 million CNY, the headway is 5.61 min and the comfort magnitude degrades to 0.83. When the headway hits the minimum value of 5.54 min, the investment is 561.30 million CNY and the comfort magnitude becomes 0.8250, which is the lowest one. This conflicting nature makes it impossible to optimize all three objectives simultaneously, yet addressing them together is essential in practical construction projects. Therefore, the trade-off solutions provided by the proposed approach serve as valuable references for project owners.

(3) A wide range of optimal trade-off solutions can provide the project owner with great flexibility for decision-making. As Fig. 10d illustrates, multiple optimal solutions address various priorities that project owners might have. While the suggested solution in this chapter is selected based on its proximity to the ideal solution, making it the optimal choice when no specific priority is given, project

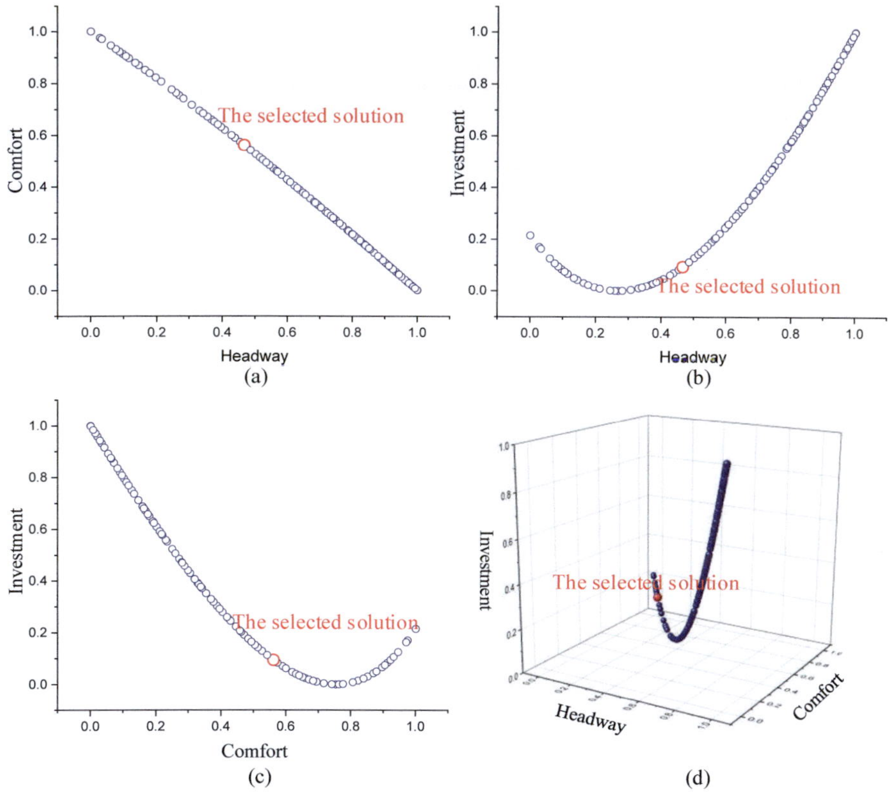

Fig. 10 Pareto front and the selected optimal solution: **a** relation between headway and comfort; **b** relation between headway and investment; **c** relation between comfort and investment; **d** Pareto front and the selected solution

owners may prioritize particular objectives, such as minimizing investment, reducing headway, or enhancing comfort. Each solution in the set is considered optimal because all are non-dominated (meaning no objective can be improved without compromising another). This flexibility allows project owners to select a solution that aligns with their preferences, ensuring an optimal decision. Thus, the proposed approach effectively supports adaptable decision-making for the tunnel line alignment project.

5 Merits Analysis

The analysis of the case study illustrates the capability of the proposed approach in generating a set of optimal solutions, which is called the Pareto front of the tunnel lines, for improved decision-making of the tunnel line alignment project. The optimization establishes optimal trade-offs between investment, headway, and comfort.

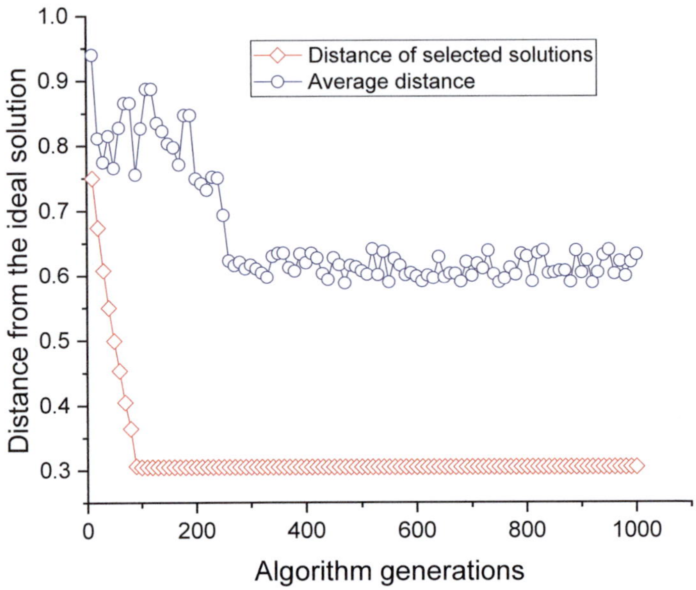

Fig. 11 Variation of distance from the optimal trade-off solutions to the ideal point in the optimization process

Table 3 Selected optimal solution and inputs from the Pareto front of the tunnel lines

Optimal solution		Objectives		
r (m)	*d* (m)	Headway (min)	Comfort	Investment (million CNY)
356.8	44.0	5.66	0.8646	559.81

A selected optimal solution can be determined from the established Pareto front, and it is suggested as the reference solution for project owners in the tunnel alignment. When designing the tunnel construction, however, priorities may be demanded because of the specific requirements of the project owner and specific environments. For example, the investment has to be within a certain budget if an upper limit exists.

To test the extendibility of the proposed approach and its application to more complex projects, additional decision variables are explored beyond the case study's variables r and d. Specifically, the speed and fleet size of trains between the two stations are considered. While subway speeds typically remain stable, fleet size can vary significantly based on operational needs, crucially affecting transportation system efficiency. Therefore, fleet size is introduced as another decision variable in this study. In addition, objectives remain focused on investment, headway, and comfort. The comfort magnitude can be kept within an ideal scope through the case study, 0.8250 to 0.8938. Thus, investment and headway objectives are now treated as constraints. This approach aims to determine whether additional improvements can be achieved using the proposed methodology.

Three scenarios can be determined according to different combinations of the two objective constraints, where the investment and/or headway are constrained. Specifically, in Scenario I, the investment is constrained to be less than 566.27 million CNY. In Scenario II, the headway is constrained to be ranging from 4.5 to 5.5 min. In Scenario III, both of them are constrained, where the investment is set as less than 559.81 million CNY, and the headway is less than 4 min. Figure 12 presents the scatter plots of the inputs for the Pareto front under three scenarios. Figure 13 shows the Pareto front and the selected optimal solution under three scenarios. Figures 11, 14, 15 and 16 present the relations among the three objectives under three scenarios. Table 4 illustrates the selected ideally optimal solutions under three scenarios.

(1) The extendibility of the proposed approach is demonstrated through its ability to adapt from simpler scenarios with fewer variables and constraints to more complex conditions involving additional decision variables and objective constraints. This flexibility allows for customization based on the decision-maker's needs, making it applicable to diverse project scenarios. Initially, the decision variables included only r and d with fewer objective constraints. In this section, fleet size (f_s) is introduced as an additional decision variable, along with more objective constraints, making the approach adaptable to changing project priorities and conditions. The proposed approach is effective in obtaining

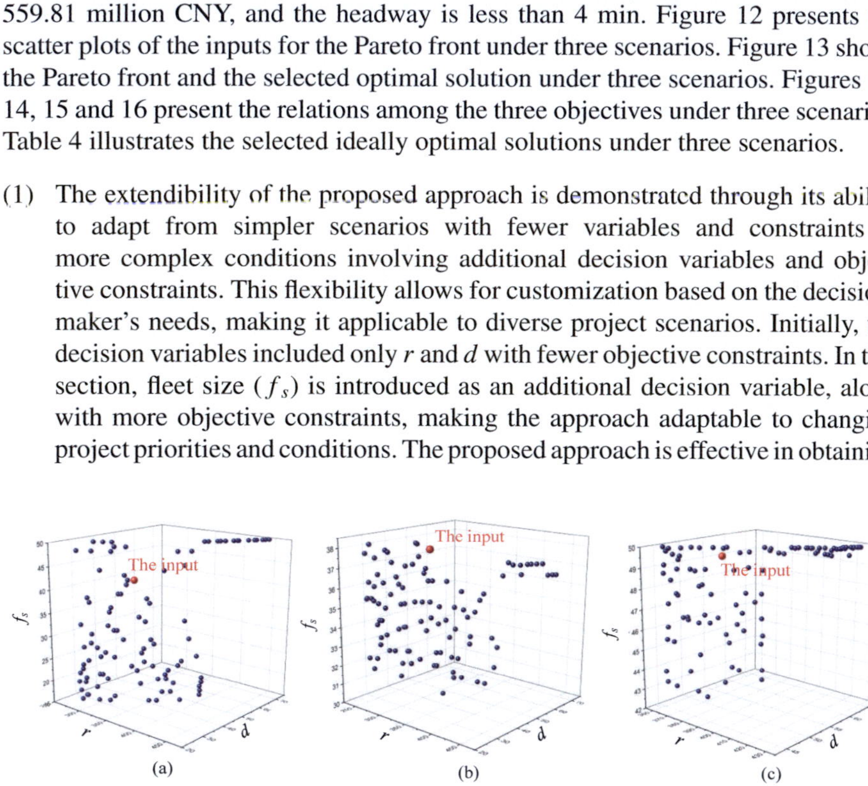

Fig. 12 Inputs for the Pareto front solutions under three scenarios: **a** Scenario I; **b** Scenario II; **c** Scenario III

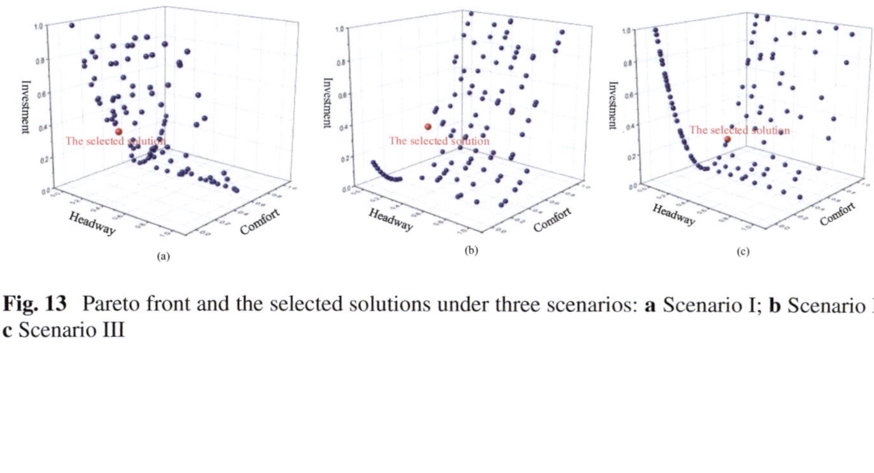

Fig. 13 Pareto front and the selected solutions under three scenarios: **a** Scenario I; **b** Scenario II; **c** Scenario III

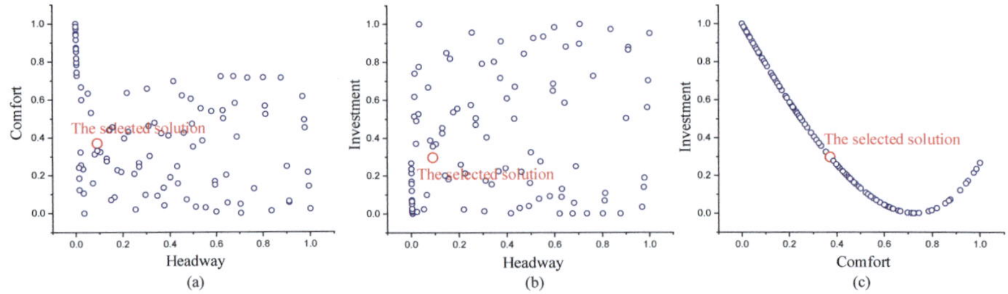

Fig. 14 Relations among three objectives under Scenario I: **a** relation between headway and comfort; **b** relation between headway and investment; **c** relation between comfort and investment

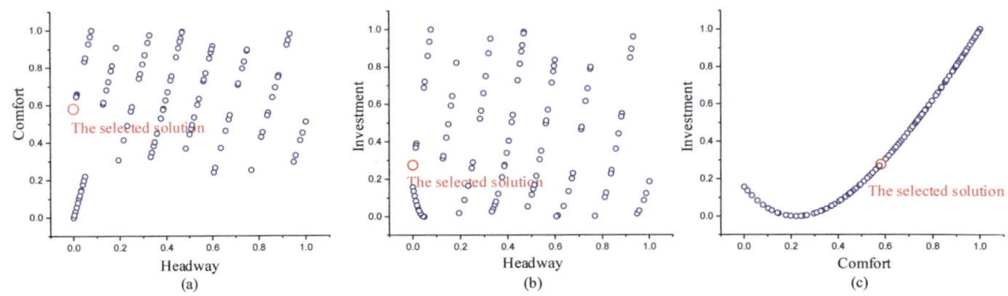

Fig. 15 Relations among three objectives under Scenario II: **a** relation between headway and comfort; **b** relation between headway and investment; **c** relation between comfort and investment

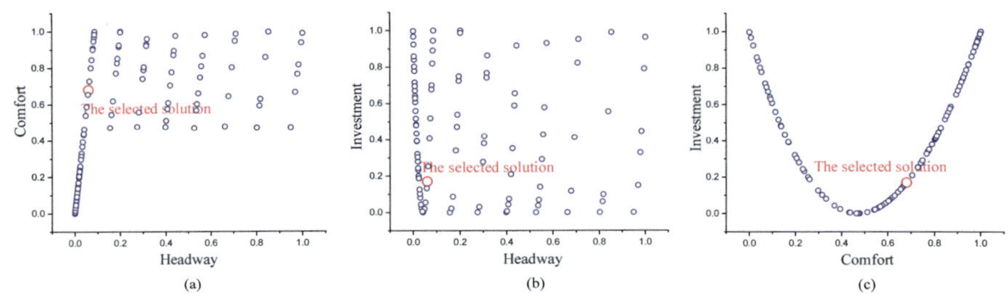

Fig. 16 Relations among three objectives under Scenario III: **a** relation between headway and comfort; **b** relation between headway and investment; **c** relation between comfort and investment

Table 4 Discovered ideally optimal solutions for three scenarios

Scenario	Optimal solution			Objectives		
	r (m)	d (m)	f_s	Headway (min)	Comfort	Investment (million CNY)
I	331.40	38.4	42	4.07	0.8763	561.30
II	326.15	37.4	38	4.50	0.8781	561.80
III	382.21	50.3	50	3.38	0.8529	559.31

the Pareto front of optimal solutions even under complex scenarios, providing the project owner with an optimal solution for decision-making. The results present that under Scenario I, the selected optimal solution includes a headway of 4.07 min, a comfort level of 0.8763, and an investment of 561.30 million CNY, with inputs of $r = 331.40$ m, $d = 38.4$ m, and $f_s = 42$. In Scenario II, the optimal solution shifts to a headway of 4.50 min, comfort of 0.8781, and investment of 561.80 million CNY, with inputs $r = 326.15$ m, $d = 37.4$ m, and $f_s = 38$. Scenario III yields a headway of 3.38 min, comfort of 0.8529, and investment of 559.31 million CNY, with $r = 382.21$ m, $d = 50.3$ m, and $f_s = 50$.

(2) More constraints typically result in less ideal solutions for tunnel line alignment, reflecting the conflicting nature of the objectives. Nonetheless, this also highlights the reliability of the proposed approach. Putting the optimized solutions together, using Eq. (8) to standardize them and Eq. (7) to calculate the distance of the three selected optimal solutions from each scenario to the ideal point, it can be seen that the distance of the optimal solution to the ideal point is 0.50 under Scenario I, 0.51 under Scenario II, and 0.69 under Scenario III. The optimal solution in Scenario III is less ideal than those in Scenarios I and II but satisfies two constraints simultaneously, making it preferable when both priorities are required for the tunnel alignment project. This demonstrates the approach's capability to provide robust optimal solutions under varying conditions. Moreover, the optimized tunnel radius ranges from 326.15 to 382.21 m across scenarios, closely matching the actual line alignment decision of 350 m in the Wuhan project, further validating the reliability of the proposed method.

6 Chapter Summary

Complex construction projects often involve conflicting objectives, making it challenging to find optimal solutions that satisfy all requirements. This chapter introduces a genetic algorithm-based approach for multi-objective optimization. The approach begins with analyzing the problem to identify the decision variables and objectives, establishing functional relationships between them. Through the optimization process, a set of trade-off solutions is generated along the Pareto front, from which the solution closest to the ideal point is determined as the reference for decision-making. The method's applicability is tested using a real-world tunnel line alignment project in China, where the tunnel's radius and depth are considered as decision variables, while investment, headway, and comfort are the optimization objectives.

Through the optimization process, an optimal solution is found with an investment of 559.81 million CNY, a headway of 5.66 min, and a comfort level of 0.8646, with corresponding decision variables of a 356.8 m radius and 44.0 m depth. To further test the approach's extendibility, three scenarios with varying priorities are considered, introducing fleet size as a third decision variable and applying constraints on investment and headway. Each scenario results in an optimal trade-off solution that meets the specific priority. It is also presented that further reducing the investment to 559.31 million CNY and the headway to 3.38 min would compromise comfort, highlighting the inherent conflicting nature. These findings indicate that the approach provides reliable optimal solutions, adaptable to complex projects. As long as decision variables, objectives, and their relationships are established, this method can effectively perform multi-objective optimization, offering practical solutions across different scenarios. The case study showcases the approach's capability to handle both simple and more complex multi-objective problems.

From the perspective of research methodology, the novelty lies in the development of holistic optimization with the capabilities of (1) assessing more than two decision variables and objectives and evaluating their combined effect in tunnel alignment projects; (2) generating the optimal solutions for the tunnel line alignment projects, which can help for improved decision making when conflicting objectives exist; (3) providing great flexibility and extendibility with the proposed approach for achieving priorities in tunnel alignment projects. From the perspective of practical implementation, the proposed approach not only overcomes the difficulty of the decision-making for conflicting objectives in construction projects, but it also provides the project owner with great flexibility and extendibility by producing a certain range of optimal trade-off solutions and allowing for further analysis in more dimensions.

References

1. Liu J, Sarker R, Elsayed S, Essam D, Siswanto N (2024) Large-scale evolutionary optimization: a review and comparative study. Swarm Evol Comput 85:101466. https://doi.org/10.1016/j.swevo.2023.101466
2. Guo C (2017) Study on optimization of construction management of large project. In: 2017 international conference on innovations in economic management and social science (IEMSS 2017). Atlantis Press, Atlantis Press. https://doi.org/10.2991/iemss-17.2017.247
3. Fu X, Wu M, Tiong RLK, Zhang L (2024) Data-driven joint multi-objective prediction and optimization for advanced control during tunnel construction. Exp Syst Appl 238:122118. https://doi.org/10.1016/j.eswa.2023.122118
4. Gunantara N (2018) A review of multi-objective optimization: methods and its applications. Cog Eng 5(1):1502242. https://doi.org/10.1080/23311916.2018.1502242
5. Koo C, Hong T, Kim S (2015) An integrated multi-objective optimization model for solving the construction time-cost trade-off problem. J Civ Eng Manage 21(3):323–333. https://doi.org/10.3846/13923730.2013.802733
6. Guo K, Zhang L (2021) Multi-source information fusion for safety risk assessment in underground tunnels. Knowl Based Syst 227:107210. https://doi.org/10.1016/j.knosys.2021.107210
7. Zhang L, Chiaradia A (2018) How to design the metro network for maximal accessibility potential? A comparative analysis of Shanghai. In: The 24th international seminar on urban form, editorial Universitat Politècnica de València, Valencia, Spain, pp 1223–1231. https://doi.org/10.4995/ISUF2017.2017.4975
8. Nian G, Chen F, Li Z, Zhu Y, Sun D (2019) Evaluating the alignment of new metro line considering network vulnerability with passenger ridership. Transp A Transp Sci 15(2):1402–1418. https://doi.org/10.1080/23249935.2019.1599080

9. Gu T, Kim I, Currie G (2019) Measuring immediate impacts of a new mass transit system on an existing bike-share system in China. Transp Res Part A Policy Pract 124:20–39. https://doi.org/10.1016/j.tra.2019.03.003
10. Fu X, Gu Y (2018) Impact of a new metro line: analysis of metro passenger flow and travel time based on smart card data. J Adv Transp 2018. https://doi.org/10.1155/2018/9247102
11. Pham VHS, Nguyen Dang NT, Nguyen VN (2024) Achieving improved performance in construction projects: advanced time and cost optimization framework. Evol Intell 17(4):2885–2897. https://doi.org/10.1007/s12065-024-00918-7
12. Neghină M, Dicoiu A-I, Chiş R, Florea A (2024) A competitive new multi-objective optimization genetic algorithm based on apparent front ranking. Eng Appl Artif Intell 132:107870. https://doi.org/10.1016/j.engappai.2024.107870
13. RazaviAlavi S, AbouRizk S (2017) Site layout and construction plan optimization using an integrated genetic algorithm simulation framework. J Comput Civ Eng 31(4):04017011. https://doi.org/10.1061/(ascc)cp.1943-5487.0000653
14. D. Schmaranzer, R. Braune, K.F. Doerner, Simulation-based headway optimization for a subway network: a performance comparison of population-based algorithms. In: 2018 winter simulation conference (WSC). IEEE, Gothenburg Sweden, pp 1957–1968. https://doi.org/10.1109/WSC.2018.8632362
15. Ding L, Zhang L, Wu X, Skibniewski MJ, Qunzhou Y (2014) Safety management in tunnel construction: case study of Wuhan metro construction in China. Saf Sci 62:8–15. https://doi.org/10.1016/j.ssci.2013.07.021
16. Hou Z, Chen D, Li Y (2012) Fuzzy reasoning based evaluation method for the comfort of the train and its application. Railway Comput Appl 21(7):4–7. https://doi.org/10.3969/j.issn.1005-8451.2012.07.002
17. Yu S, Zhao Y (2001) JERK (the time rate of change of acceleration)—impact, passenger's comfortability, transition curve. Phys Eng 11(3):7–22. https://doi.org/10.3969/j.issn.1009-7104.2001.03.002

Safety Risk Assessment

1 Introduction

With the rapid growth of cities worldwide, the demand for tunneling projects has surged, characterized by their large scale and significant uncertainty. As project scale and complexity increase, risk control becomes crucial for ensuring underground construction safety [1]. However, while complex projects generate vast amounts of information, the risk assessment often relies on static data (missing the dynamic nature), and current risk management practices are still largely manual and depend on experience and mathematical analysis. Therefore, enhancing existing risk management practices with a system that can perceive risk probabilities in real time is an urgent and valuable step forward [2].

The rapid advancement and widespread use of Building Information Modeling (BIM) offers a promising solution for risk management in construction. BIM is considered the next-generation information technology, surpassing traditional 2D drawing-focused methods like computer-aided design (CAD). Since its introduction, BIM has quickly become a primary tool in the construction industry due to its advantages, such as multi-dimensional capabilities, interactive functions, and information-sharing features. BIM platforms facilitate the extraction of engineering information from dynamic environments, and this ability of handling dynamic data enhances construction management to a great extent. Additionally, BIM's visualization capabilities enhance risk perception by making it easier to identify risk distributions and implement appropriate measures.

Despite that BIM has already been incorporated into studies for risk management, few of the studies have emphasized risk assessment under fuzziness and uncertainty. Most current studies primarily focus on quantitative risk assessment, often overlooking the uncertainty and the utility of fuzzy information, which are inherent and unavoidable during construction phases. It is important to note that expert knowledge and experience will continue to play a crucial role for the foreseeable future due to their effectiveness in addressing the subtle, complex, and uncertain aspects of

L. Zhang et al., *Intelligent Construction in Tunnels*, Lecture Notes in Civil Engineering 679, https://doi.org/10.1007/978-981-96-8785-5_12

underground construction. Thus, a hybrid model combining the advantages of BIM with uncertainty reasoning technologies is supposed to greatly enrich and enhance risk management, especially in the field of underground construction.

To deal with uncertainty, the Dempster Shafer (D-S) evidence theory, an evidence fusion theory based on the study of Dempster [3] and further developed by Shafer [4], is proved as an effective tool. Different from the traditional probability theory where evidence is only associated with one possible piece of evidence, the evidence in the D-S evidence theory can be associated with multiple pieces of evidence, i.e., sets of evidence. It owns the advantage of dealing with uncertainty by treating imprecise and incomplete information as a scope of probabilistic uncertainty and fuzziness. With this advantage, the D-S evidence theory nowadays has been incorporated into many research fields. Zhang et al. [1] incorporated the D-S evidence theory into a model for assessing the risk status of tunnel-induced buildings. Pan et al. [5] developed a model based on the combination of D-S evidence theory with interval-valued fuzzy sets and Bayesian networks to cope with the problems of insufficient and imprecise data collection in risk management.

To develop an effective approach for the risk assessment of underground tunnel constructions, the key questions in this chapter are: (1) How to take advantages of BIM and the D-S evidence theory, and integrate them into one model; (2) How to visualize the risk perception for better risk comprehension and management; and (3) How to capture dynamic features of tunnel construction. For solving these questions, a hybrid model is proposed, in which the application programming interface (API) is developed to extract necessary information from the BIM platform, and the D-S evidence theory is incorporated to fuse the information and perceive the risk. Dynamo within BIM is used to realize the visualization, and an evidence updating rule is proposed to complete the approach. By integrating these tools, the approach outlined in this chapter serves as an effective risk evaluation tool for underground tunnel construction. This method leverages dynamic data to visualize risk likelihood in real time, providing an immediate sense of potential risks in tunneling. Such capabilities support timely decision-making, enhancing proactive risk management in underground construction projects.

2 BIM Modelling for Safety

BIM and related technologies have become the most promising developments in the Architecture, Engineering, and Construction (AEC) industries. With the incorporation of BIM technologies, a virtual environment of construction projects can be digitally built, and dynamic data, relevant information, and precise geometry are included for the procurement activities and the implementation of construction. With the utility of BIM, engineering information can be easily extracted, the scope of works can be readily defined, and different aspects of works can be promptly interrelated. Dynamo, a plug-in tool within BIM, can be used to set up automated platforms through constructing a node-based visual compilation interface, in order to realize data processing and visualization of structural parameters. This advantage of BIM in visualization could better help engineers effectively perform data analysis, detect potential risks, and make decisions in time.

Risk management has been a global issue for years, especially in the AEC industry. Risks in AEC projects could be caused by many possible factors, such as design deficiency, material defects, inexperienced operations, and lack of proper management [6]. The risk could occur in various phases of construction projects, such as the planning and designing phase, construction stage, operation and maintenance, and demolition phases. Especially in recent years, risks in AEC projects are rapidly growing due to the increasing project size and construction complexity. Because of the rapid development and advantages in BIM, the incorporation of BIM into risk management has become a research trend.

Studies show that BIM can not only serve as a systematic platform but also can be used as a data generator to allow other BIM-related tools to perform further analysis. However, even though the project data is exploding, the information closely related to the risk problems of specific projects could still be incomplete and vague, and it probably has a great level of uncertainty due to the high complexity of construction projects. This inevitably leads to inefficiency and even mistakes in the risk management of projects, especially when the project information could come from different sources. How to extract useful information and accurately fuse the information becomes a challenge.

To deal with vagueness and uncertainty, methods, such as fuzzy theory, rough set theory, and D-S evidence theories, are developed and proven as effective tools in problems of knowledge discovery, pattern recognition, and decision support. The effectiveness of these methods has been proved in many studies, especially the D-S evidence theory stands out for effectively handling information vagueness and uncertainty in engineering. Integrating BIM with D-S evidence theory can significantly enhance risk management in underground construction by enabling targeted data collection, information fusion under uncertainty, and visualization for more effective risk control. Building on the effectiveness of D-S evidence theory and BIM integration in managing uncertainty, this chapter puts effort into developing an approach that enables visualized risk management by leveraging the BIM platform's capabilities alongside the D-S evidence theory's information fusion. Relevant engineering information is extracted from BIM models, with analysis and risk perception conducted using D-S evidence theory. Finally, the perceived risk results are reintegrated into the BIM model to achieve effective risk visualization.

3 Information Fusion for Safety Risk Assessment

This section introduces the detailed steps of how to blend implementations of BIM, D-S evidence theory, and visualization technology. Figure 1 presents the conceptual risk perception framework. First, a risk problem is analyzed and the risk influencing factors are identified. Then, the related information and data can be extracted from the built database through the developed API from the constructed BIM model. The extracted information then will be treated as evidence in the D-S evidence theory for further information fusion. Afterward, the perceived risk grades will be put back into the BIM model to realize the visualization. The final output is to present the visualization of the risk probability, which could be used to provide proactive preventive suggestions.

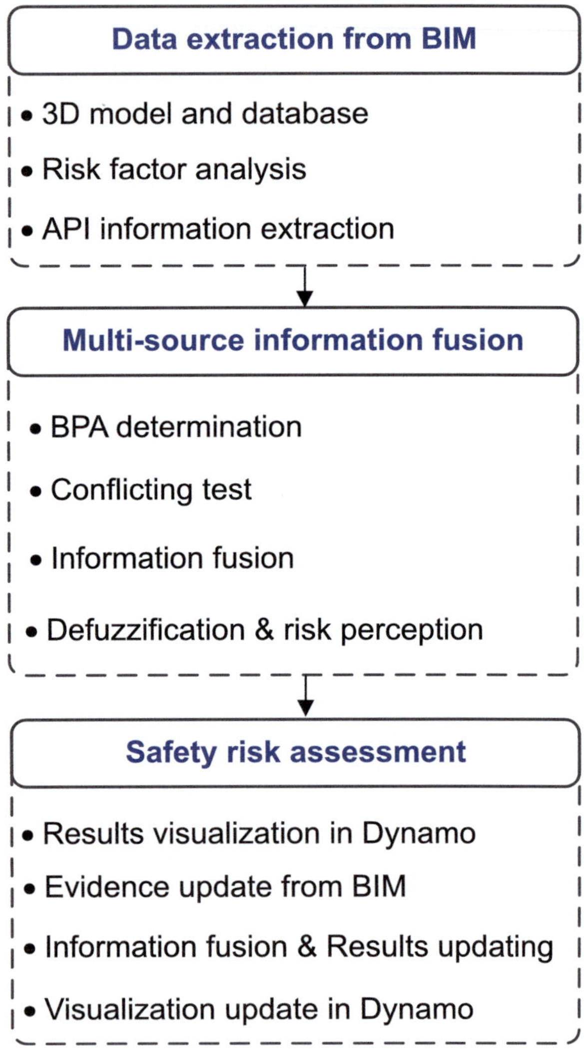

Fig. 1 Workflow of the proposed BIM-based risk perception model

3.1 Data Extraction from BIM

API in BIM is a very useful tool for realizing BIM-related research in constructions, including process modeling, algorithms, and object-based instantiations. API is particularly helpful when it comes to how different software tools interact with each other. In BIM-related research, many applications rely on API programming to build communications with models of other existing BIM platforms. The specific information on construction projects can be extracted from BIM by the implementation of API, such as project attributes, geometry, and spatial data. In this chapter, the needed information on influencing factors of risk management is extracted by API and transferred into the D-S evidence theory, where the information will be processed by the D-S fusion rule, and the risk status is perceived from the fused results.

To assess the leakage risk in tunneling projects, four steps are needed to extract the required information: (1) Snoop objects; (2) Filter element information; (3) Extract information by programming; (4) Save information as the required data type. Step 1 is to investigate the target project and gain needed information from the constructed BIM model, which can be realized by using Revit add-ins, such as *Revit Lookup*. With *Revit Lookup*, users can select any particular segments, and all engineering information of the segments can be queried, as shown in Fig. 2b. Parameters, such as design and environmental variables, have been embedded into the model, and different segments of the model carry different engineering information.

Next, Step 2 is to filter the huge amount of information and to collect those needed for leakage risk evaluation. All the information could be extracted by users according to the requirements. Step 3 is to extract the information by using Revit add-ins, such as *pyRevit*. With *pyRevit*, users can filter all the information embedded in the BIM model and extract useful information by programming. An example algorithm of the API information extraction and transaction is shown in Table 1. With the programming, step 4 also can be realized, i.e., saving the extracted information in a needed format, such as Excel files for the later use of the assessment.

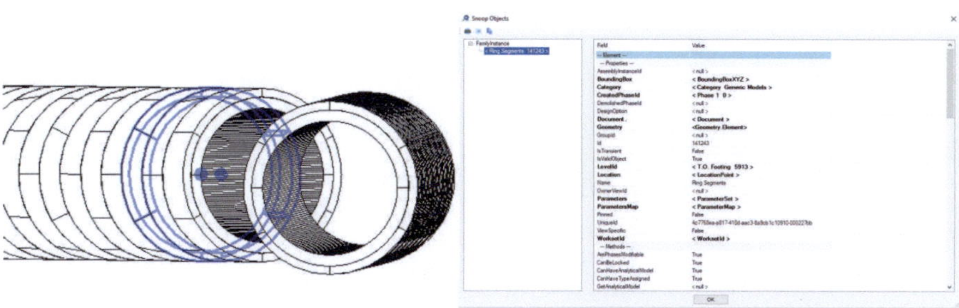

(a) (b)

Fig. 2 Example of API programming in Revit to filter elements: **a** constructed tunnel section model; **b** extracted information of the selected ring segment

Table 1 API algorithm for extracting the needed information

Algorithm: API information extraction

```
Import Revit database
_revit_.ActiveUIDocument.Document
  DB.FilteredElementCollector(doc)
     OfCategory(DB.BuiltInCategory.OST_MechanicalEquipment)
     WhereElmentIsNotElementType()
     for eq in all_element:
        print(eq)
  Information Transaction
Save element information
```

3.2 Multi-source Information Fusion

The risk perception process mainly consists of three steps: the basic probability assignment (*BPA*) determination, the improved information fusion, and the risk recognition by the defuzzification. The basic probability of factors belonging to certain intervals could be determined through the *BPA* calculation process. Based on Dempster's rule, an improved information fusion rule is proposed to fuse the information gained from different factors. A crisp number of the risk status is obtained by the defuzzification process.

(1) *BPA* determination

In the D-S evidence theory, the evidence in its frame is used to assign probabilities to events, without having to rely on the assumptions of the events. It regards all pieces of evidence as equally important and allows the direct representation of uncertainty, where imprecise information can be coped with as a set. Three critical functions are developed in D-S evidence theory: the basic probability assignment (*BPA*), the belief function (*Bel*), and the plausibility function (*Pl*). *BPA*, often represented by *m*, is a primitive concept in D-S evidence theory. It defines a mapping from a power set to an interval between 0 and 1. The *BPA* value for a particular set *A* ($m(A)$) indicates the certainty of all available evidence that supports the claim. Given the intervals of different states of measured factors, *BPA* can be represented by the following functions.

$$\begin{cases} f_{ij}(x_i) = exp\left(-\frac{(x_i - a_{ij})^2}{2{b_{ij}}^2}\right) \\ a_{ij} = \frac{x_{ij}(L) + x_{ij}(R)}{2} \\ b_{ij} = \frac{x_{ij}(L) - x_{ij}(R)}{3} \end{cases} \quad (1)$$

where x_i ($i = 1, 2, \ldots .N$) is the observed value of a factor, $f_{ij}(x_i)$ is the basic probability of sample *i* over attribute *j*, $x_{ij}(R)$ is the upper bound of the interval of the attribute, and $x_{ij}(L)$ is the lower bound of the interval of the attribute, the divided coefficients 2 and 3 are derived from the study of Zhang et al. [1], where they are reasoned in detail and demonstrated that the risk perception results of D-S evidence theory can reach up to the highest reliability level.

As the *BPA* has been determined, the bounds, upper and lower, of an interval of certain evidence can be defined. The precise probability of the evidence set is bounded by this interval, and the upper and lower bounds are called *Pl* and *Bel*, respectively. The lower bound *Bel* is determined by the sum of *BPAs* of the subsets B of the set of interest A ($B \in A$). The upper bound *Pl* is the sum of *BPAs* of the subsets B that intersect the set of interest A ($B \cap A \neq \varnothing$). *Bel* and *Pl* can be defined by the following Eqs. (2) and (3), respectively.

$$Bel(A) = \sum_{B|B \in A} m(B) \tag{2}$$

$$Pl(A) = \sum_{B|B \cap A \neq \varnothing} m(B) \tag{3}$$

(2) Improved information fusion

In order to fuse the information extracted from BIM models, the Dempster rule of combination is incorporated. The process of the Dempster's rule is to combine multiple belief functions through their *BPAs*, which are defined from the same evidence frame. The result of the combined evidence can be calculated from the sum of two *BPAs*, m_1 and m_2, by the following Eqs. (4)–(6) [1].

$$m_{12}(A) = \frac{\sum_{B \cap C = A} m_1(B) m_2(C)}{1 - K} \tag{4}$$

$$m_{12}(\varnothing) = 0 \tag{5}$$

$$K = \sum_{B \cap C = \varnothing} m_1(B) m_2(C) \tag{6}$$

where K is the conflicting coefficient of the associated *BPAs*. The greater the K value is, the greater the conflicting degree of the evidence is [7].

One problem existing in the Dempster's combination rule is that it could yield counterintuitive results when a high conflicting coefficient K exists between different sources of information. Unreasonable results would be produced by using Dempster's fusion rule when K is close to 1. To solve this problem, Zhang et al. [1] proposed an improved information rule, in which the weighted fusion rule was incorporated to gain the fused results when a high conflicting coefficient ($K > 0.95$) exists. Following the work, the process of evidence fusion by the weighted rule is defined by the following Eq. (7). The distance between the two pieces of evidence can be calculated. Then, a matrix of the distance between evidence in the evidence frame can be formed. Then, the weight of the evidence is determined. With the obtained weight of every evidence, the new evidence is calculated, and then, the fusion is conducted by using this weighted evidence. In this chapter, the evidence fusion is processed through the Dempster's rule when K is smaller than 0.95, and the weighted fusion rule when K is greater than 0.95.

$$\begin{cases} m(A) = \sum_{i=1}^{o}(w_i * m_i(A)) \\ w_i = 1/d_i/\sum_{i=0}^{o}(1/d_i) \\ d_i = \sum_{j=1}^{o}\sqrt{\sum_{k=1}^{k=n} m_i(A_k) - m_j(A_k)} \end{cases} \tag{7}$$

where w_i is the weight for the piece of evidence m_i, and d_i is the Euclid distance in total between the ith piece of evidence and other pieces of evidence.

(3) Risk perception

To gain a crisp value of the final risk level, a defuzzification process is needed to defuzzy the information obtained from the D-S fusion rule above. Through the defuzzification process, quantifiable results from fuzzy sets and corresponding membership degrees can be produced. Commonly used defuzzification methods include the Centroid method, Bisector method, Middle of maximum, Smallest of maximum, and Largest of maximum. However, there are some defects existing in these methods, such as a time-consuming process or information loss. To improve the defuzzification approach, Zhang et al. [1] proposed a novel defuzzification method, Center of Distribution (COD), which was demonstrated to be a more effective defuzzification method, and its calculation is realized by using Eq. (8). A crisp number could be gained to represent the center of the result distribution. This crisp number is regarded as the perceived final risk level.

$$\begin{cases} T = 0.5 * \frac{\varphi(A_j)-1}{\varphi(A_j)+1} + j \\ \varphi(A_j) = \frac{\sum_{i=j}^{i=m} m(A_j)}{\sum_{i=1}^{i=j} m(A_j)} \end{cases} \tag{8}$$

where T represents the defuzzied value of the final risk perception, $m(A_i)$ is the membership degree of the ith grade, and j is the jth grade which has the largest membership degree. With this defuzzification process, the value T can be seen as the evaluative result of the risk perception. The scale of the risk levels can be determined by the value of T, which has N levels, starting from [0.5, 1.5] to [N − 0.5, N + 0.5].

3.3 Safety Risk Assessment and Visualization

The next step is to visualize the results to facilitate decision making. The visualization of perceived risk levels is achieved through Dynamo, a visual programming environment for Revit. As a free Revit plug-in, Dynamo serves as a parametric design tool that enables users to conduct parametric operations on individual objects or object families using APIs. Built on an object-based parametric model, which is

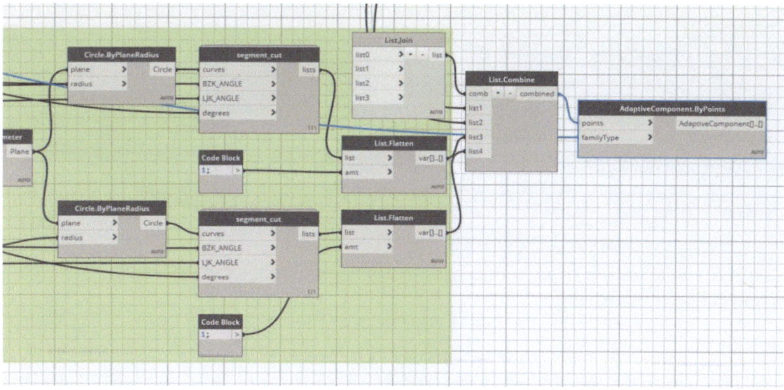

(a)

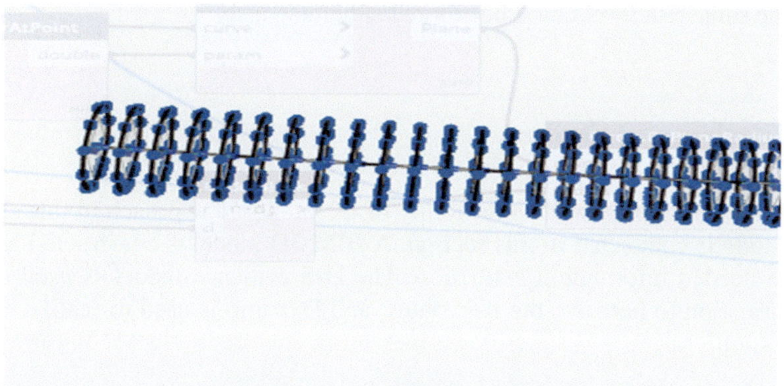

(b)

Fig. 3 Example of visualization of Dynamo programming: **a** part of the node-based programming; **b** produced tunnel rings of the programming

fundamental to BIM, Dynamo integrates procedural information into BIM models, allowing users to define construction processes and modify architectural components within BIM platforms. Through Dynamo's node-based visual interface, users can automate data processing, perform parameter operations, and ultimately visualize the risk assessment results effectively.

Figure 3 shows an example of node-based Dynamo programming for the visualization of BIM models. The frame of a tunnel could be constructed in Dynamo by setting up the starting point, the position of the central axis, and the width and diameters of the rings. Figure 4 presents a Dynamo programming example for the colored visualization, which could make the visualization richer and more discernible. Specific colors could be used to represent different safety risk grades. Due to the advantage of Dynamo object-based programming, every single segment could be automatically colored for its risk corresponding level color, and a lot of repetition work could be saved. With the determined risk level, the corresponding color could be programmed

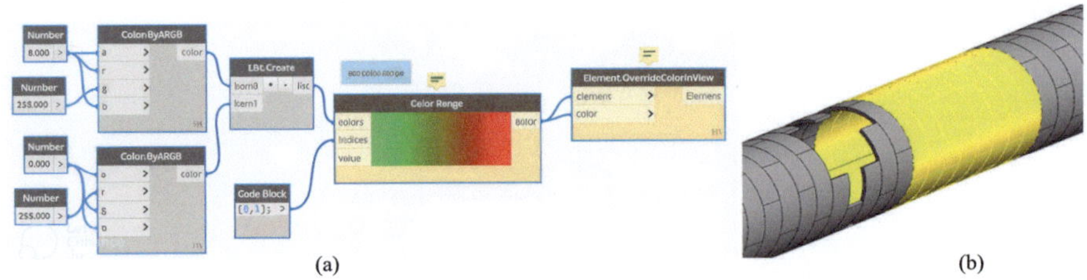

Fig. 4 Colored visualization of Dynamo programming: **a** coding for the visualization; **b** colored segments and rings of the model

in Dynamo to override the color of the selected segments, which could be chosen by using the built-in function in Dynamo. As presented in Fig. 4b, segments and rings under the same risk level could be colored simultaneously.

4 Experimental Studies

To test the effectiveness and applicability of the proposed approach in this chapter, a case study is conducted in this section. A BIM 3D model is constructed to facilitate the needed information extraction. The D-S evidence theory is used to fuse the information to perceive the risk status, and Dynamo is used to realize the risk visualization.

4.1 Data and Resources

Cross-river tunnel in Wuhan Sanyang Road, as shown in Fig. 5, connecting Wuchang and Hankou, is the cross-river section of the Wuhan Metro Line 7 Yangtze River tunnel. The Sanyang road cross-river tunnel is the first tunnel in China that links together the highway and railway. It is the cross-river shield tunnel with the largest diameter in China to date. The tunnel adopts two mega-mix shield TBMs with a super large diameter of 15.76 m, which is the first largest in China and the third-largest in the world in terms of the diameter. The slurry pressure balance shield tunneling machine is the largest in China at present and is nearly 4000 tons in total weight, 167 m in length, and 20 thousand tons in total thrust.

This tunnel includes two lines, the left line and the right line, and each line is 2590 m in length. It consists of three layers. The upper layer will be the discharged flue of the highway tunnel; the middle will be the running lane of the highway tunnel and will be equipped with unidirectional three lanes with a width of 3.5 m; the lower layer will be the running lane of the subway line 7 and its escape way, cable tunnel, and discharge flue.

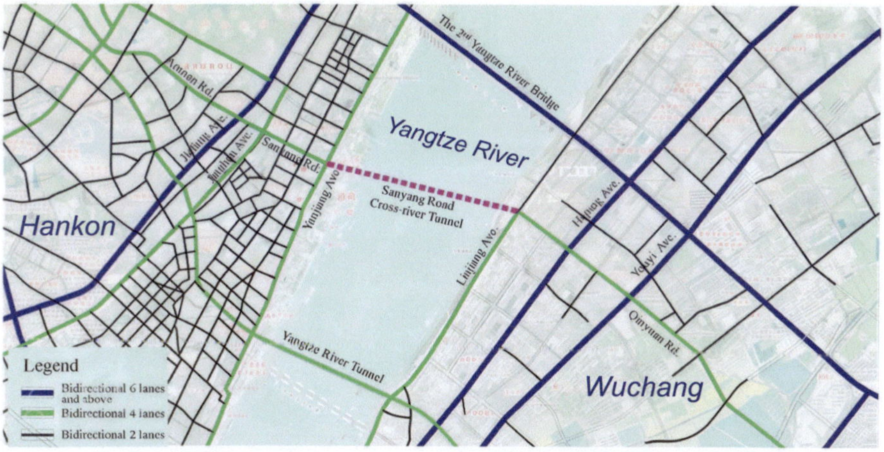

Fig. 5 Layout of the cross-river tunnel in the Wuhan Sanyang Road

The geological conditions of the Sanyang road cross-river tunnel are very complicated. Figure 6 presents the longitudinal section of the Sanyang road cross-river tunnel [8]. The stratum where the tunnel passes through mainly consists of a medium dense to dense fine sand layer, which contains a local undercutting strongly weathered silty mudstone, weakly weathered silty mudstone, and weakly cemented conglomerate. The cut-in bed rock segment's total length and maximum depth are approximately 1100 and 12.6 m, respectively. The bedrock consists of soft and extremely soft rock. The uniaxial compressive strength of the weakly weathered silty mudstone and the weakly cemented conglomerate are 0.6–5.4 and 0.1–12.2 MPa, respectively. The super-large tunnel section's monoline in the river is 2590 m long. Nearly all of the line is full-face sandy strata except for approximately 1200 m upper-soft and lower-hard composite stratum in the river [9]. Figure 7 demonstrates a simulated 3D BIM model of the tunnel. It consists of 1295 rings, and each ring is made of six segments.

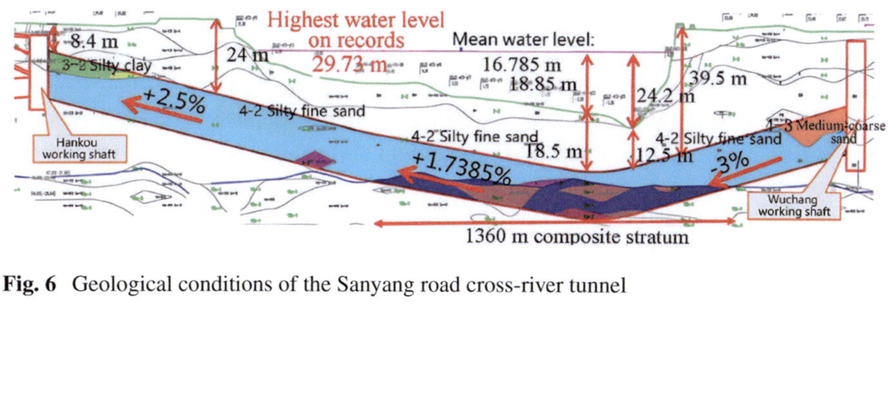

Fig. 6 Geological conditions of the Sanyang road cross-river tunnel

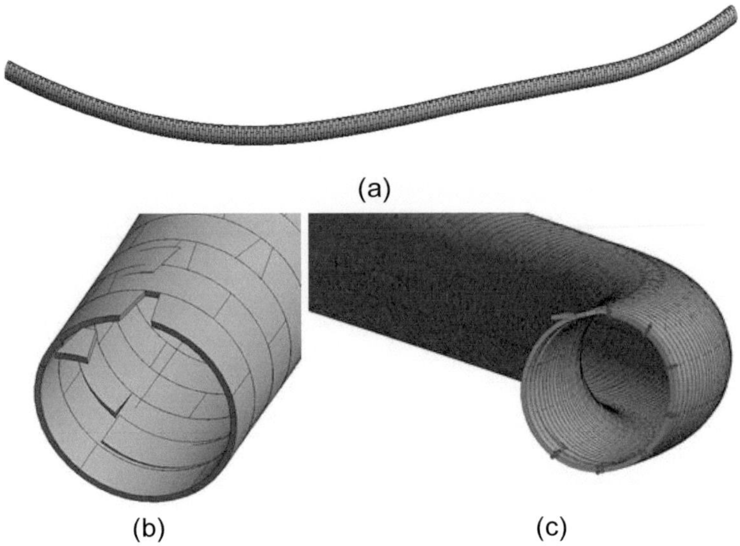

Fig. 7 Constructed tunnel model from BIM: **a** whole tunnel section; **b** ring segments; **c** part of the tunnel with slope

4.2 Safety Risk Analysis in Tunnel Construction

The leakage risk during the tunnel construction could be caused by many factors, such as the design of the tunnel, the groundwater depth, differential settlement of the tunnel, and construction techniques. The evaluation of the tunnel leakage risk status should consider all possible influencing factors and select the most important factors to be analyzed. Most risks are directly related to the engineering information, which is embedded in BIM platforms, such as construction techniques, structural attributes, and dynamic environmental data. The influencing factors are identified by reviewing engineering documents and literature reviews. The intervals corresponding to the risk magnitude of specific factors are recognized later.

Due to the complexity of the tunneling environment, the identification of influencing factors could be fuzzy, and the reasonable identification could be conducted from a lot of practical engineering experience and reviews of related studies. Considering the cross-river tunneling condition and study review in the related field, two key leakage risk indicators are identified, geological and design variables [10]. Design variables are often determined in the pre-construction stage, which is before the actual excavation activity. Geological conditions play an important role during the tunnel construction phase. Geological uncertainty is often the primary reason causing risk in underground projects. Table 2 describes the details of the identified influencing factors, and Fig. 8 presents the leakage risk index system adopted in this chapter.

Table 2 Descriptions of influencing factors in tunnel leakage

Code	Factor	Description	References
C_1	Soil cohesion	Referring to the component of shear strength of the soil	[1]
C_2	Soil friction angle	A shear strength parameter of soils, used to describe the friction shear resistance of soils with normal stress	[1]
C_3	Soil compression modulus	The ratio of the vertical stress to the corresponding vertical strain under full confinement conditions (without lateral deformation). Its size reflects the resistance of the soil to compression deformation under uniaxial compression	[1]
C_4	Soil permeability	Referring to the property of the soil to transmit water and air, indicating the capability of the soil to allow fluids to pass through	[14]
C_5	Water level	The depth from the water surface to the tunnel	[19]
C_6	Cover depth	The distance from the tunnel to the ground surface	[1]
C_7	Cover-span ratio	The ratio of the cover depth and the tunnel span	[1]
C_8	Inclination angle	The slope of the tunnel sections under the river	[20]

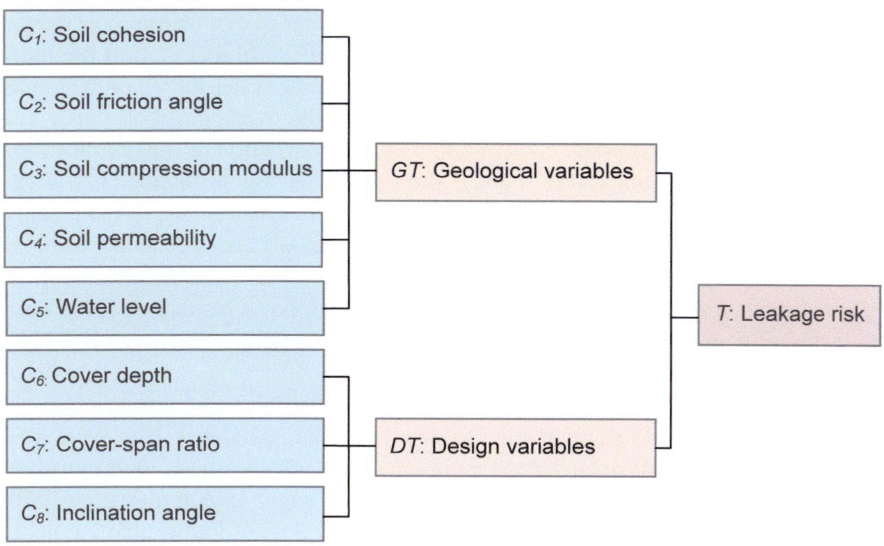

Fig. 8 Leakage risk index system in the tunnel project

(1) Geological variables

The geological condition is a critical factor influencing the uncertainty and construction risk in tunnel construction. Building the tunnel in accordance with the geological conditions can effectively reduce the construction risk [11]. Important geological factors include soil cohesion (C_1), friction angle (C_2), and compression modulus (C_3) [12]. Soil cohesion and friction are the influencing factors of the shear strength of the soil, which indicates the ability of the soil to resist the movement along internal surfaces. The soil compression modulus refers to the ratio of the vertical stress to the corresponding vertical strain under full confinement conditions (without lateral deformation). These are three basic parameters that are often used to test the leakage pressure on the soil [13]. The soil permeability (C_4) of the surrounding environment plays an important role in the tunnel lining behavior related to the leakage risk [14]. It indicates the capability of the soil to allow fluids to pass through, i.e., impermeable soil loses little water and permeable soil is prone to lose much water. The permeability of soil is often closely related to seepage problems [15]. Another important factor is the water level (C_5). Water pressure will increase along with the depth of the water level, and a high water pressure may accelerate the rock damage, cause the leakage hazard, and cause damage to the tunnel support structure.

(2) Design variables

The rationality of the design in tunnel construction exerts significant impacts on the surrounding environment. Poor design could lead to construction problems causing safety issues, such as foundation deformation [16]. The cover depth (C_6) and the cover-span ratio (C_7) are used as critical parameters for the safety assessment in tunnel construction, and they are identified as influencing design factors. The deeper the tunnel is buried, the larger the catchment area of the tunnel is, and the more sufficient the groundwater supply is. The total and specific water inflow of the tunnel is characterized by the increase of the thickness of the overlying rock mass. The increase of the cover-span ratio could strengthen the joint action of surrounding rock and support, which significantly reduces the risk of leakage problems [17]. Thus, the cover depth and cover-span ratio are regarded as a critical factor in the leakage analysis of tunnels. The supporting structure of the tunnel carries eccentric loads in the inclined double-deck mined-out area. The bigger the inclination angle, the more the eccentric load, which causes the instability of the supporting structure [18]. Thus, the inclination angle (C_8) is set as another factor influencing the leakage risk level in the tunnel construction.

To gain the probability of the leakage risk severity in tunnel construction, an interval index needs to be constructed to determine the risk status. However, great fuzziness and uncertainty exist due to the complexity of underground tunnel construction, and the recognition of the risk intervals is often vague in determination. With the massive construction of tunnel projects, large amounts of knowledge are being accumulated from the actual construction sites, such as the monitoring records, technical manuals, and research reports, and numerous researchers have built simulation models for safety risk control, which provide valid references for the determination and classification of influencing factors [1]. Based on the related studies, three safety risk intervals are constructed, i.e., safe, low risk, and high risk. The safe interval refers to the fact that it is unlikely to cause risk when the value of parameters stays within this interval, the low risk level means that there is a low probability of causing the risk, and the high risk refers to the high probability. The classification of intervals for influencing factors is presented in Table 3. To realize the visualization of the perceived risk status, three colors are used to represent the three different grades, as shown in Table 4.

Engineering information has been embedded into the model, and the needed information of three tunnel sections, from the cover depth of 8.4 m, 24 m, and 36.5 m, respectively, can be extracted by the application of API as stated in Sect. 3.1. A total of five experts from the tunnel construction field are invited to check the data, in order to ensure that the information is consistent with the condition of the actual construction site. The information of the seven influencing factors is presented in Table 5.

Table 3 Risk grade intervals of identified risk influential factors

Code	Unit	States		
		I (safe)	II (low risk)	III (high risk)
C_1	(KPa)	[20, 25]	[10, 20]	[0, 10]
C_2	(°)	[25, 45]	[10, 25]	[0, 10]
C_3	(MPa)	[40, 60]	[10, 40]	[0, 10]
C_4	(1E−8/cm/s)	[0, 0.6]	[0.6, 1.2]	[1.2, 1.5]
C_5	(m)	[0, 10]	[10, 20]	[20, 30]
C_6	(m)	[0, 12]	[12, 28]	[28, 40]
C_7		[3, 5]	[1, 3]	[0, 1]
C_8	(%)	[0, 1]	[1, 3]	[3, 5]

Table 4 Colors representing each level of risk perceptions

States	Color	RGB values	Color
I	Green	(0, 255, 0)	
II	Yellow	(255, 255, 0)	
III	Red	(255, 0, 0)	

Table 5 Extracted values of different variables of the three tunnel sections from the BIM model

Variables	Unit	Extracted values of different variables at three tunnel sections		
		Tunnel section 1	Tunnel section 2	Tunnel section 3
C_1	(KPa)	24.9	23.1	19.9
C_2	(°)	26.2	23.2	22.3
C_3	(MPa)	41.3	38.9	36.7
C_4	(1E−8/cm/s)	0.53	0.63	0.83
C_5	(m)	2.0	12.0	23.0
C_6	(m)	0.55	1.57	2.5
C_7		2.5	2.5	3

4.3 Analysis of Results

With the extracted values of the seven influencing factors for the three sections, *BPAs* for every factor can be calculated by using Eq. (1). The results are presented in Table 6. Setting these *BPAs* as evidence frames and applying the information fusion rule, [i.e., Eqs. (2)–(5)], the fused result of the risk status of the target tunnel sections can be gained. To gain crisp values of the risk levels, the defuzzification rule in Sect. 3.2 is applied. The final risk status of the tunnel section 1 is determined as under grade I (safe level), tunnel section 2 as under II (low risk level), and tunnel section 3 as under II (low risk level). The results are presented in Table 7. With the determined risk levels, Dynamo is further used to realize the visualization as stated in Sect. 3.3. As shown in Fig. 9, (a) presents the visualized part of tunnel section 1, (b) is the part of tunnel section 2, (c) is the part of tunnel section 3, and (d) is a holistic perspective of the tunnel with the visualized tunnel sections. According to this process, the risk status of the whole tunnel could be visualized, when the measured data is obtained.

(1) The proposed approach can effectively perceive the risk status of the target tunnel section and realize the visualization. As presented in Tables 6 and 7, along with the increasing cover depth of the tunnel, there is a trend of the magnitude developing from a safe level to a risk level. The risk status of the tunnel section under the lowest cover depth is safe, and the tunnel sections with deeper cover depths are under low risk levels. Putting these perceived risk levels of the sections into Dynamo and applying the colour rule in Table 4, the visualization can be realized. Figure 9 presents the visualized results of three tunnel sections. To validate the accuracy of the perceived results, five experts who participated in the construction site investigation are invited again to evaluate the risk status of the tunnel sections. The ranking results from these experts are in accordance with the perceived risk results by the proposed approach. This indicates that the perceived risk status of the three sections is in line with the actual construction condition, and it again demonstrates the effectiveness and accuracy of the proposed approach.

Table 6 *BPAs* of influential factors

Variables	*BPAs*			
	m(I)	m(II)	m(III)	$m(\Theta)$
Tunnel section 1				
C_1	0.3546	0.0121	1.82E−08	0.6332
C_2	0.4184	0.2201	1.64E−09	0.3615
C_3	0.4268	3.03E−14	1.77E−52	0.5732
C_4	0.5162	0.1806	2.50E−15	0.3032
C_5	0.8353	0.0939	1.63E−09	0.0708
C_6	0.9888	0.0939	1.52E−06	0.0827
C_7	1.52E−08	0.7548	0.0796	0.1656
Tunnel section 2				
C_1	0.9373	0.0522	3.96E−07	0.0105
C_2	0.2088	0.5221	3.36E−07	0.2690
C_3	0.2500	6.87E−14	3.47E−23	0.7499
C_4	0.2563	0.4020	5.53E−12	0.3416
C_5	4.00E−05	0.7548	0.0439	0.2011
C_6	0.0058	0.8122	0.0013	0.1807
C_7	1.52E−08	0.7548	0.0796	0.1656
Tunnel section 3				
C_1	0.2962	0.3394	4.58E−05	0.3643
C_2	0.1629	0.6308	1.41E−09	0.2063
C_3	0.1367	6.27E−10	2.29E−20	0.8633
C_4	0.0298	0.9406	1.34E−06	0.0296
C_5	5.87E−16	0.0013	0.3886	0.6102
C_6	1.52E−08	0.7548	0.0796	0.1656
C_7	6.10E−13	0.3247	0.3247	0.3506

(2) The unsafety caused by design variable defect is identified to be an important reason leading to an increase in tunnel risk levels. From Table 7, the probabilities of higher risk levels of both geological variables and design variables are increasing, and this leads the tunnel sections 2 and 3 into the low risk levels. It should be further noticed that the probabilities of geological variables of tunnel section 2 are roughly the same under grade I and II, and no chance under grade III. The probability of geological variables of the tunnel section 3 has a higher degree under grade II, but there is nearly no probability of geological variables of tunnel section 3 under grade III. However, there is an evident increase of the probability of design variables under grade II for the tunnel section 2, and there

is even a noteworthy probability of the design variables under grade III for the tunnel section 3. This indicates that the unsafety caused by design variables is the main factor leading the status of the tunnel sections to unsafe levels.

Table 7 Fusion results of *BPAs* for the observed tunnel point

Result	*BPAs*				Risk level
	m(I)	m(II)	m(III)	$m(\Theta)$	
Tunnel section 1					
GT	0.4923	0.1489	2.31E−09	0.3586	
DT	0.6075	0.3101	0.0260	0.0563	
T	0.8206	0.1320	9.50E−08	0.0473	
Defuzzied result	1.03				I
Tunnel section 2					
GT	0.3078	0.4324	0.0001	0.2595	
DT	0.0022	0.7764	0.0393	0.1820	
T	0.0018	0.5720	0.0197	0.2271	
Defuzzied result	1.99				II
Tunnel section 3					
GT	0.1292	0.4825	0.1676	0.2205	
DT	4.91E−09	0.3661	0.2648	0.3690	
T	2.09E−09	0.5841	0.1468	0.2691	
Defuzzied result	2.05				II

Note GT represents the fusion result of geological variables, and DT is the fusion result of design variables

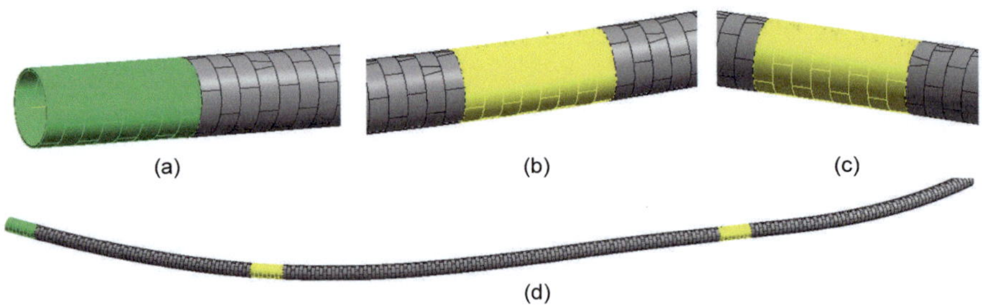

Fig. 9 Visualized risk perception of tunnel sections: **a** screenshot of the visualized tunnel section 1; **b** screenshot of the visualized tunnel section 2; **c** screenshot of the visualized tunnel section 3; **d** holistic perspective of the risk visualization of the tunnel

5 Merits Analysis

Nowadays, various kinds of monitoring systems have been developed to capture the dynamics of underground construction projects, and real-time data is available for use. This presents the opportunity that the risk perception could be able to reflect the dynamic feature of the tunnel projects for timely decision making. The risk condition should be perceived through the dynamic data measured from the tunnel condition, i.e., the existing risk status can be updated timely with the newly gained information. The aim of this section is to find out how to update the risk status when new evidence is available.

In the conventional probability theory, the conditional probability of an event A is the probability that the event will occur given the knowledge that an event B has already occurred. This probability is written as $P(A|B)$, and it can be calculated from the following Eq. (9). Based on this, a Bayesian conditional rule is proposed to realize the probability redistribution and update under uncertainty, when the new information is obtained. Let α and $1 - \alpha$ are the probability of events B and $\overline{B}$, respectively. The probability update can be realized from the following Eq. (10).

$$P(A|B) = P(A \cap B)/P(B) \tag{9}$$

$$P'(A) = \alpha * P(A|B) + (1 - \alpha) * P(A|\overline{B}) \tag{10}$$

where P is a prior probability and $P(A|B)$ is a posterior probability when B occurs, α is the probability of event B, and $P'(A)$ is the updated probability. The updating rule equation can be reduced to the following one:

$$P'(A) = \sum_{i=1}^{s} \alpha_i * P(A|B_i) \tag{11}$$

where B_i (i = 1, 2, ..., s) represents all the possibilities of the events, α_i is the possibility of B_i, and it satisfies $\sum_{i=1}^{s} \alpha_i = 1$.

Established on Jeffrey's rule, an information fusion rule can be used to update the fusion results when obtaining new information. Let $X_1 = \{B_i|i = 1,2, \ldots, N_1\}$, $X_2 = \{A_j|j = 1,2, \ldots, N_2\}$ are the frames of two pieces of evidence, and m_1, m_2, represent their *BPAs*, respectively. The evidence m_2 is the newly gained information, and *BPAs* of m_2 are the α_i in Eq. (11), which can be used to update the existing information m_1. The evidence updating can be processed by using the following Eqs. (12) and (13).

$$m_1(C|A_j) = \sum_{\varnothing = C = B_i \cap A_j} m_1(B_i)/Pl_1(A_j) \tag{12}$$

$$m(C|m_2) = \sum_{A_j \in X_2} m_2(A_j) m_1(C|A_j) \tag{13}$$

where $m(C|m_2)$ is the information updated result, $C \in X_1 \cap X_2$.

To capture the dynamic feature of tunnel construction, the above evidence update approach is used to perceive the instant risk status. Design variables are determined within the design phase, meaning that they would not change during the construction. Geological conditions could be changed along with the construction process of the tunnel. Among the variables identified in Sect. 4.1, the value of C_1, C_2, and C_3 would be reduced, and the value of C_4 would be raised during the construction phase, as shown in Table 8.

The values of the three tunnel sections are captured along with the excavation. The percentages next to the values in Table 8 are the variation degree compared to the original values of the geological variables. The water level is not changing during the excavation process, only factors from C_1 to C_4 varied. Accordingly, the frames of the evidence of the risk status for that moment can be obtained by putting these sets of data into the D-S evidence theory. The results are presented in Table 9. These frames of the evidence are then set as new pieces of evidence to update the existing evidence. Table 10 presents the *Pl* degree of the existing evidence calculated from the extracted data value by using Eq. (3). Gaining these values and applying the above information updating through Eqs. (12) and (13), the updated evidence information of the influencing factors at the timely measured moment could be gained. The results are presented in Table 11. Applying the information fusion rule in Sect. 3.2, the corresponding results are determined. The tunnel section 1 is under I, tunnel section 2 under II, and tunnel section 3 under III. The defuzzied results are presented in Table 12. Figures 10, 11 and 12 demonstrate the comparisons of risk levels with and without information updating.

(1) The proposed information updating rule effectively captures the dynamic nature of tunnel construction. As shown in Table 12, the risk status of tunnel section 1 remains at grade I but slightly increases from 1.04 to 1.06 during construction. Tunnel section 2's risk level rises from 1.93 to 1.99, staying within grade II, while section 3 also sees a slight increase yet remains under grade II. Since water levels

Table 8 Observed values of geological variables for three tunnel sections at certain moments

Variables	Observed values of geological variables		
	Tunnel section 1	Tunnel section 2	Tunnel section 3
C_1	23.7 (– 4.8%)	21.1 (– 7.8%)	17.7 (– 11.2%)
C_2	24.2 (– 7.6%)	20.8 (– 10.2%)	20.5 (– 8.2%)
C_3	39.5 (– 4.3%)	34.9 (– 10.2%)	33.1 (– 9.7%)
C_4	0.58 (+9.7%)	0.70 (+11.3%)	0.93 (+12.2%)

Note percentages in the parentheses represent the variation degree compared to the original values of the geological variables

Table 9 *BPAs* of the measured data of geological variables of the three tunnel sections

Variables	*BPAs*			
	m(I)	m(II)	m(III)	$m(\Theta)$
Tunnel section 1				
C_1	0.7717	0.0331	1.46E−07	0.1952
C_2	0.2692	0.4075	6.24E−08	0.3233
C_3	0.2892	0.3495	5.47E−24	0.3612
C_4	0.3753	0.2780	1.33E−13	0.3467
C_5	0.6669	0.0004	4.58E−11	0.3325
Tunnel section 2				
C_1	0.7027	0.1874	8.59E−06	0.1099
C_2	0.1035	0.8043	1.32E−05	0.0922
C_3	0.0769	0.6126	3.37E−18	0.3105
C_4	0.1353	0.6065	6.69E−10	0.2581
C_5	0.1102	0.6669	0.0004	0.2222
Tunnel section 3				
C_1	0.0158	0.7203	0.0007	0.26315
C_2	0.0939	0.8353	2.01E−05	0.0707
C_3	0.0402	0.7203	3.70E−16	0.2394
C_4	0.0070	0.9888	0.0001	0.0040
C_5	4.65E−07	0.0561	0.8352	0.1085

Table 10 The *Pl* degree of the existing evidence of factors C_1 to C_4 from extracted data in Sect. 5.3

Variables	*Pl* degree		
	Pl(I)	*Pl*(II)	*Pl*(III)
Tunnel section 1			
C_1	0.9878	0.6453	0.6332
C_2	0.7799	0.5816	0.3615
C_3	1	0.5732	0.5732
C_4	0.8194	0.4838	0.3032
Tunnel section 2			
C_1	0.9478	0.0627	0.0105
C_2	0.4778	0.7911	0.2690
C_3	0.9999	0.7499	0.7499
C_4	0.5979	0.7436	0.3416
Tunnel section 3			
C_1	0.6605	0.7037	0.3643
C_2	0.3692	0.8371	0.2063
C_3	1	0.8633	0.8633
C_4	0.0594	0.9702	0.0296

Table 11 Updated evidence of geological variables C_1 to C_4

Variables	*BPAs*			
	m(I)	m(II)	m(III)	$m(\Theta)$
Tunnel section 1				
C_1	0.2769	0.0006	4.21E−14	0.7223
C_2	0.1444	0.1541	2.84E−16	0.7013
C_3	0.1679	0.1615	3.14E−32	0.6705
C_4	0.2364	0.1038	1.10E−22	0.6597
Tunnel section 2				
C_1	0.7265	0.0127	5.28E−12	0.2608
C_2	0.0452	0.5307	1.65E−12	0.4240
C_3	0.0310	0.3108	3.17E−24	0.6580
C_4	0.0580	0.3278	1.08E−16	0.6140
Tunnel section 3				
C_1	0.0070	0.3474	8.86E−08	0.6454
C_2	0.0414	0.6294	1.38E−11	0.3291
C_3	0.0110	0.4208	2.36E−22	0.5680
C_4	0.0035	0.6102	6.71E−10	0.3863

and design parameters are fixed at the pre-construction stage, these risk changes mainly result from evolving geological conditions, specifically variables C_1, C_2, C_3, and C_4, as reflected in Table 11. This perceived risk varying trend is in line with the changing trend of the influencing factors. These findings are verified by the experts from the tunnel projects.

(2) Incorporating the information updating rule enhances the certainty and identifiability of risk assessments. The uncertainty degree is decreasing along with the information fusion process. m(Θ) presents the uncertainty degree, and as indicated in Tables 11 and 12, $m(\Theta)$ has a decreasing trend. Especially for the results gained by the information fusion rule, m(Θ) is decreasing to the minimum, indicating that the results are reliable. It was concluded that the results gained from the evidence updating rule were more reliable because of the comprehensive consideration of the existing data and the timely captured data. Besides, the risk level in the D-S evidence theory is decided by the greatest value among the intervals of risk levels. A better identifiable degree means that there is a bigger difference between the highest value and the second highest value of the intervals. As illustrated by Figs. 10, 11, and 12, the results gained by the information updating rule in the figure (c) have better identifiable degrees than the other two.

Table 12 Comparison of risk levels of the three tunnel sections perceived from the extracted data, dynamic data with information updating, and dynamic data without information updating

Results		*BPAs*				Defuzzied result	Risk level
		m(I)	m(II)	m(III)	$m(\Theta)$		
(a)	Tunnel section 1	0.8206	0.1320	9.50E−8	0.0473	1.03	I
	Tunnel section 2	0.1812	0.5720	0.0197	0.2271	1.99	II
	Tunnel section 3	2.09E−09	0.5841	0.1467	0.2691	2.05	II
(b)	Tunnel section 1	0.7549	0.1946	2.22E−09	0.0504	1.06	I
	Tunnel section 2	0.0010	0.8240	9.21E−06	0.1749	1.99	II
	Tunnel section 3	4.74E−10	0.7044	0.1452	0.1503	2.05	II
(c)	Tunnel section 1	0.8307	0.1692	1.30E−16	8.32E−18	1.05	I
	Tunnel section 2	0.0008	0.9991	3.10E−12	1.78E−16	1.99	II
	Tunnel section 3	3.15E−10	0.7985	0.2014	8.32E−17	2.06	II

Note (a) is the result gained from the extracted data in Sect. 5.3, (b) is the result from the observed data without considering the existing status, (c) is the result by updating the existing results with the newly observed data

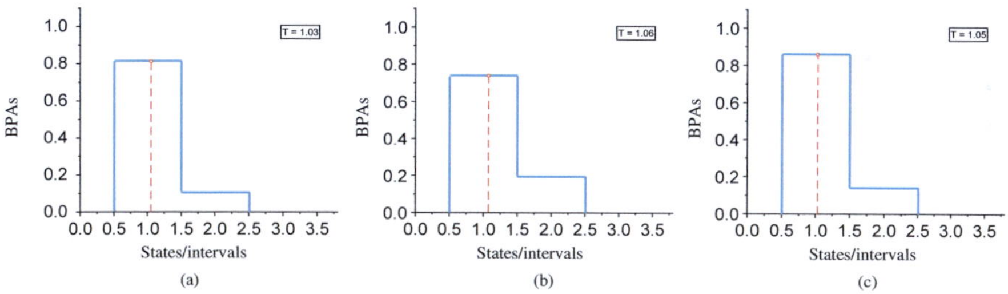

Fig. 10 Risk levels of three tunnel sections with and without information updating for the tunnel section 1: **a** perceived risk level with the extracted data; **b** perceived dynamic risk level without information updating; **c** perceived risk level with information updating

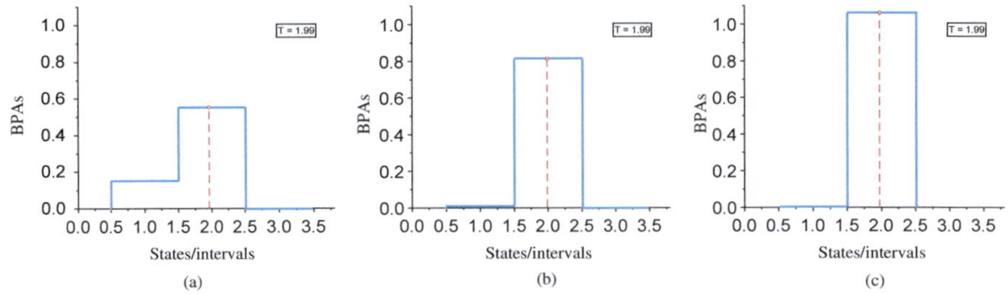

Fig. 11 Risk levels of three sections with and without information updating for the tunnel section 2: **a** perceived risk level with the extracted data; **b** perceived dynamic risk level without information updating; **c** perceived risk level with information updating

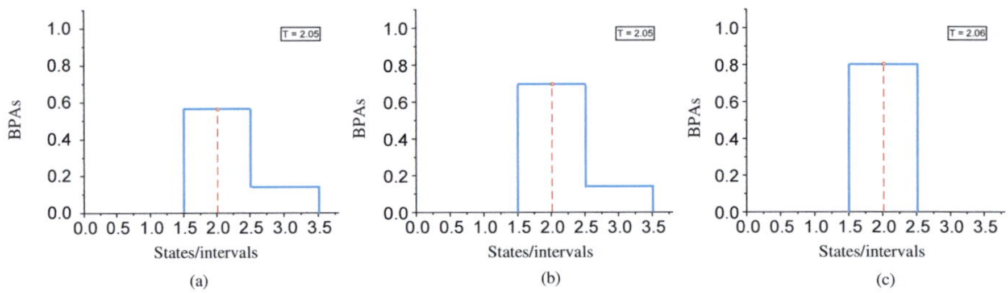

Fig. 12 Risk levels of three sections with and without information updating for the tunnel section 3: **a** perceived risk level with the extracted data; **b** perceived dynamic risk level without information updating; **c** perceived risk level with information updating

6 Chapter Summary

The proposed approach in this chapter enriches the risk management field for complex underground projects by combining the advantages of BIM in data processing and visualization and D-S evidence theory in dealing with the vagueness and uncertainty of information. The complexity and uncertainty of underground constructions make risk assessment challenging. This issue is compounded by the fact that current risk management heavily relies on expert knowledge and experience, while the growing volume of information and the need for timely processing exceed manual handling capabilities. To solve this problem, a hybrid approach combining BIM, BIM-related technologies, and the D-S evidence theory is proposed in this chapter. A case study is performed to demonstrate the effectiveness of the proposed hybrid approach for assessing and managing risk in tunnel construction.

Through integration with Dynamo, the proposed approach facilitates direct visualization of risk levels, which enables users to quickly perceive the current risk status and implement timely measures. Furthermore, the dynamic tracking of risk status in underground construction is achieved by applying an information updating rule, enhancing the responsiveness of the approach to changing conditions. By incorporating an information fusion rule, this approach also increases the certainty and accuracy of risk identification. In sum, this hybrid approach allows for real-time risk perception, visualization, and data updates, empowering users to take proactive steps that significantly reduce potential risks in tunnel construction projects.

References

1. Zhang L, Ding L, Wu X, Skibniewski MJ (2017) An improved Dempster-Shafer approach to construction safety risk perception. Knowl Based Syst 132:30–46
2. Guo K, Zhang L (2021) Multi-source information fusion for safety risk assessment in underground tunnels. Knowl Based Syst 227:107210
3. Dempster AP (1967) Upper and lower probability inferences based on a sample from a finite univariate population. Biometrika 54:515–528
4. Shafer G (1976) A mathematical theory of evidence. Princeton University Press
5. Pan Y, Zhang L, Li Z, Ding L (2019) Improved fuzzy bayesian network-based risk analysis with interval-valued fuzzy sets and D-S evidence theory. IEEE Trans Fuzzy Syst 28(9):2063–2077
6. Getuli V, Ventura SM, Capone P, Ciribini ALC (2017) BIM-based code checking for construction health and safety. Proc Eng 196:454–461
7. Pan Y, Zhang L, Wu X, Skibniewski MJ (2020) Multi-classifier information fusion in risk analysis. Inf Fusion 60:121–136
8. Zhong Y (2018) Technical innovation for Sanyang road cross-river tunnel project. Fron. Eng Manage 5:411–415
9. Li W (2017) The design of horizontal and vertical sections of the cross-river tunnel of Wuhan Sanyang Road. Transp Technol 83–85
10. Zhang L, Wu X, Chen Q, Skibniewski MJ, Zhong J (2014) Developing a cloud model based risk assessment methodology for tunnel-induced damage to existing pipelines. Stoch Env Res Risk Assess 29:513–526
11. Xie X, Wu Q, Liu K (2017) Analysis on excavation face stability of deep shield tunnel considering effect of pore water pressure. J Highw Transp Res Dev 13–23
12. Likitlersuang S, Surarak C, Wanatowski D, Oh E, Balasubramaniam A (2013) Geotechnical parameters from pressuremeter tests for MRT blue line extension in Bangkok. Geomech Eng Int J 5:99–118
13. Fu W (2013) Mechanism analysis and treatment countermeasures of highway tunnel leakage, in, Southwest University
14. Soga K, Laver RG, Li Z (2017) Long-term tunnel behaviour and ground movements after tunnelling in clayey soils. Undergr Space 2:149–167
15. Zhang D, Huang Z, Yin Z, Ran L, Huang H (2017) Predicting the grouting effect on leakage-induced tunnels and ground response in saturated soils. Tunn Undergr Space Technol 65:76–90
16. Shi J, Liu X, Cao W, Chen C, Yang Y (2013) The influential factors of evaluation of leakage of water diseases by improved AHP on multi-arch tunnel. Prog Geophys 28(1):0482–0487
17. Wang X, Zhou S, Chi J, Wu X (2011) Study on overburden to span ratio of shallow tunnel in upper soft and lower hard stratum. J Undergr Space Eng 7:700–705
18. Fang Y, Fu Y, Zhou C, Yang Z (2014) Model test of the excavation process of the double goaf under the highway tunnel. J Rock Mech Eng 33:2247–2257
19. Zhou H, Zhao Y, Shen Q, Yang L, Cai H (2020) Risk assessment and management via multi-source information fusion for undersea tunnel construction. Autom Constr 111
20. Li Z, Liu P (2016) Cause analysis and prevention technology of water leakage in loucun tunnel of Hefei-Fuzhou railway. China Harbour Eng 36(4):62–66

Risk Mitigation Planning

1 Introduction

With urbanization accelerating, metro systems have become vital components of public transport infrastructure, providing significant long-term environmental, social, and economic benefits. Urban metro systems typically rely on tunnels. However, the complexity and uncertainty of underground environments introduce various risks, complicating the tunnel excavation process [1]. Consequently, tunnel construction is inherently high-risk, potentially leading to severe safety issues. In China, the rapid expansion of urban rail transit has correlated with a rise in construction accidents, with 246 incidents recorded between 2002 and 2016 [2]. These accidents can result in cost overruns, project delays, and significant casualties. Despite growing attention to tunneling safety management, effective risk control remains elusive due to reliance on empirical methods that depend heavily on prior engineering experience and documentation.

Given insufficient and unreliable resources, it is challenging to assess complex tunneling environments fully. Basic analytical methods often fail to accurately represent the relationships between influential factors and potential risks, leading to significant decision-making errors [3]. To address this gap, there is an urgent need for a systematic decision support framework that intelligently analyzes tunnel-induced risks and informs appropriate control measures [4]. This approach aims to evaluate intrinsic risk factors and underground conditions accurately and automatically, offering objective guidelines for preventing risk-prone events during shield-based tunnel excavation.

To mitigate the negative impacts of underground tunnel construction, the tunnel boring machine (TBM) is widely used due to its high speed and minimal disturbance. However, shield tunneling in soft ground at shallow depths in densely populated areas raises significant safety concerns [5]. On the one hand, deformation of the surrounding ground will inevitably occur and then possibly cause considerable tunnel-induced settlement in response to shield penetration. Ground settlement is

L. Zhang et al., *Intelligent Construction in Tunnels*, Lecture Notes in Civil Engineering 679, https://doi.org/10.1007/978-981-96-8785-5_13

heavily influenced by soil properties and shield operational parameters, making it a critical factor in assessing potential damage from vertical and lateral surface movements [6]. Thus, analyzing ground settlement is particularly important in shallow tunnel projects and is a key step in tunneling risk control. Numerous studies have focused on evaluating and predicting ground settlement to enhance tunneling safety and develop strategies for minimizing settlement within acceptable limits [7].

On the other hand, since the planned route of a new tunnel is often in the area with the existing surface buildings and underground pipelines in a short distance, excessive settlement is bound to trigger some kinds of structural damage to these adjacent structures and facilities. To be more specific, too large structural deformation tends to incur cascading structural diseases, like distortion, displacement, crack, leakage, and others, which can even degrade the operational performance of the facility system, especially founded on shallow foundations [3]. Due to the obvious interaction between the tunneling activities and adjacent buildings, the effect of tunneling on nearby facilities has become another key focus [8]. Hence, it is necessary to deeply investigate the tunnel-soil-building interaction under consideration of various influencing factors, which can potentially pave an effective way to minimize the uncertainty and randomness in safety management of the tunnel construction and protect both the underground and surface facilities [9]. However, it is not a straightforward task to monitor, estimate, and control both the ground settlement and building tile rate carefully at an early stage, which remains a great challenge.

In short, to realize the great reduction of casualties and properties during shield-driven tunnel construction, an important problem to be solved in this research is how to reliably build a model to capture the complicated tunnel-soil-building interaction. Expectantly, the intelligent model is beneficial in forecasting the potential tunnel-induced risk and then informing corresponding countermeasures to mitigate the magnitude of targeted risk in a data-driven manner. This chapter presents a deep neural network-based model equipped with gradient descent optimization (DNN-GDO), which has shown promising performance in both multi-output prediction and multi-objective optimization. For one thing, the deep neural network (DNN) is superior in well handling the non-linear regression end to end to deliver high-quality prediction results. For another, the gradient descent optimization (GDO) technique is understandable and easy-to-use to increase the interpretability of the optimization process. Therefore, the contribution of each component to risk reduction can be numerically uncovered and key influential features can be easily identified. Actional suggestions that can fit the practical tunneling conditions can be provided for running safety improvement.

2 Risk Control in Tunnel Construction

It is well-known that proper safety management can play a critical role in making sure the project of metro tunnel construction proceeds smoothly under the complicated and volatile underground environment. Since subjective judgment and limited prior knowledge tend to mistakenly evaluate the risk of excessive deformation and the potential damage to adjacent structures, the traditional manual method is hardly a guarantee of project success [10]. To address this issue, several efforts have been made to develop some data analytics models for mitigating the inherited risk closely associated with metro tunnel excavation, which can systematically investigate the shield tunneling and inform protective and preventive measures. As reviewed, the typical data-driven methods in response to the risk analysis of tunnel engineering can be divided into two major categories, including statistical and machine learning methods.

Most existing studies have employed statistical models for risk identification and assessment in tunnel excavation. However, their application is limited, as the success of such models relies heavily on extensive theoretical and practical knowledge to establish relationships between variables. The challenge of obtaining accurate prior knowledge from real-world engineering projects can undermine model performance. With the rapid advancement of artificial intelligence, machine learning algorithms have garnered increasing attention for their ability to autonomously process large volumes of multi-dimensional data, uncover hidden patterns, and enhance prediction accuracy. For instance,

Zhang et al. [11] built the least-squares support vector machine (LSSVM) model integrating with wavelet packet transformation (WPT) to accurately predict the daily tunnel-induced ground settlement, which had proven to be a useful prerequisite for improving safety assurance in the tunneling progress. However, SVMs often suffer from low computational efficiency and limited interpretability. Zhou et al. [12] employed a random forest (RF) model to predict excavation risk levels and identify key predictive variables. While these models achieve satisfactory accuracy, they require labor-intensive manual feature extraction, with improperly selected features potentially degrading prediction quality.

Differently, DNN offers a significant advantage by performing feature engineering with minimal human intervention [13]. Unlike traditional methods, DNN integrates feature extraction and risk prediction into a single learning block, automatically identifying and optimizing features from raw data [14]. What's more, DNN exhibits a potential capacity of fully approximating the complicated non-linear relationship in the tunnel excavation system especially when the collected data is growing into a large volume. According to a comparative study for tunneling-induced settlement prediction by Zhang et al. [15], neural networks exhibited superior extrapolation capabilities compared to SVM and RF models. While most studies focus on single-output predictions, DNNs are inherently suited for multi-output regression, allowing multiple risk factors to be forecasted simultaneously. Despite these advantages, DNNs have seen limited application in predicting risks in shield tunneling projects. Therefore, we propose constructing a DNN meta-model capable of intelligently learning input variables related to tunneling safety and accurately predicting multiple tunnel-induced risks simultaneously.

Moreover, much research on tunnel excavation primarily focuses on prediction, leaving the challenge of using these predictive results for risk reduction unsolved. While identifying risk factors and issuing early warnings is common, effective risk control still relies heavily on expert interpretation of meta-model predictions, which can be unreliable under varying geological conditions. Setting optimal input variables can mitigate risks, but most studies fail to generate automatic, objective solutions. This issue aligns with a multi-objective optimization (MOO) problem [16]. Studies about the use of MOO-related algorithms in tunnel projects are still rare, most of which mainly put efforts into determining a set of optimal trade-off solutions in the tunnel planning stage. For example, Guo et al. [17] performed the non-dominated sorting genetic algorithm (NSGA-II) for tunnel line alignment under a great balance among investment amount, headway length, and comfort degree. Feng et al. [18] implemented the particle swarm optimization algorithm (PSO) to control the two objectives of limit support pressure and ground surface deformation for the determination of newly-built tunnel location under uncertainty. Another popular usage of the MOO problem in tunnel construction is to help tune the parameters of TBM to improve tunneling operation performance. For example, Wang et al. [19] conducted NSGA-II to search for the most satisfying value of TBM control variables, such as the advanced speed and rotation speed, to reach the desired target of high efficiency, safety, and stability, which can even increase the automation of TBM. Nikakhtar et al. [20] applied PSO to identify optimal operational parameters in the mechanized tunneling rather than the determined value based on engineering experience, which had the potential to decrease the surface settlement.

While some studies explore multi-objective optimization (MOO) for tunnel design and TBM operations, applying MOO to tunneling-induced risk mitigation remains in its early stages. Our work aims to optimize decision variables during tunneling for enhanced safety control, addressing the complex interaction between tunnels and their surroundings. Herein, we propose a novel approach, DNN-GDO, which combines the deep neural network (DNN) with gradient descent-based optimization (GDO). This method leverages the computational efficiency and stable convergence of gradient descent to find optimal solutions within the decision space. DNN-GDO offers flexibility in setting objective function weights based on real-world importance, leading to more practical solutions. Additionally, GDO enhances explainability by calculating the contribution of each variable to risk mitigation, a feature often lacking in current MOO studies. The DNN-GDO algorithm minimizes multiple risks simultaneously under complex tunneling conditions, while also identifying critical variables for adjustment. The optimization results from the novel decision-making tool based on the proposed DNN-GDO can therefore serve as references for managers to work out precautionary safety measures for risk management in tunnel excavation.

3 Risk Mitigation Planning Framework

The proposed decision support framework is illustrated in Fig. 1. Such an end-to-end learning pipeline relies on a hybrid algorithm named DNN-GDO, which takes full advantage of deep learning and the first-order iterative optimization method. There are three major steps incorporated, including data preparation, DNN-based

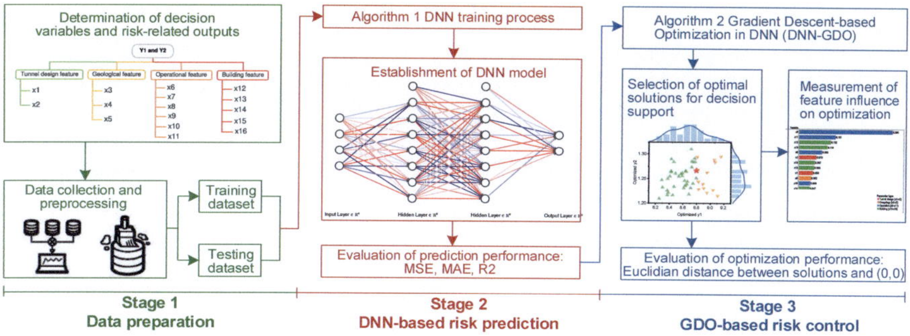

Fig. 1 Proposed DNN-GDO-based framework for decision support in controlling risk in tunnel excavation

risk prediction, and GDO-based risk control. This systematic approach serves two key functions: it enables early detection of potential risks in complex tunneling environments with high accuracy, and it employs gradient descent optimization to provide solutions for managing tunnel-induced risks, helping to prevent accidents.

3.1 DNN Architecture and Training Process

DNN, also known as multilayer perceptron, excels at approximating continuous multivariate functions in high-dimensional spaces, making them ideal for modeling complex patterns and non-linear problems. We adopt DNN as the meta-model for accurate predictions. A typical DNN architecture consists of a feedforward neural network with fully connected layers. For example, Fig. 2a illustrates a DNN with an input layer $\in \mathcal{R}^4$, three hidden layers $\in \mathcal{R}^{12}$, and one output layer $\in \mathcal{R}^2$. As shown in Fig. 2b, each hidden layer owns the rectified linear activation function (ReLU) as a nonlinear approximation, defined in Eq. (1). It returns a linear output for positive values and zero for negatives. ReLU's monotonic properties help address vanishing and exploding gradients, offering low computational complexity and fast convergence. In addition, residual connections (or skip connections) are introduced between hidden layers and earlier layers, allowing the network to learn residuals. These residual blocks enable deeper networks, improving the modeling of complex relationships. They also allow faster feedforward propagation, making residual mappings easier to learn than original mappings. No non-linear activation function is applied in the output layer, ensuring effective risk prediction.

$$\sigma(y) = \max(0, y) \tag{1}$$

$$\sigma\prime(z) = \begin{cases} 0 \; if \; y \leq 0 \\ 1 \; if \; y > 0 \end{cases} \tag{2}$$

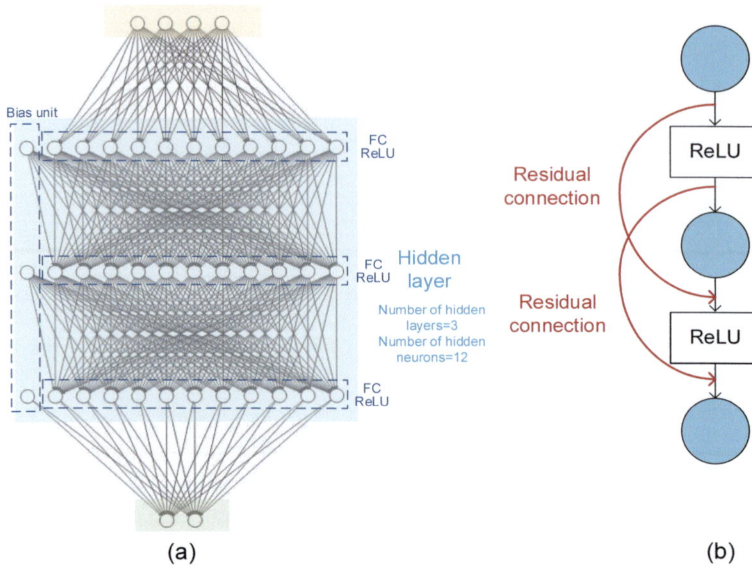

Fig. 2 Neural network structure: **a** an example of DNN architecture; **b** an example of connections between neurons in neighboring hidden layers (*Note* FC is the abbreviation of fully connected layer. ReLU activation function is utilized in the input layer and hidden layer, which is computationally efficient)

where y is the input.

The training process for a L-hidden layer DNN model begins by randomly selecting the model parameters $\Theta = \{w^{(l)}, b^{(l)}\}, l \in \{1,2, \ldots, L\}$. When the input vector X passes to the hidden layer, it is mapped by the weight and bias. As a result, the output of each hidden layer can be calculated according to Eq. (3). The goal of DNN is to minimize the objective function in the form of mean-square error (MSE) given by Eq. (4). Backpropagation is used to fine-tune the DNN's parameters, where gradient descent backpropagates the error from the current layer to previous layers, updating the weights to reduce the prediction error. The negative gradient indicates the direction for weight adjustment, with the update calculated as the product of the learning rate and the partial derivative of the error. The updated weight is derived using Eq. (5) to optimize the loss. Algorithm 1 provides a detailed breakdown of the DNN training process.

$$\widehat{Y_i^l} = \sigma\left(\sum_{j=1}^{n} w_{ij}^{(l)} Y_j^{(l-1)} + b_{ij}^{(l)}\right) \tag{3}$$

$$\Theta \leftarrow \arg\min_{\Theta} \frac{1}{N} \sum_{i=1}^{N} \left(\widehat{Y_i^l} - Y_i^l\right)^2 \tag{4}$$

$$w = w + \eta \times \frac{\partial E}{\partial w} \tag{5}$$

where $\widehat{Y_i^l}$ stands for the prediction of the ith neuron at the lth layer. Y_i^l is the actual value of the ith neuron at the lth layer. $w_{ij}^{(l)}$ and $b_{ij}^{(l)}$ are the weight and bias of the ith neuron at the lth layer, respectively. w is the previous weight, η is the learning rate, and E is the error between the prediction and actual value.

Algorithm 1 Pseudocode for the DNN training process

```
Input: number of samples n, feature dimension d, training data
  X ∈ R^{n×d}, training labels Y^1 ∈ R^{n×1}, Y^2 ∈ R^{n×1}, maximum number
  of epoch E, learning rate η
Output: model parameters w
randomly initialize the model parameters w
initialize number of epoch e = 0
while e < E do
    shuffle the training data into K mini-batches;
    for mini-batch (X_i, Y_i^1, Y_i^2) in K mini-batches do
        Ŷ_i^1, Ŷ_i^2 = Infer Deep Learning Model with parameter w(X_i)
        objective function O = MSE(Ŷ_i^1, Ŷ_i^2, Y_i^1, Y_i^2);
        ∇_w = backpropagation(O);
        w = w + η∇_w;
    end
    e += 1
end
```

3.2 Gradient Descent-Based Optimization (DNN-GDO)

To tackle the MOO problem, we directly equip the DNN model with a novel gradient descent-based optimization (GDO) algorithm to create an end-to-end learning pipeline named DNN-GDO. Backpropagation based on gradient descent is used to optimize multiple objective functions formulated by DNN. In general, the goal of the MOO problem is to minimize the vector of k objective functions simultaneously. The DNN-GDO algorithm aims to find diverse optimal solutions that closely approach the Pareto-optimal front. This optimization relies on a gradient descent process, which numerically estimates the optimal direction for minimizing the objective functions.

$$\begin{aligned} &\min O(x) = [o_1(x), o_2(x), \ldots, o_k(x)]^T \\ &s.t. x \in \chi, x = (x_1, x_2, \ldots, x_n)^T \end{aligned} \tag{6}$$

where $k \geq 2$ is the number of objective functions, x is the vector of decision variables that needs to satisfy some constraints, $\chi \in \mathcal{R}^n$ is the feasible decision space, and n is the number of variables.

The proposed DNN-GDO algorithm addresses a bi-objective optimization problem, detailed in Algorithm 2. To begin with, the usage of the weighted Tchebycheff scalarization given by Eq. (7) [21] can combine the multiple objectives into one aggregated objective function for further GDO implementation. Particularly, the weight can be easily set in proportion to the relative importance of every objective function $O_k(x)$. When the values of weights are varied, it is possible to obtain all Pareto points. For simplicity, both objectives are assigned an equal weight (1.0), with the ideal solution at (0, 0), and p set to 2. The decision variable fluctuates within 5% of its value, so the solution lies in $0.95x \leq x \leq 1.05x$. Later, an initial point is randomly selected from the defined range, where the gradient descent starts to seek the minima of a function. Following the negative of the gradient, the value can go in the direction of steepest descent to rapidly move towards a local minimum.

$$\min\left(\sum_{k=1}^{K} w_k^p \left|O_k(x) - O_k^{ideal}\right|^p\right)^{\frac{1}{p}} \tag{7}$$

where $w_k \in (0,2]$ is the weight of kth objective function $O_k(x)$, $\sum w_k = 2$, O_k^{ideal} is the ideal solution for the kth objective function $O_k(x)$.

During iterations, the value of the variable will be updated according to Eq. (8) until the termination criterion is met. In short, the updated x can be calculated by subtracting the first-order derivative of the objective function multiplied by the learning rate from itself. With the help of the partial derivative, the variation of the objective function can be precisely measured as the variable changes. One thing to be noted is that gradient descent is unable to guarantee a global minimum. To avoid the situation where the optimal result is stuck in the local minima, it is necessary to repeat the optimization process at a certain point several times.

$$\widehat{x_i} = \widehat{x_i} + \eta \nabla O(x_i) = x_i + \eta \frac{\partial O(x)}{x_i} \tag{8}$$

where η is the learning rate to define the step size of gradient descent to control the convergence speed of the algorithm.

Remarkably, we also propose a gradient descent-based method to quantitatively explain the impact of a feature on the optimization problem. This can be considered as a kind of sensitivity analysis, which has been well incorporated into the DNN-GDO algorithm outlined in Algorithm 2. This method improves the interpretability of the black-box DNN model by calculating an influence degree (or sensitivity value) β for each decision variable, defined as the average gradient descent across all points (Eq. 9). A larger influence degree indicates a sharper gradient decline, suggesting that the variable significantly decreases the output. That is to say, the defined index offers an appropriate insight into the contribution of the variable to optimization, which can further make sense to identify key variables that have a major effect on optimization. Regarding the important application of feature influence measurement, it allows for informing evidence-based decisions to better control the optimized output in a data-driven manner instead of expert expertise. Therefore, we can reasonably give priority to adjust features with a relatively larger influence degree due to their higher contributions in significantly optimizing the problem.

$$\beta(x_i) = \frac{1}{N} \nabla O(x_i) \tag{9}$$

where N is the number of data points.

Algorithm 2 Pseudocode for the optimization process in DNN-GDO

Input: model parameters w, initial samples $X \in R^{n \times d}$, maximum gradient descent steps G, maximum optimization steps L, gradient descent learning rate η
Output: optimized input $X^* \in R^{n \times d}$, feature influence degree β
initialize feature influence degree list = empty list
for *sample* $X_i \in X$ **do**
  Y_i^1, Y_i^2 = Infer Deep Learning Model with parameter $w(X^i)$
  objective function $O = abs(Y_i^1) + abs(Y_i^2)$
  $\nabla_X = backpropagation(O)$
  feature influence degree list append $abs(\nabla_X) \in R^d$
  initialize optimization step l = 0, best objective O^* = inf, best features = None
  for *optimization steps* $l < L$ **do**
    sample r = Uniform(0.95, 1.05) $\in R^d$
    $\hat{X}_i = r * X_i$
    initialize gradient descent step g = 0
    while $abs((\hat{X}_i - X_i)/X_i) < 0.05$ *and* $g < G$ **do**
      $\hat{Y}_i^1, \hat{Y}_i^2$ = Infer Deep Learning Model with parameter $w(\hat{X}^i)$
      objective function $O = abs(\hat{Y}_i^1) + abs(\hat{Y}_i^2)$
      if $O < O^*$ **then**
        $O^* = O, X_i^* = \hat{X}$
      end
      $\nabla_{\hat{X}} = backpropagation(O)$
      $\hat{X} = \hat{X} + \eta \nabla_{\hat{X}}$
      $g = g + 1$
    end
    $l = l + 1$
  end
  feature influence degree β =average (feature influence degree list)
end

3.3 Risk Modeling and Mitigation Performance

The performance of the proposed DNN-GDO algorithm is evaluated in two aspects. First, the accuracy of DNN predictions is assessed using mean-square error (MSE) and mean absolute error (MAE), defined in Eqs. (10) and (11). These metrics measure the absolute and squared differences between predicted and true values, with values closer to 0 indicating a more accurate model. Additionally, the coefficient of determination R^2 expressed by Eq. (12) is used to evaluate how well the regression model explains the variance in the dependent variable. An R^2 value close to 1 indicates a strong fit between the predicted and observed results.

$$MSE = \frac{1}{N}\sum_{i=1}^{N}\left(\widehat{Y_i} - Y_i\right)^2 \tag{10}$$

$$MAE = \frac{1}{N}\sum_{i=1}^{N}\left|\widehat{Y_i} - Y_i\right| \tag{11}$$

$$R^2 = 1 - \frac{SS_{RES}}{SS_{TOT}} = 1 - \frac{\sum_i\left(Y_i - \widehat{Y_i}\right)^2}{\sum_i\left(Y_i - \overline{Y_i}\right)^2} \tag{12}$$

where SS_{RES} is the residual sum of squared error, SS_{TOT} is the total sum of squared error, $\widehat{Y_i}$ is the predicted value, Y_i is the actual value, and $\overline{Y_i}$ is the mean value of Y_i.

On the other hand, how to help the decision-maker to select the most optimal solution among all possible trade-off candidates remains another question. According to Eq. (13), we calculate the Euclidean distance of the optimized value from the reference point that is the original point (0, 0) in this research. The shorter the distance is, the more ideal the solution is to minimize the two objective functions.

$$d = \sqrt{\sum_{k=1}^{K}\left|O_k(x) - O_k^{ideal}\right|^2} \tag{13}$$

where $O_k(x)$ is the optimized value for the kth objective function, and O_k^{ideal} is the ideal solution for the kth objective function $O_k(x)$.

4 Experimental Studies

To validate this approach, we implement the DNN-GDO framework in a practical engineering project for tunnel construction in the Wuhan metro system, China, leveraging hundreds of measurement records collected from a web-based early-warning system during excavation.

4.1 Data and Resources

The proposed decision support approach begins with carefully identifying the potential risk in the tunneling procedure along with a series of relevant variables. Expectedly, a high-quality dataset can be well prepared for deep learning model development. During shield tunnel excavation in soft ground, two significant risks must be monitored: accumulative ground settlement (Y1) and building tilt rate (Y2). Y1 tends to increase as the shield machine advances, posing threats to the stability of the TBM's excavation face and the integrity of the tunnel and nearby structures. Similarly, adjacent surface facilities, particularly aging buildings on shallow foundations, may experience deformation and damage due to the tunnel–soil–building interaction. Building tilt, in particular, is a common safety concern during nearby tunneling operations, making Y2 a critical indicator of potential deterioration. In summary, we emphasize the negative impacts of tunneling on both accumulative ground settlement (Y1) and building tilt rate (Y2), as controlling these factors within acceptable limits (according to standards such as GB 50007-2011 and GB 50715-2011) is essential for ensuring tunneling safety [22].

It is common knowledge that the target risk can be greatly affected by a lot of influencing factors. Based upon the expert judgment and technical documents, we build an evaluation system involving 16 decision variables and 2 outputs toward an engineering application of modeling the tunnel-induced risk, whose statistical characteristics have been outlined in Table 1. To facilitate a clear understanding, these 16 inputs of interest (x1–x16) that are highly relevant to outputs of tunnel-induced risk (Y1 and Y2) can be divided into the following four categories.

(1) Tunnel design features: Improper design of the tunnel structure is prone to cause too large disturbances in underground environments under a high degree of complexity and uncertainty. Tunnel cover depth (x1) about the depth of the tunnel and cover-span ratio (x2) concerning both the tunnel depth and span are two critical variables in estimating the ground movement, which then are possible to pose adverse influences on foundation subsidence and deformation of adjacent buildings.
(2) Geological features: The actual geological condition plays a strong part in bridging intermediate connections between the tunnel and its surrounding environment. Aiming to describe the soil properties in this tunneling environment, we refer to three key variables. More specifically, soil friction angle (x3) acts as a shear strength parameter of soil for evaluating soil stability. The soil compression module (x4) measures the soil resistance to the deformation caused by compression. Soil cohesion (x5) stands for the component of shear strength of soil for revealing the soil consistency.
(3) Operational features: Operational features are crucial for enhancing TBM adaptability. Adjusting operational parameters effectively reduces disturbances in the surrounding rock mass and limits ground surface settlement, ensuring safety during tunnel excavation. Based on the installation of sensors in the TBM cutter head, data about six popular variables can be easily obtained for a deeper investigation, including the excavation speed (x6), cutter torque (x7), trust force (x8), cutter rotating speed (x9), soil warehouse pressure (x10), and grouting volume (x11).
(4) Building features: The shield tunneling will impose a stronger influence on buildings closer to it. Three kinds of distance are measured to reveal the spatial relationship between the tunnel and adjacent buildings, which are the relative

Table 1 Statistical description of determined inputs and outputs

Category	Variable	Name	Unit	Mean	Std	[Min, Max]
Tunnel design feature	x1	Tunnel cover depth	m	16.783	7.191	[5.025, 29.921]
	x2	Cover-span ratio	%	2.630	1.360	[0.222, 4.992]
Geological feature	x3	Soil friction angle	°	16.241	3.378	[10.047, 21.999]
	x4	Soil compression modules	MPa	24.817	11.421	[5.114, 44.965]
	x5	Soil cohesion	MPa	20.086	5.824	[10.006, 29.978]
Operational feature	x6	Excavation speed	mm/min	44.913	14.700	[20.050, 69.939]
	x7	Cutter torque	bar	2541.683	804.058	[1204.032, 3892.574]
	x8	Trust force	kN	10,895.800	3282.631	[5026.006, 15,992.480]
	x9	Cutter rotating speed	r/min	1.998	0.857	[0.506, 3.490]
	x10	Soil warehouse pressure	bar	1.976	0.823	[0.505, 3.480]
	x11	Grouting volume	m^3	5.742	2.468	[1.505, 9.967]
Building feature	x12	Relative horizontal distance	%	2.470	1.487	[0.011, 4.992]
	x13	Relative vertical distance	%	1.226	1.302	[− 0.998, 3.486]
	x14	Relative longitudinal distance	%	73.240	44.034	[0.020, 149.898]
	x15	Foundation integrity	score	79.924	8.761	[65.010, 94.968]
	x16	Building intact status	score	78.824	8.395	[65.071, 94.970]
Objective 1	Y1	Accumulative ground settlement	mm	8.715	2.247	[1.604, 18.021]
Objective 2	Y2	Building tilt rate	%	1.684	0.479	[0.435, 2.757]

horizontal distance (x12), relative vertical distance (x13), and relative longitudinal distance (x14). Besides, the structural health of the building itself is a key factor to consider, which can be represented by two additional parameters, namely the foundation integrity (x15) and building intact status (x16).

Another step of data preparation is data cleaning. Removal of missing values and correction of inaccurate data help to convert the raw data into high-quality information, enabling more sound decision-making in mitigating tunnel-induced damage. After that, a reliable dataset in a dimension of 500 × 16 can be prepared, which owns 500 lines of valid records and 16 important features in the numerical form. For further improving the prediction performance and training speed of DNN, it is necessary to perform feature normalization to change all features in different units to the same scale. Herein, we refer to the min–max scaling technique to normalize the range of features of data before model fitting. According to the equation $x_{scaled} = (x - \min(x))/(\max(x) - \min(x))$, all features can be eventually transformed and rescaled into a fixed range between 0 and 1.

4.2 *Intelligent Risk Prediction*

We build a robust DNN-based model for supporting the multi-output regression task following Algorithm 1, which can intelligently make accurate predictions about two kinds of tunnel-induced risk represented by Y1 and Y2. The DNN excels at automatically extracting important features from input data without manual intervention, and the inclusion of residual connections facilitates deeper network architectures. MSE is chosen as the loss function to quantify the error between predicted and actual values, with the objective of minimizing this loss during training. To optimize model performance, a grid search strategy is employed to identify the best values for three critical parameters. The learning rate is tested at {0.01, 0.001, 0.0001}, while the number of hidden neurons and hidden layers are assessed at {64, 128, 256} and {3, 4, 5}, respectively. The optimal configuration discovered is {learning rate, hidden neurons, hidden layers} = {0.001, 128, 4}. Figure 3 provides some examples in the grid-search progress to display the effects of the three targeted parameters during the model testing, and their corresponding prediction in the two risk-related outputs have been evaluated in Table 2. Especially, DNN with 5 hidden layers or 256 hidden neurons tends to demonstrate worse performance than the best model. This suggests that blindly increasing model complexity does not necessarily improve prediction accuracy and may instead increase computational demands.

Based upon grid search-based parameter tuning, it is appropriate to build a DNN model owning one input layer with 16 features, four hidden layers with 128 neurons and ReLU activation function in each hidden layer, and one output layer with two nodes, and train it at a learning rate of 0.001. The prepared dataset is randomly split into a training set and a testing set in an 80–20% ratio, which is then fed into DNN to fit the model and predict two outputs. To further improve the training efficiency, an

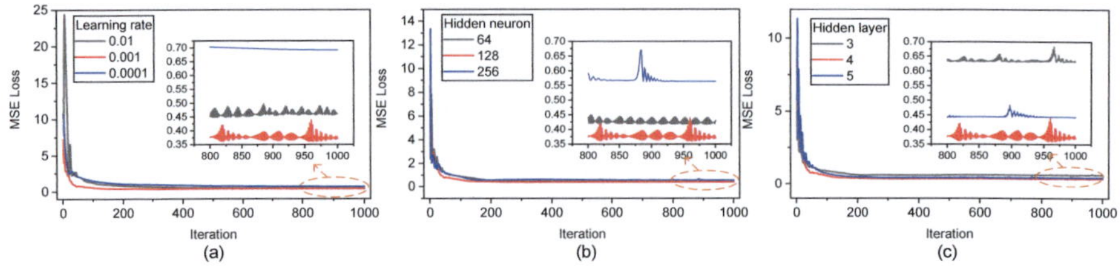

Fig. 3 Example of parameter combinations from grid search: **a** Different learning rate under 128 hidden neurons and 4 hidden layers; **b** Different hidden neurons under the learning rate 0.001 and 4 hidden layers; **c** Different hidden layers under the learning rate 0.001 and 128 hidden neurons

Table 2 Evaluation of DNN prediction performance under different parameter settings

Learning rate	Hidden neuron	Hidden layer	MSE			MAE			R^2		
			Y1	Y2	Mean	Y1	Y2	Mean	Y1	Y2	Mean
0.01	128	4	0.924	0.002	0.463	0.689	0.036	0.363	0.792	0.988	0.890
0.001	**128**	**4**	**0.737**	**0.002**	**0.370**	**0.626**	**0.033**	**0.330**	**0.821**	**0.99**	**0.906**
0.0001	128	4	1.379	0.004	0.692	0.884	0.047	0.466	0.689	0.981	0.835
0.001	64	4	0.858	0.002	0.430	0.744	0.037	0.3905	0.807	0.989	0.898
0.001	**128**	**4**	**0.737**	**0.002**	**0.370**	**0.626**	**0.033**	**0.3295**	**0.821**	**0.990**	**0.906**
0.001	256	4	1.124	0.002	0.563	0.772	0.034	0.403	0.747	0.990	0.869
0.001	128	3	1.260	0.002	0.631	0.869	0.038	0.4535	0.716	0.983	0.850
0.001	**128**	**4**	**0.737**	**0.002**	**0.370**	**0.626**	**0.033**	**0.3295**	**0.821**	**0.990**	**0.906**
0.001	128	5	0.884	0.002	0.443	0.658	0.033	0.3455	0.801	0.990	0.896

The bold values represent the best prediction performance under different evaluation metrics

optimization technique for stochastic gradient descent called the Adam is deployed to automatically adjust DNN weights in response to errors. Since the Adam optimizer can take advantage of both the Adaptive Gradient Algorithm (AdaGrad) and Root Mean Square Propagation (RMSProp), it is outperforming in driving DNN to converge faster and better handling sparse gradients. Figure 4 shows the training loss and testing loss curve to demonstrate the validity of our established DNN model, where the loss quickly decreases at the first 200 iterations of optimization. After that, the loss can remain stable under a relatively small oscillation, leading to a plateau status in the end. It is observed that DNN performance in the training set and testing set is very close, whose MSE loss converges around 0.201 and 0.370, respectively.

As a result, this defined DNN model can fully learn 16 input variables and return two numerical outputs Y1 and Y2 to meet the accuracy requirements. Its prediction performance under different parameter configurations has been evaluated and compared in Fig. 3 and Table 2. It is evident in Fig. 5a and c that the predictive results on the testing dataset can almost match the observed value. For all the 100 testing points, their relative error, a ratio of the absolute value to the ground truth, is in the range of [− 0.378, 0.373] on Y1 and [− 0.050, 0.140] on Y2, quantitatively indicating that the magnitude of the discrepancy between the exact value and

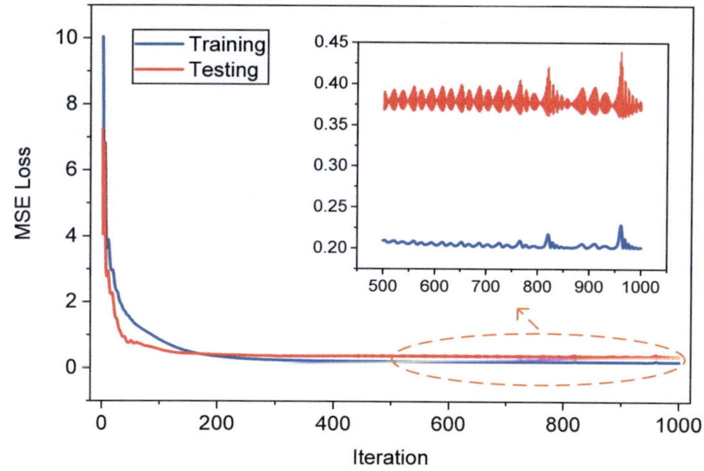

Fig. 4 The training and testing loss of our established DNN model over iterations

the DNN-based approximation is quite small. Regarding the standardized residual in Fig. 5b and d, only four predictions on Y1 and five predictions on Y2 go out of the acceptable range [− 2, − 2]. That is to say, over 95% of testing data can obtain reliable forecasts closer to the actual data. Overall, the lowest MSE, MAE, and R^2 of Y1 and Y2 in Table 2 also validate that the DNN model can well fit the observed data and then precisely estimate both the tunnel-induced ground settlement and building title rate under the underground environments with a high degree of complexity and uncertainty.

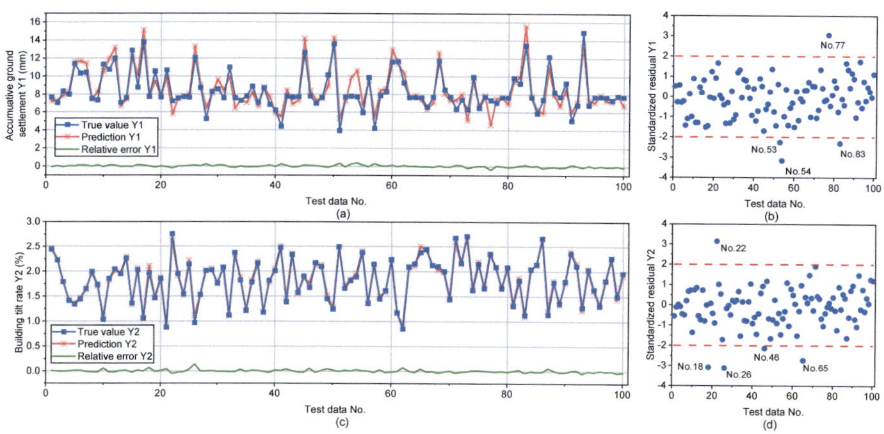

Fig. 5 Visualization of DNN prediction performance on 100 output points in the testing dataset: **a** true value and prediction of Y1 along with the relative error; **b** standard residual of Y1 prediction; **c** true value and prediction of Y2 along with the relative error; **d** standard residual of Y2 prediction

4.3 Optimal Solution Analysis

With the constructed DNN architecture achieving satisfactory prediction accuracy, we apply the GDO algorithm to the DNN-based multi-output prediction model to tackle the MOO problem. This method, inspired by gradient descent, searches a large solution space to minimize the two objectives, Y1 and Y2, derived from the DNN model. Consequently, optimal or near-optimal candidates along the Pareto frontier are generated, aiding project managers in implementing protective measures for tunnel excavation and adjacent buildings. Our experiment restricts decision variables to a $\pm$ 5% adjustment range of their original values. Following the DNN-GDO algorithm outlined in Algorithm 2, we analyze the optimized results for effective tunneling risk control as follows:

(1) The optimization on output data points #121 and #126 are taken as an example to demonstrate the effectiveness of the GDO algorithm in searching optimal solutions for mitigating the tunnel-induced risks simultaneously. The gradient descent procedure is repeated 50 times for each target point to enhance the search capability and avoid local minima, yielding 50 potential solutions per target, as shown in Fig. 6. Solutions are deemed acceptable if their Euclidean distance to the ideal point (0, 0) is less than the original distance from the initial point. For data point #121, with coordinates (Y1, Y2) = (12.612, 1.781), the GDO algorithm produced 34 optimal solutions, reducing the distance to (0, 0). Specifically, 80% of optimized Y1 and 76% of optimized Y2 values are lower than their original values, leading to an average distance improvement of 0.139%. Depending on the improvement shown in Fig. 6b, we can easily select the top three optimal solutions for minimizing two objectives, which are (12.373, 1.762), (12.374, 1.762), and (12.461, 1.756) derived from the 3rd, 27th, and 29th repetition of GDO, respectively. Similarly, data point #126 showed an average distance reduction of 1.492%, with 74% of results closer to (0, 0) than the initial point. Table 3 indicates that the 16 decision variables can be reasonably adjusted to rapidly achieve the top three optimal solutions for both points. This highlights how the DNN-based GDO algorithm offers an intuitive decision-making framework for risk control in tunneling environments, allowing for data-driven strategies that reduce reliance on expert evaluations and prior knowledge.

(2) We adopt the top 3 most optimal solutions under the three largest distance improvements as a meaningful reference in effectively guiding the risk control during the tunnel construction. The GDO-based optimal solutions that are ranked at the top three for all 500 output points have been provided in Fig. 7. As shown in the linear regression along with the 95% confidence interval in Fig. 7b, there is a negative linear relationship between the original distance and the improvement of distance after optimization. The absolute value of the slope shows a decreasing tendency from the top 1 optimal solution (0.0049) to the top 3 (0.0039), indicating that the top 1 solutions can drive a larger reduction in lowering the targeted tunnel-induced risks. This conclusion can also be drawn

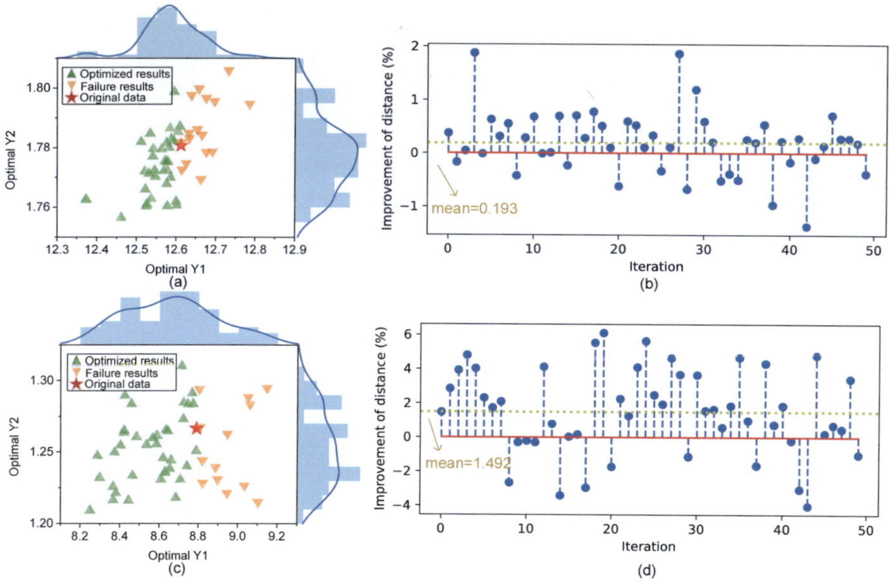

Fig. 6 Example of generated results when GDO algorithm is repeated 50 times: **a** Joint plot of two objectives with bivariate and univariate graphs for the data point #121; **b** Stem plot of distance improvement for all the 50 results after optimization for the data point #121; **c** pair plot of two objectives with bivariate and univariate graphs for the data point #126; **d** stem plot of distance improvement for all the 50 results after optimization for the data point #126

from the distribution of distance improvement in Fig. 7c, where the peak of the top 1–3 damage mitigation degree gradually moves towards 0. Besides, the slope with a negative sign signifies that initial points with a smaller distance to the ideal point (0, 0) tend to experience a larger improvement in risk alleviation. The top 3 optimal solutions rather than the best one can increase the diversity of feasible alternatives in decision support, contributing to further boosting the application potential in tunneling safety management under complicated tunnel-soil-building interaction.

(3) Apart from the equal weight for two objectives, their individual weights in the definition of the objective function can be flexibly adjusted, and thus the most optimal solutions for a target point can change accordingly. Additional experiments have been conducted under the eight alternative weight combinations (w1, w2) as (0.2, 1.8), (0.4, 1.6), (0.6, 1.4), (0.8, 1.2), (1.0, 1.0), (1.2, 0.8), (1.4, 0.6), (1.6, 0.4), and (1.8, 0.2), respectively. Figure 8 takes two output points #6 and #72 as an example to show the most optimal solutions from different weight settings. Through calculating the distance improvement, it can be found that the optimization can result in a reduction of two risks in the range of [0.996%, 3.944%] for #6 and [3.974%, 25.458%] for #72. It is observed that these eight possible solutions exhibit an obvious trade-off in decreasing the accumulative ground settlement and building tilt rate. In other words, the representative set

Table 3 Profile of decision variables x1–x16 to achieve the top 3 most optimal solutions

Input	Data point #121				Data point #126			
	Original value	Top 1	Top 2	Top 3	Original value	Top 1	Top 2	Top 3
x1	15.962	15.469	15.944	15.560	20.520	19.804	20.238	20.114
x2	2.981	3.020	3.118	2.853	1.044	1.007	1.077	1.007
x3	16.442	16.478	16.472	16.371	12.500	12.607	12.579	12.502
x4	11.784	11.522	11.531	11.688	30.939	30.667	30.756	31.669
x5	21.643	21.936	22.041	21.328	29.296	29.176	28.611	28.365
x6	33.872	33.968	34.166	34.321	64.146	65.560	65.367	64.145
x7	1434.530	1437.590	1429.689	1438.124	1700.475	1713.130	1691.228	1717.069
x8	12,099.780	12,344.050	12,241.970	12,275.590	11,162.380	11,110.910	11,335.940	11,442.670
x9	2.270	2.182	2.194	2.293	1.930	1.920	1.859	1.883
x10	1.155	1.154	1.130	1.125	2.430	2.500	2.400	2.413
x11	3.809	3.902	3.921	3.905	6.689	6.942	6.797	6.828
x12	0.511	0.498	0.524	0.498	2.234	2.203	2.307	2.243
x13	− 0.704	− 0.697	− 0.708	− 0.692	0.222	0.272	0.200	0.259
x14	143.126	140.336	149.162	136.582	106.737	105.544	102.032	102.324
x15	76.728	76.860	76.246	76.211	78.130	77.941	77.668	78.605
x16	78.761	79.321	78.962	78.129	81.950	82.687	81.697	82.686
Y1	12.612	12.373	12.374	12.461	8.790	8.251	8.287	8.299
Y2	1.781	1.762	1.762	1.756	1.266	1.209	1.247	1.232

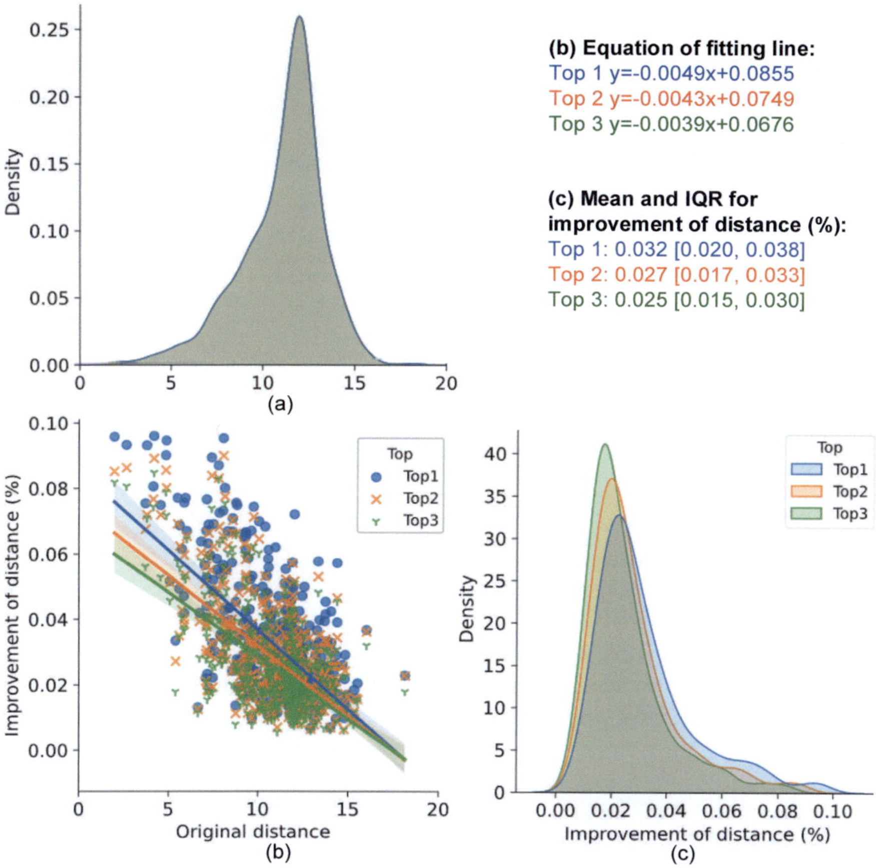

Fig. 7 Pair plot of original distance and improvement of distance (%) from the top 3 most optimal solutions for all 500 output points: **a** Kernal density estimation of the original distance; **b** Scatter plot for the relationship between the original distance and the improvement of distance (%); **c** Kernal density estimation of the improvement of distance (%) from the top 3 most optimal solutions

of optimal choices that are determined by the GDO algorithm under different combined weights can be reasonably regarded as a good approximation towards the Pareto-optimal front following the risk mitigation goal.

(4) Adjusting only six operational variables instead of all 16 is sufficient to effectively reduce risks associated with ground settlement and building tilt. Operational features of the TBM can be modified more easily and flexibly during tunneling, making it practical to focus solely on these parameters in the optimization phase. Figure 9 illustrates the optimization improvements of the top three solutions for all 500 output points, comparing two scenarios: one using all parameters (x1–x16) and another using only operational parameters (x6–x11). By adjusting operational parameters, all top three solutions achieved similar risk

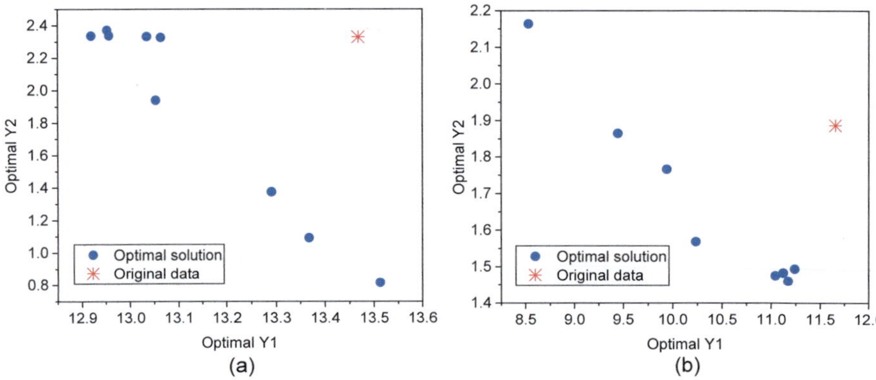

Fig. 8 The most optimal solutions under alternatives of weight setting including (0.2, 1.8), (0.4, 1.6), (0.6, 1.4), (0.8, 1.2), (1.0, 1.0), (1.2, 0.8), (1.4, 0.6), (1.6, 0.4), and (1.8, 0.2): **a** example of output point #6; **b** example of output point #72

mitigation, with an average distance improvement of around 0.021%. The corresponding interquartile range (IQR) of top 1–3 optimal solution improvement is also similar, which is [0.014%, 0.025%], [0.014%, 0.026%], and [0.013%, 0.025%], respectively. Comparatively, the discrepancy between the shortened distance in the two scenarios is gradually narrowed from top 1 to top 3. Especially for the optimal solutions ranking at the third place, their mean and median of distance improvement based on adjustment of operational parameters is only 0.004% and 0.002% lower than that from all parameters, which further confirms the availability of tuning operational parameters. As shown in Fig. 9b and c, the mean of Y1 and Y2 can be averagely decreased by 2.586%–3.275% and 2.106%–2.146%, respectively, under all variable adjustments, while the six operational parameters can also bring a reduction in the range of [2.106%, 2.146%] for Y1 and [0.310%, 0.380%] for Y2. This indicates that intelligent control of these six operational parameters provides a straightforward yet effective method for ensuring satisfactory TBM performance, positively impacting the mitigation of damage from shield-driven tunnel excavation.

5 Merits Analysis

When the proposed GDO algorithm has solved the optimization task on the DNN model for tunneling risk control, another issue worthy of discussion is how to estimate feature influence on optimization. For this purpose, we define a metric still relying on the core idea of gradient descent to calculate the influence degree. Accordingly, the interpretability of the developed DNN-GDO algorithm can be significantly improved. According to Eq. (9), an input variable experiencing a larger gradient descent during the optimization procedure tends to cause a larger decrease in two objectives, which

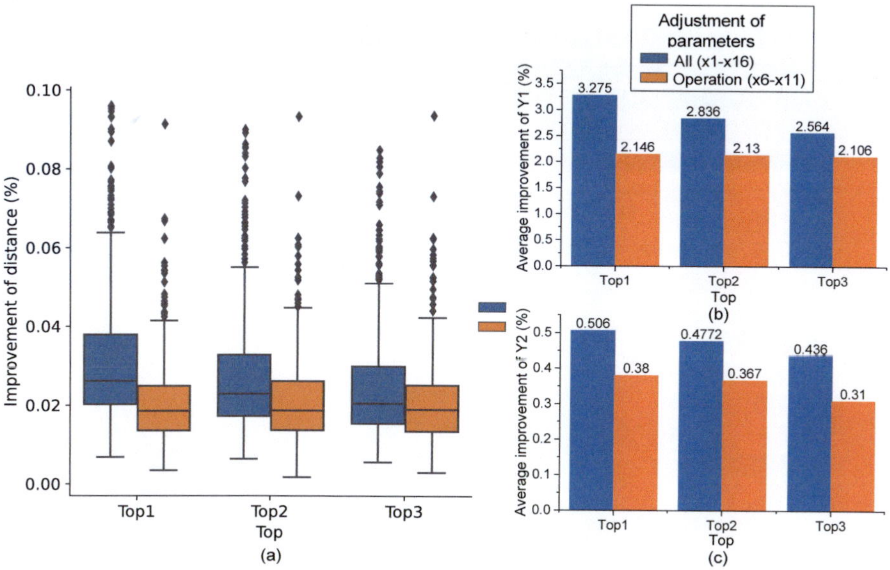

Fig. 9 Comparison of results in all 500 output points after GDO-based optimization: **a** box plot about the improvement of distance from the top 1–3 optimal solutions; **b** bar plot about the average improvement of Y1 from the top 1–3 optimal solutions; **c** bar plot about the average improvement of Y2 from the top 1–3 optimal solutions

can be identified as a more influential feature. Detailed analysis about the feature influence measurement is presented as follows.

(1) The proposed metric allows quantitatively interpreting the contribution of each feature in minimizing two targeted risks caused by the shield-driven tunnel construction, which is helpful in resulting in a better understanding of the GDO-based optimization. Figure 10 ranks the influence degree calculated by Eq. (9) from largest to smallest. It is obvious that the most influential variables ranking at the first five belong to the operational parameters or the building parameters, which means that these two types of parameters are prone to play a more prominent role in mitigating the potential risk in tunneling than the other two types named tunnel design and geology. Especially for the cutter rotating speed (x9), grouting volume (x11), and relative horizontal distance (x12), they are recognized as the top three most influential variables with the influence degree of 0.390, 0.151, and 0.132, respectively. Their influence degree can surpass the variable ranking at the bottom (x4) by 0.092–0.350. That is to say, these three identified key variables can actually bring the greatest contribution to the optimal solution. In this sense, we'd better pay more attention to these identified top three variables and give top priority to properly adjusting them. This suggestion seems to be practical, since x9 and x11 that are relevant to the configuration of TBM are highly adjustable.

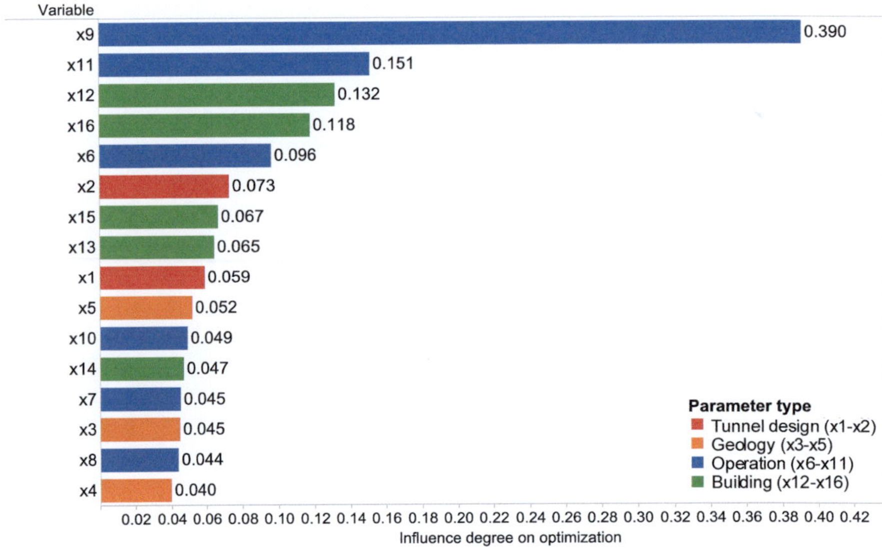

Fig. 10 Influence degree of 16 input variables on optimization

(2) For simplicity and efficiency, adjusting the top three most influential variables (x9, x11, and x12) is sufficient to significantly improve optimization quality and lower tunnel-induced risks. Comparative experiments are conducted using various combinations of these variables, with results summarized in Fig. 11 and Table 4. The ranking of average distance improvement is: all parameters (0.032%) > x9, x11, x12 (0.018%) > x9, x11 (0.016%) > x9, x12 (0.012%) > x11, x12 (0.008%) > x9 (0.008%) > x11 (0.002%) > x12 (0.002%). Noticeably, the involvement of the three identified key variables in the optimization process can gain the second-highest improvement in reducing two objectives, whose mean and medium of improvement degree are about 56.3% and 61.5% of those from the adjustment of all variables, respectively. When only one variable is allowed to be adjusted, x9 is more effective in shortening the distance between the objectives and the ideal point (0, 0), which can raise the average improvement of distance by around 0.006% in comparison with x11 and x12. Besides, optimal results based on two combinations of (x9, x11) and (x9, x12) can be averagely increased by 0.008% and 0.004%, respectively, than the consideration of (x11 and x12). Hence, it can be assumed that the operational feature named the cutter rotating speed (x9) contributes more than other variables in the optimization task, which deserves more attention. Thus, prioritizing x9 adjustments, followed by x11 and x12, offers a practical approach to quickly reduce tunnel-induced risks.

(3) Apart from focusing on all parameters or operational parameters, we formulate three additional feasible plans based on three adjustment schemes: the single key feature (x9), two key features (x9, x11), and three key features (x9, x11,

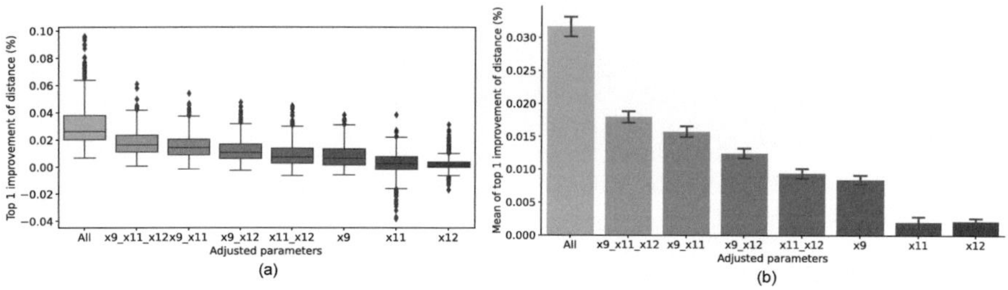

Fig. 11 Improvement of distance from the top 1 most optimal solution through adjusting different parameters: **a** boxplot about the distance improvement in all 500 output points; **b** bar plot about the mean of the distance improvement in all 500 output points (*Note* The error bar stands for the 95% confidence interval)

Table 4 Improvement of optimization when the key variables are adjusted

Adjusted variable	Mean (Std)	95% CI for mean	Medium	Minimum	Maximum
x9	0.008 (0.008)	[0.008, 0.009]	0.007	− 0.006	0.038
x11	0.002 (0.010)	[0.001, 0.003]	0.003	− 0.038	0.038
x12	0.002 (0.005)	[0.002, 0.003]	0.000	− 0.017	0.031
x9, x11	0.016 (0.009)	[0.015, 0.017]	0.014	0.015	0.017
x9, x12	0.012 (0.009)	[0.012, 0.013]	0.011	− 0.003	0.047
x11, x12	0.009 (0.008)	[0.009, 0.010]	0.007	− 0.006	0.045
x9, x11, x12	0.018 (0.010)	[0.017, 0.019]	0.016	0.001	0.061
x1–x16	0.032 (0.018)	[0.030, 0.033]	0.026	0.007	0.096

x12), which are expected to increase engineering efficiency and reduce computation complexity. Relying on the GDO-based optimization, Figs. 12, 13 and 14 visualize the desired value setting of the targeted features across all the 500 output points to reach the top 1 most optimal solution under the three different combinations of key features. That is to say, Figs. 12, 13 and 14 can be considered as the proper measures for safety control. To strictly follow the value shown in Figs. 12, 13 and 14 provides a great opportunity to diminish the possibility of risk occurrence during the process of shield tunneling. For example, it is preferred to adjust x9 due to its superiority in minimizing two objectives than x11 and x12. According to the determined profile of x9 in Fig. 12, we'd better decrease the cutter rotating speed, and the proportional reduction is relatively concentrated on the [− 5%, − 3%]. When two critical features are taken into account, the combination of x9 and x11 is a more suitable choice, whose value can be reasonably set as Fig. 13. Around 57.4% of optimal solutions of (Y1, Y2) need x9 under a decline over 4% and x11 under an increase of more than 2%. Besides, Fig. 13 provides the referable solutions to adjust the three most critical features simultaneously. It is required to reduce the value of x9 and x12 for 82.2% (411/500) and 70.4% (352/500) of output points, while in contrast,

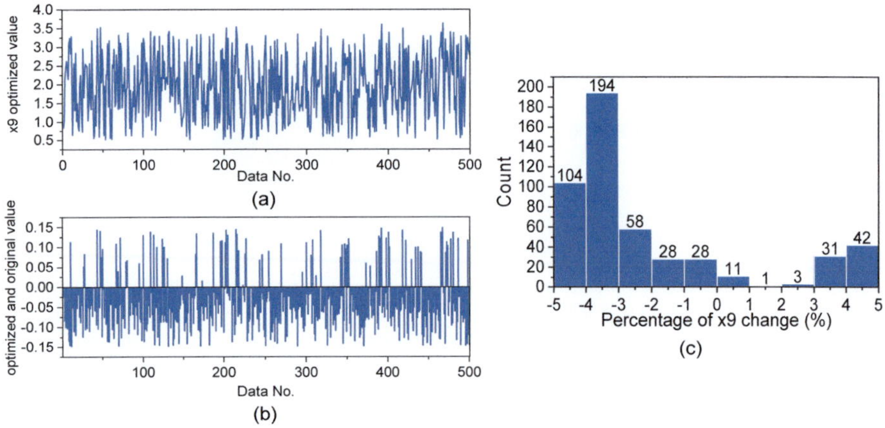

Fig. 12 Adjustment of x9 to obtain the top 1 optimal solution using the GDO algorithm: **a** value setting of x9 across all 55 output points; **b** difference between the optimized value and original value of x9 across all 55 output points; **c** histogram of percentage of x9 change after optimization

x11 for 70.5% (353/500) of output points need to experience a proper increase. In short, it can be inferred that a lower cutter rotating speed and relative horizontal distance along with a higher grouting volume are likely to lead to lower tunnel-induced risk.

6 Chapter Summary

To conclude, this paper presents an intelligent decision support framework using a novel deep neural network-based method for both risk prediction and optimization during shield tunnel excavation. By following three major stages, including data preparation, DNN-based risk prediction, and GDO-based risk control, the proposed decision tool can deeply explore the complicated tunnel-soil-building interaction, resulting in some positive guidelines to assure the structural safety of the tunnel being built and its adjacent buildings. The contribution of this study lies in: (1) the state of knowledge by developing a hybrid algorithm named DNN-GDO under the integration of the DNN prediction model owning the residual connection for multi-output non-linear regression task and the gradient descent technique for multi-objective optimization, which serves as an end-to-end learning pipeline. Moreover, a self-defined metric named influence degree can fully explain the influence of each feature on the optimization process, which can make the DNN-GDO model more interpretable; (2) the state of practice by taking a variety of influential factors into account to not only accurately predict the potential risk caused by the shield tunneling in advance, but

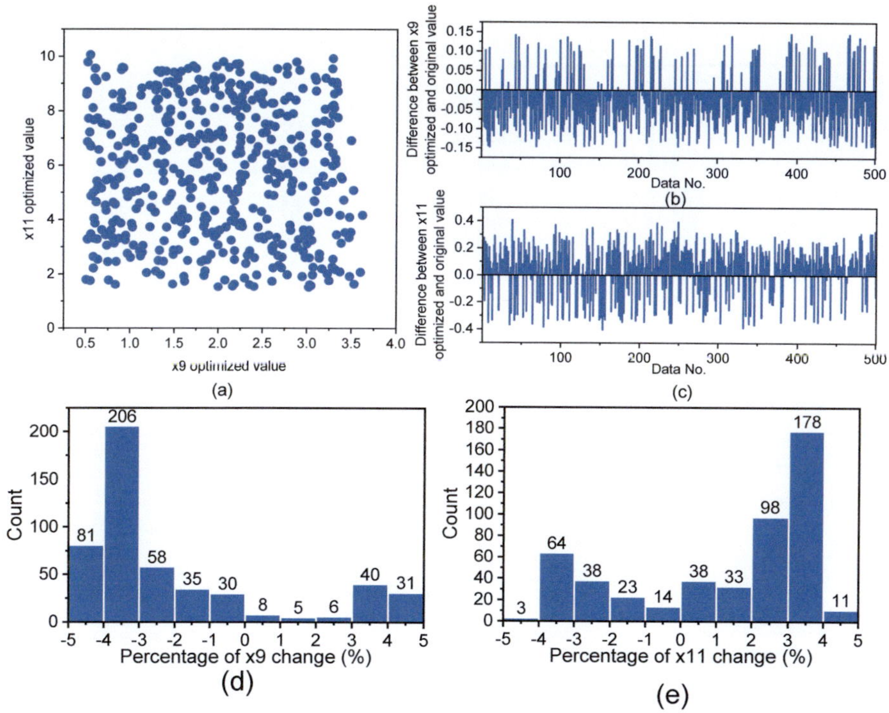

Fig. 13 Adjustment of x9 and x11 to obtain the top 1 optimal solution using the GDO algorithm: **a** value setting of x9 and x11 for all 55 output points; **b** difference between the optimized value and original value of x9 across all 55 output points; **c** difference between the optimized value and original value of x11 across all 55 output points; **d** histogram of percentage of x9 change after optimization; **e** histogram of percentage of x11 change after optimization

also objectively formulate protective measures for enhancing safety improvement relying on the optimization results instead of expertise-based estimation.

The proposed framework has been validated in a case study involving the Wuhan metro tunnel project. To be more specific, DNN is built to model the complex relationships between 16 inputs in terms of tunnel design features (x1, x2), geological features (x3–x5), operational features (x6–x11), and building features (x12–15) and two outputs associated with the important tunnel-induced risk (Y1, Y2). Framed as a multi-objective optimization (MOO) problem, the GDO algorithm was applied to the DNN-based model to minimize Y1 and Y2. The DNN accurately predicted these risks, with mean squared errors of 0.737 for Y1 and 0.002 for Y2. The GDO algorithm optimized input variables within a $\pm$ 5% range, reducing Y1 by 3.275% and Y2 by 0.506%. Adjusting only six operational variables further lowered Y1 by 2.146% and Y2 by 0.38%. The DNN-GDO approach provided three optimal solutions, offering practical guidance for risk mitigation. Moreover, the calculation of

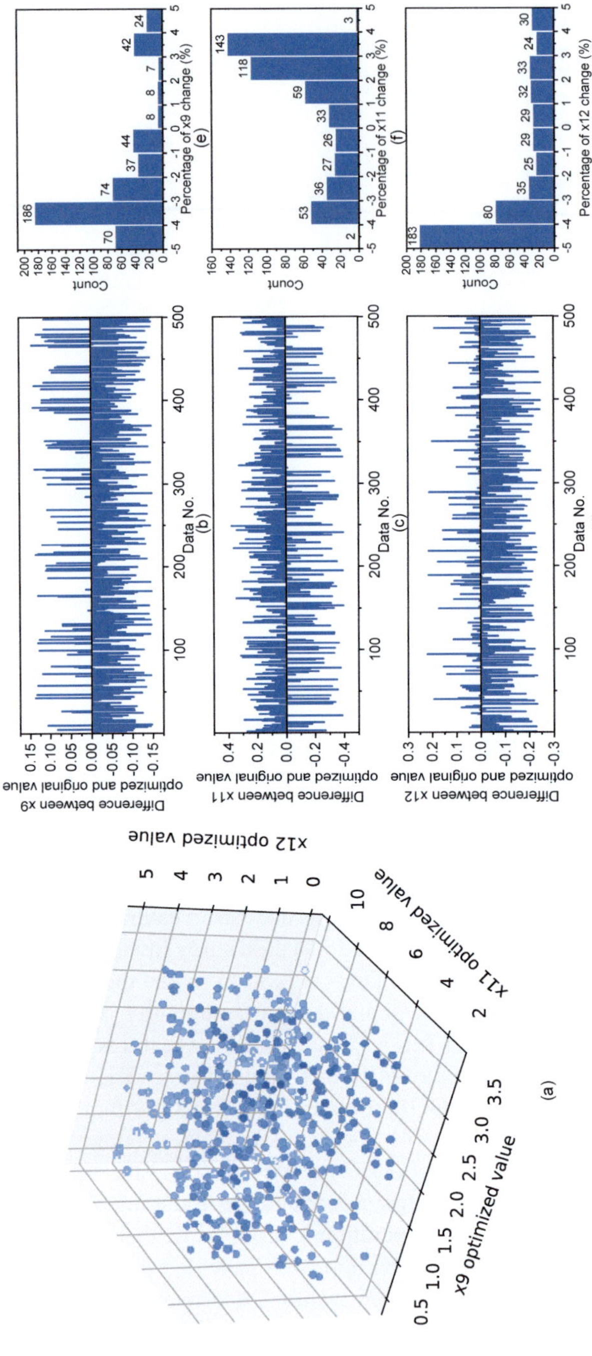

Fig. 14 Adjustment of x9, x11, and x12 to obtain the top 1 optimal solution using the GDO algorithm: **a** value setting of x9, x11, and x12 for all 55 output points; **b** difference between the optimized value and original value of x9 across all 55 output points; **c** difference between the optimized value and original value of x11 across all 55 output points; **d** difference between the optimized value and original value of x12 across all 55 output points; **e** histogram of percentage of x9 change after optimization; **f** histogram of percentage of x11 change after optimization; **g** histogram of percentage of x12 change after optimization

the influence degree metric opens a unique way to clearly explain the mechanism behind the GDO-based MOO. It is evident that features belonging to the category of operation and building herein are found to contribute more to mitigating risks represented by the two objectives. Notably, cutter rotating speed (x9), grouting volume (x11), and horizontal distance (x12) emerged as the key factors in minimizing risks, prioritizing their adjustment for effective tunneling safety control.

References

1. Guo K, Zhang L (2021) Multi-source information fusion for safety risk assessment in underground tunnels. Knowl Based Syst 227:107210
2. Luo Z, Zeng L, Pan H, Hu Q, Liang B, Han J (2019) Research on construction safety risk assessment of new subway station close-attached undercrossing the existing operating station. Math Probl Eng 2019
3. Zhang L, Wu X, Zhu H, AbouRizk SM (2017) Perceiving safety risk of buildings adjacent to tunneling excavation: an information fusion approach. Autom Constr 73:88–101
4. Pan Y, Zhang L (2022) Mitigating tunnel-induced damages using deep neural networks. Autom Constr 138:104219
5. Fu X, Zhang L (2021) Spatio-temporal feature fusion for real-time prediction of TBM operating parameters: a deep learning approach. Autom Constr 132:103937
6. Wang F, Gou B, Zhang Q, Qin Y, Li B (2016) Evaluation of ground settlement in response to shield penetration using numerical and statistical methods: a metro tunnel construction case. Struct Infrastruct Eng 12:1024–1037
7. Zhang L, Wu X, Liu H (2016) Strategies to reduce ground settlement from shallow tunnel excavation: a case study in China. J Constr Eng Manag 142:04016001
8. Zhang L, Wu X, Chen Q, Skibniewski MJ, Hsu S-C (2014) Towards a safety management approach for adjacent buildings in tunneling environments: case study in China. Build Environ 75:222–235
9. Zhou C, Kong T, Zhou Y, Zhang H, Ding L (2019) Unsupervised spectral clustering for shield tunneling machine monitoring data with complex network theory. Autom Constr 107:102924
10. Wang F, Ding L, Love PE, Edwards DJ (2016) Modeling tunnel construction risk dynamics: addressing the production versus protection problem. Saf Sci 87:101–115
11. Zhang L, Wu X, Ji W, AbouRizk SM (2017) Intelligent approach to estimation of tunnel-induced ground settlement using wavelet packet and support vector machines. J Comput Civ Eng 31:04016053
12. Zhou Y, Li S, Zhou C, Luo H (2019) Intelligent approach based on random forest for safety risk prediction of deep foundation pit in subway stations. J Comput Civ Eng 33:05018004
13. Huang X, Kroening D, Ruan W, Sharp J, Sun Y, Thamo E, Wu M, Yi X (2020) A survey of safety and trustworthiness of deep neural networks: verification, testing, adversarial attack and defence, and interpretability. Comput Sci Rev 37:100270
14. Abdeljaber O, Avci O, Kiranyaz MS, Boashash B, Sodano H, Inman DJ (2018) 1-D CNNs for structural damage detection: verification on a structural health monitoring benchmark data. Neurocomputing 275:1308–1317
15. Zhang P, Wu H-N, Chen R-P, Chan TH (2020) Hybrid meta-heuristic and machine learning algorithms for tunneling-induced settlement prediction: a comparative study. Tunn Undergr Space Technol 99:103383
16. Liu R, Yang P, Liu J (2021) A dynamic multi-objective optimization evolutionary algorithm for complex environmental changes. Knowl Based Syst 216:106612
17. Guo K, Yuan Y (2021) Geographic distribution and influencing factor analysis of green residential buildings in China. Sustainability 13:12060

18. Feng L, Zhang L (2021) Reliability-based multi-objective optimization in tunneling alignment under uncertainty. Struct Multidisc Optim 1–19
19. Wang H, Wang J, Zhao Y, Xu H (2021) Tunneling parameters optimization based on multi-objective differential evolution algorithm. Soft Comput 25:3637–3656
20. Nikakhtar L, Zare S, Nasirabad HM, Ferdosi B (2020) Application of ANN-PSO algorithm based on FDM numerical modelling for back analysis of EPB TBM tunneling parameters. Eur J Environ Civil Eng 1–18
21. Hunter SR, Applegate EA, Arora V, Chong B, Cooper K, Rincón-Guevara O, Vivas-Valencia C (2019) An introduction to multiobjective simulation optimization. ACM Trans Model Comput Simul 29:1–36
22. Zhang L, Wu X, Zhu H, AbouRizk SM (2017) Performing global uncertainty and sensitivity analysis from given data in tunnel construction. J Comput Civ Eng 31:04017065

Digital Twin Platform Development

1 Introduction

Exploration of intelligent tunnel excavation has consistently attracted continuous research over the past decades to improve efficiency, reliability, and safety for urban tunnel construction projects [1]. Generally, the success of tunnel projects highly relies on the performance of tunnel boring machines (TBM). Due to complex geological conditions and knowledge gaps, the current industry practice for TBM operation mainly relies on human experience which may result in compromised TBM performance or safety hazards if there is a lack of experience or human errors [2]. Thanks to the continuous advancement in sensor technologies, sensors that are equipped around TBM can generate big data that records the working status in real time. Such data provides unprecedented research opportunities to upgrade TBM with a higher level of intelligence and better performance.

A bulk research effort has been spent on the accurate prediction of TBM's performances using the operating parameters. For instance, Gao et al. [3] predicted TBM's key operating parameters using various recurrent neural networks. Wang et al. [4] developed a non-parametric Bayesian network that evaluates tunnel risks based on various failure modes. The success of these studies has proven the hidden relationships between various key operating parameters and TBM's performance indicators. However, there are some limitations for continuous studies. Firstly, previous studies normally extract spatial features which are TBM's operational parameters of the same timestep to predict the performance. As a result, there is a lack of consideration of the temporal dependencies for performance estimation and subsequent optimization. Secondly, the majority of current studies focus on proposing some reliable approaches to predict TBM's performances. However, there is limited effort to further guide TBM operation according to the prediction results.

L. Zhang et al., *Intelligent Construction in Tunnels*, Lecture Notes in Civil Engineering 679, https://doi.org/10.1007/978-981-96-8785-5_14

Successful tunnel excavation has multiple objectives, such that the project must be delivered with high excavation efficiency, low risk on safety, and low cost. Following this idea, some studies attempt to convert the problem of TBM control into a mathematical multi-objective optimization problem. In particular, TBM's operating parameters are treated as the adjustable inputs while the performance indicators are the objectives to be optimized. Among various algorithms, non-dominated sorting genetic algorithm II (NSGA-II) [5] was widely adopted for its low computational cost, well spread of solutions, and better convergence near the true Pareto-optimal. The method has been adopted and successfully resolved some tunneling-related problems including tunnel alignment design and construction optimization. However, there are some limitations for further improvement. Firstly, there is limited consideration of the temporal factors. Since TBM's performances of past time steps may potentially influence the performance at the current time step, the optimized results may not be practical, as it may deviate from the actual performance during tunneling. Secondly, previous studies on TBM optimization normally engage 2 performance indicators, which do not cover all factors that need to be considered during tunnel excavation. Additionally, most existing researches focus on developing the algorithm for MOO, while there is limited effort in establishing an integrated platform that enables real-time interaction between digital algorithms and physical objects for better TBM control and monitoring [6].

Digital Twin (DT) enables the concept of physical-cyber interaction and has been widely implemented in many industries for process monitoring and advanced control. There are some research efforts for DT implementations in the Architecture, Engineering, and Construction (AEC) industry [7, 8]. These studies have proven the potential benefits of using DT for the construction of civil products. However, the research and implementation of DT is generally at the prototype stage. Current research mainly focuses on the construction and facility maintenance of building projects. There are very limited efforts in exploring intelligent TBM control and tunnel excavation using DT-based methods. As a matter of fact, the sensors equipped by TBM generate a considerable amount of real-time data, which provides great opportunities for intelligent DT implementation and subsequent intelligent TBM operation.

Motivated by the successful estimation of TBM performances by the deep learning method and the strong potential of NSGA-II in resolving the MOO problems, this study intends to hybrid these two methods and propose a novel online multi-objective optimization method for advanced control of TBM's operational parameters and enhance TBM's performance in real-time. Additionally, practical objectives during tunnel excavation including progress, safety, energy consumption, and tool wear are incorporated in the proposed intelligent method for realistic TBM operation optimization during tunnel excavation. A digital twin platform will be established that provides data transmission for physical-cyber interaction, such that advanced control of TBM can be practically achieved.

This chapter presents a digital twin model of TBM that will be established for real-time monitoring of the major TBM operating parameters. Advanced control of TBM will be achieved by resolving the MOO problem with optimum solutions obtained by adopting the NSGA-II method in this study. Deep learning models that map the complex relationship between operational parameters and TBM performances are established, and an online MOO process is proposed by integrating NSGA-II and TOPSIS to determine the most suitable operating parameters for TBM under localized conditions.

2 Digital Twin in the AEC Industry

DTs generate a realistic virtual representation of the physical assets which can monitor the performance and working status continuously through bi-directional data exchange. The concept of DT has drawn considerable attention over the past decades since it was introduced in 2002. Although the applications in the AEC industry are mainly at the prototype stage [8], there are a large number of conceptual frameworks that have been proposed for above-ground structures. The current research efforts can be briefly summarized according to the supporting technologies for DT implementations for virtual model creation, bi-directional data exchange, and intelligent decision-making, respectively [9].

Building information modeling (BIM) is widely investigated in DT-related studies for virtual model creation. Nadarri and Shojaei [9] pointed out in their recent review paper that around 53% of research implements BIM models as the virtual model. Many studies adopted the BIM model as the as-built condition of buildings and integrated it with DTs to optimize building operation and maintenance. For instance, Shim et al. [10] integrated the BIM model with DT to improve the maintenance works for a bridge. An inspection model was generated by collecting data from light detection and ranging (Lidar) and unmanned aerial vehicle (UAV) images. Lu et al. [11] designed an architecture that integrated DT and BIM that aimed to assist in the operation and maintenance of the buildings. Jiang et al. [12] introduced a DT-enabled smart modular integrated construction framework that improved the productivity for the on-site assembly of the prefabricated units. In addition to BIM, simulation models, geographic information systems (GIS) and finite element methods are other approaches for model creation.

Compared to conventional 3D computer-generated models, the main advantage of DTs is the real-time monitoring and interaction between the virtual model and physical objects. One of the key contributing factors to achieving this feature is thorough smooth bi-directional data flow. The data from physical objectives can be utilized to update the virtual model, and vice versa to provide guidance for actual construction. To achieve bi-directional data exchange, current studies mainly investigate IoT sensors as the main approaches for data acquisition and collection. The adoption of IoT sensors has been successfully applied in structural maintenance, structural health monitoring, and construction safety management. For instance, Lin et al. [13] proposed a DT model that assisted bridge design by providing pre-construction seismic collapse assessment and fragility analysis. Yu et al. [14] proposed a DT-based framework for the operation and maintenance of tunnels by fusing the monitoring and inspection data from multiple sources. Park et al. [15] investigated Bluetooth low energy (BLE) sensors for safety management in construction sites. In addition to IoT sensors, LiDAR and UAV are other major technologies that have been implemented to provide real-time data for DTs implementation.

Machine learning methods and computer-vision techniques are widely explored to assist the DT models for data analysis and intelligent decision-making. The machine learning methods are mainly investigated for performance prediction and maintenance works, while computer-vision methods are employed for quality control during construction and maintenance works. For instance, Cheng et al. [16] established a data-driven predictive maintenance framework for the mechanical and electrical (M&E) components with the assistance of artificial neural network (ANN) and support vector machine (SVM). Lee et al. [17] integrated DT with reinforced learning to optimize the task allocation for the robotic arm. A virtual experiment was conducted and resulted in a 36% improvement in terms of construction efficiency. Hamledari et al. [18] proposed a DT-based method to improve quality control during construction with the aid of UAVs and computer vision methods for image data analysis.

Intelligent tunnel excavation has been an interesting topic and it has attracted a considerable amount of research attention. With the continuous development of machine learning and deep learning algorithms, numerous methods have been investigated for TBM performance prediction and optimization. Although these studies demonstrated the powerful capabilities of DT by integrating with BIM, IoT, and AI technologies, there are some research gaps for further improvement. Firstly, the current studies mainly focus on using AI methods for prediction tasks but lack investigation on optimization. Secondly, the literature is more focused on facility maintenance and structural health monitoring, but there is a lack of exploration on how DTs can improve underground construction. To address these limitations, this study is motivated to integrate MOO with DTs to optimize TBM operation with improved performance.

3 Digital Twin Platform for Enhanced TBM Control

To establish the proposed digital twin model that enables real-time monitoring of TBM parameters through IoT sensors, a 3D TBM model with the key components is established. In particular, a deep learning model that incorporates graph convolutional network (GCN) and long short-term memory (LSTM) neural networks is developed to map the non-linear relationships between the key operating parameters and TBM performance indicators. The developed models are then utilized as the fit function to optimize TBM operation by using the NSGA-II method with online updating. Detailed elaboration on digital twin model establishment, intelligent module integration, and advanced TBM control through MOO are shown in Sects. 3.1–3.3, respectively. An overview of the proposed framework is shown in Fig. 1.

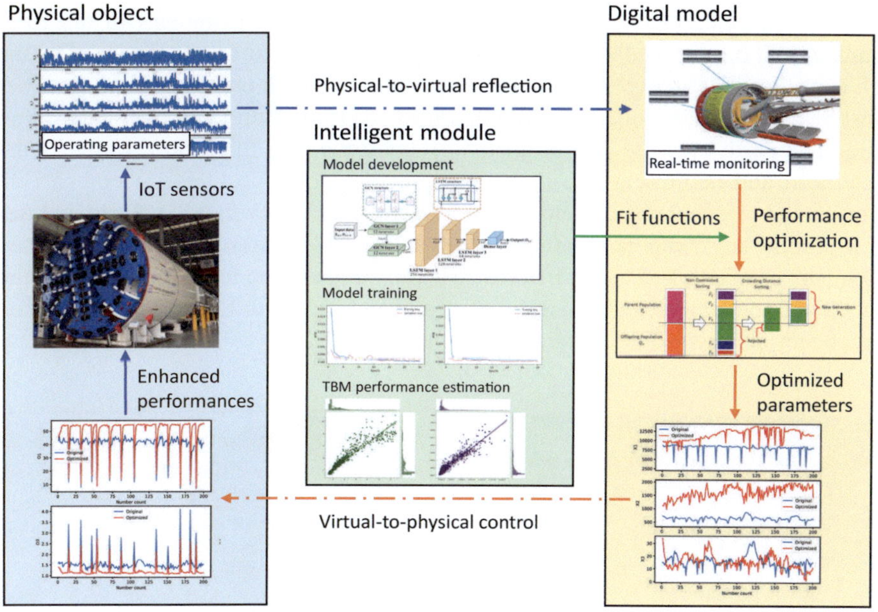

Fig. 1 Framework of the proposed advanced TBM control method

3.1 Physical-to-Virtual Reflection

Modern TBMs are equipped with sensors that closely monitor and record the operation status during tunnel excavation. The information recorded by the sensors will be transferred and processed by TBM's data acquisition system. These sensors record useful information involving the excavation status and the TBM's operational parameters. In particular, the date and time of excavation, cutter head's thrust and torque, pressure in the chamber, trajectory deviations, status of grouting, etc. can be extracted. Currently, information on the key operational parameters will be presented through the screens installed in the control chamber of TBM and above ground for remote monitoring. However, there is a lack of intuitive presentations that can directly reflect the operation status of TBM for ease of analysis. As a result, massive data are generated during TBM excavation, yet there is a limited exploration of the knowledge that lies in the large amount of the data.

As an improvement, a 3D digital twin platform can be established to directly transform the physical equipment into a digital model. The model shall truthfully reflect the actual condition of TBM including the length and diameter, depth of excavation, the number of tunnel rings that it is excavating, and trajectory deviations, etc. The key components within TBM shall also be presented, where the essential ones, including cutterhead, earth chamber, thrust system, screw and belt conveyor, etc. In particular, TBM's cutterhead is one of the most important components as it is in charge of soil breaking and excavation. The excavated soil will then be transferred into the earth chamber to balance the earth pressure. Next, the soil in the earth chamber will be removed through the screw conveyor and further transfer out by the belt conveyor. For an experienced TBM operator, the earth pressure in the chamber must be carefully controlled as it directly affects face stability and over-excavation ratio. The earth pressure can be generally controlled by the inflow which is the cutterhead and outflow, which is the screw conveyor, respectively. A comparison between the physical objects and digital reflection for the overall TBM and a closer view of some key components are presented in Fig. 2a–c, respectively. The developed 3D model can be linked with TBM's data acquisition system for visualization of the key operational parameters and excavation status. The sensor location is highlighted for a better understanding of the operating parameters.

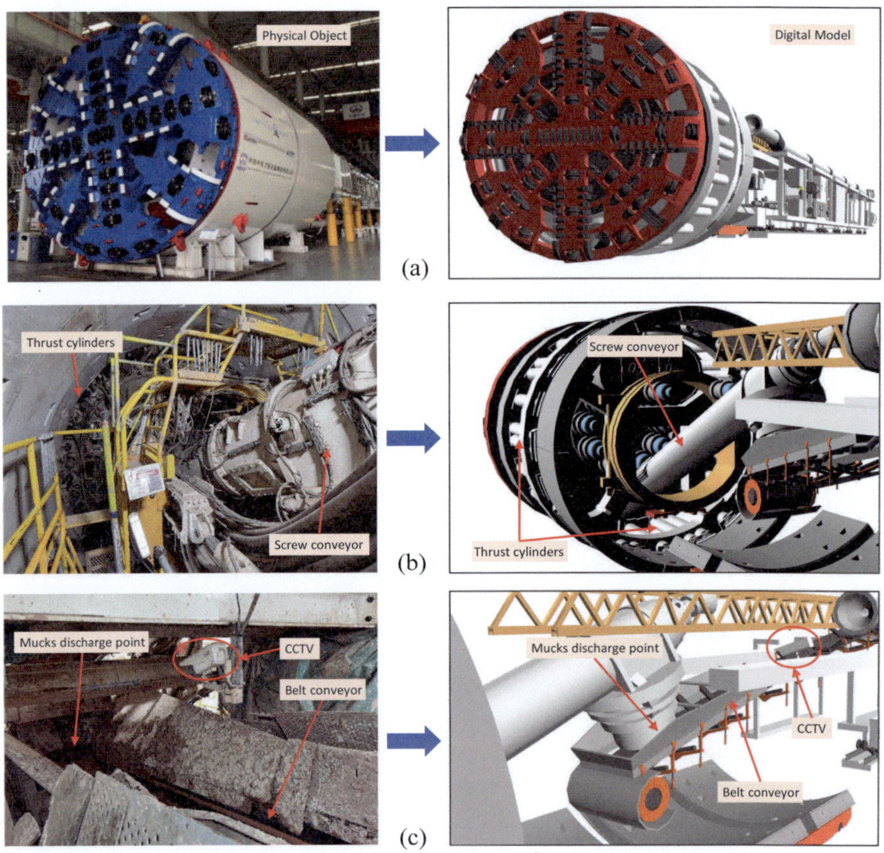

Fig. 2 Comparison of the physical and digital objects for **a** overview of TBM, **b** close view of thrust cylinders and screw conveyor, and **c** close view of the muck transport system

3.2 Intelligent Module Integration

With the real-time monitoring data collected from TBM, the digital twin model can integrate an intelligent module and provide reliable estimation for TBM's key performance indicators. To achieve this purpose, deep learning models can be established and trained that aim for performance prediction through the key operating parameters. With the success of our previous study that demonstrates the strong capacity to extract both spatial and temporal features, we implement a deep neural network that integrates GCN and LSTM.

Prior to establishing the intelligent model for performance estimation and optimization, data pre-processing is necessary to reduce the noisy data and transform them into a suitable format for model training. Following the standard practice, data preparation is conducted with the process of noise removal, normalization, and data reframing. One row of obtained TBM operating parameters be denoted as $x_n = \{x_1, x_2, x_3, \ldots, x_i\}$, where x_i represents one type of input parameters X_i. Those cells with null or outliers are regarded as noises and are removed accordingly. Outliers are identified by using the Z-score method. Next, all data is rescaled into the range of [0,1] for normalization. To extract both spatial and temporal features, the dataset is then reframed into the format of a matrix, where the rows represent different TBM parameters and columns represent the timesteps. A more detailed explanation of data pre-processing can be found at [19], and mathematical expressions for z-score, data normalization and rescale are shown in Eqs. (1)–(3), accordingly.

$$z = \frac{x_i - \tilde{x}_i}{S_i} \tag{1}$$

$$x_k = \frac{x_i - x_{min}}{x_{max} - x_{min}} \tag{2}$$

$$data = \begin{bmatrix} x_{1,t-n} & x_{1,t-n+1} & \cdots & x_{1,t-1} & x_{1,t} \\ x_{2,t-n} & x_{2,t-n+1} & \cdots & x_{2,t-1} & x_{2,t} \\ \vdots & \vdots & \ddots & \vdots & \vdots \\ x_{k-1,t-n} & x_{k-1,t-n+1} & \cdots & x_{k-1,t-1} & x_{k-1,t} \\ x_{k,t-n} & x_{k,t-n+1} & \cdots & x_{k,t-1} & x_{k,t} \end{bmatrix} \tag{3}$$

where $\tilde{x}_i$ is the mean value for the input parameter X_i, S_i represents the respective standard deviation, x_{min} and x_{max} are the minimum and maximum values of the variable X_i, and *data* involves both inputs and output for deep learning model's training

With the data being pre-processed and transformed into a favorable format, the necessary parameters can be further extracted for deep learning model training. In addition, a brief description of the performance indicators is presented prior to intelligent module development. The structure of the proposed intelligent module as well as the model training and evaluation procedures are elaborated subsequently.

3.2.1 Problem Statement

Identification of the TBM performance indicators is critical for advanced TBM control. In this study, penetration rate, energy consumption, over-excavation ratio, and cutting disc tool wear are selected as the key TBM performance indicators. Penetration rate (P_r) represents the speed of excavation and can be calculated as the ratio of excavation distance over the time spent for the excavation without stopping. It is one of the critical performance indicators as it reflects the excavation speed and construction productivity. A higher penetration rate indicates better excavation efficiency and will result in faster construction progress. Therefore, the penetration rate is identified as the first objective for TBM performance optimization, and it is denoted as O_1 in this research. The mathematical expression of the penetration rate for one tunnel ring can be denoted as Eq. (4).

$$O_1 = P_r = N_s/t \tag{4}$$

where N_s denotes the net stroke of excavation, and t is the time spent on the continuous excavation.

The over-excavation ratio (O_{ev}) is another critical performance indicator that reflects the safety and reliability aspects during tunnel excavation. It is defined by the ratio of actual soil mass excavated by TBM against the theoretical value. With certain underground space created, a smaller over-excavation ratio indicates the excavation has lesser disturbance to the soil with a lower risk of damage to the above-ground structures. O_{ev} is identified as the second objective and denoted as O_2 in this research. It can be estimated by Eqs. (5) and (6).

$$O_2 = O_{ev} = M_{S;ac}/M_{S;theo} \tag{5}$$

$$M_{S;theo} = 0.25D^2\pi \cdot \gamma \cdot N_s \tag{6}$$

where $M_{S;ac}$ is the actual soil mass removed by TBM in one ring, $M_{S;theo}$ is the theoretical value of soil mass for one ring, D is the excavation diameter for TBM, and γ is the average unit weight of the soil.

Energy consumption (E_T) of TBM is another critical indicator that affects the total project cost, and it is identified as the third objective O_3 in this study. In addition, previous studies have shown that the cutterhead contributes to at least 60–80% of the total energy consumption for the entire TBM [20]. In this study, the total energy consumption at the cutterhead is estimated based on Eqs. (7)–(9).

$$O_3 = E_T = E_1 + E_2 \tag{7}$$

$$E_1 = \sum_{i=1}^{n} N_{si} \cdot F_i \tag{8}$$

$$E_2 = \sum_{i=1}^{n} \omega_i \cdot T_i \cdot t_i \tag{9}$$

where E_T is the total energy consumption for the cutterhead, E_1 is the energy consumed by the hydraulic thrust system, E_2 is the energy consumed by the cutter system, F_i and T_i are thrust force and cutterhead torque, respectively, and ω_i is the cutterhead rotational speed.

The last objective, O_4 is the tool wear of TBM's cutter head, denoted as T_w. Excessive tool wear will result in less excavation efficiency and even damage the TBM's cutterhead. Therefore, it is another important indicator that has been monitored during tunnel excavation. In current industry practice, tool wear is generally obtained through site measurement during the cutterhead intervention at certain intervals. The obtained data can be further decomposed to obtain the tool wear for each portion during tunnel excavation. This process can be realized through Eq. (10).

$$O_4 = T_w = \frac{k\pi D\omega_i N_{si}}{P_r} \tag{10}$$

where k is the wear coefficient that depends on the geological condition and cutter disc material.

3.2.2 Deep Learning Model for Performance Estimation

To construct the proposed deep learning model and capture the spatial and temporal features from the input time-series data, GCN and LSTM neural networks are employed. Considering the TBM operating parameters as inputs and the performance indicators as the output, the variables can be denoted as the nodes and their spatial relationships are the links. That is to say, each TBM operating parameters $x_{n,t} = \{x_{1,t}, x_{2,t}, x_{3,t}, \ldots, x_{k,t}\}$ can be viewed as nodes, while the potential relationships between these parameters are the links between the nodes. In this case, the entire system with TBM operating parameters from a few time steps can be viewed as a graph of structured data. GCN cells are introduced to extract the spatial features from the variables, the architecture can capture information flow from the nodes in a convolutional fashion and supports graph structure. Detailed elaboration of GCN structure and the algorithm can be found in some earlier studies [21].

In addition to the spatial relationships, the TBM performance indicators are affected by their previous states. Therefore, LSTM cells are introduced to extract the temporal relationships of the variables. A detailed explanation of LSTM structure and the respective algorithm can be found in many related studies. To establish the model, multiple GCN and LSTM layers are stacked and compiled. In particular, the input data first passes through 2 layers of GCN neurons, followed by 3 layers of

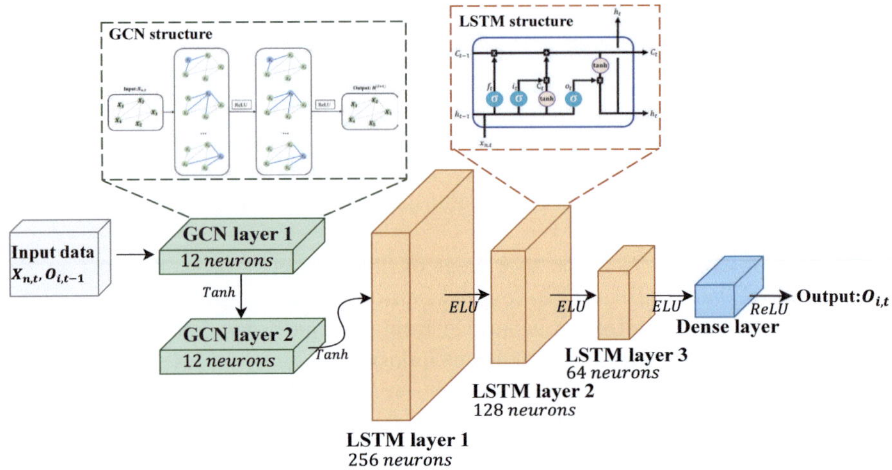

Fig. 3 Structure of the proposed deep learning model

LSTM neurons. Lastly, a single dense layer is created to output the predicted result. Each GCN layer has 12 neurons and the LSTM layers have 256, 128, and 64 neurons, respectively. The structure of the proposed deep learning model is shown in Fig. 3.

Tensorflow Keras and Stellargraph frameworks are utilized to implement the proposed deep learning model. In this study, the conventional "80–20 rule" is employed for model training, where the first 80% of the dataset is utilized for model training and the subsequent 20% dataset is for testing purposes. The input and output of the training data and testing data are denoted as X_{train} and y_{train}, and X_{test} and y_{test}, respectively. To capture the spatio-temporal features from the parameters, the input data includes both TBM's key operational parameters $\{x_1, x_2, x_3, \ldots, x_{k-1}\}$ from timesteps $t - n$ to t as well as the objectives O_i from previous timesteps from $t - n$ to $t - 1$. Outputs are the objectives O_i at the current timestep t. Therefore, the input and output parameters can be elaborated by Eqs. (11) and (12), respectively.

$$X_{train}, X_{test} = \begin{bmatrix} x_{1,t-n} & x_{1,t-n+1} & \cdots & x_{1,t-1} & x_{1,t} \\ x_{2,t-n} & x_{2,t-n+1} & \cdots & x_{2,t-1} & x_{2,t} \\ \vdots & \vdots & \ddots & \vdots & \vdots \\ x_{k,t-n} & x_{k,t-n+1} & \cdots & x_{k,t-1} & x_{k,t} \\ O_{i,t-n} & O_{i,t-n+1} & \cdots & O_{i,t-1} & - \end{bmatrix} \tag{11}$$

$$y_{train}, y_{test} = O_{i,t} \tag{12}$$

where X_{train} and X_{test} are the input parameters. y_{train} and y_{test} are the output for prediction.

During model training, *tanh*, exponential linear unit (*ELU*), and rectified linear activation unit (*ReLU*) are employed as the activation functions for the GCN, LSTM, and dense layers, respectively. Mean squared error (*MSE*) is utilized as the loss function and *Adam* is adopted as the optimizer during model training. The activation functions and the loss function can be calculated by Eqs. (13)–(16), respectively.

$$\tanh(x) = \frac{e^x - e^{-x}}{e^x + e^{-x}} \tag{13}$$

$$RELU(x) = \begin{cases} x, & if\ x > 0 \\ 0, & if\ x \leq 0 \end{cases} \tag{14}$$

$$ELU(x) = \begin{cases} x, & if\ x < 0 \\ \alpha(\exp(x) - 1) & if\ x \leq 0 \end{cases} \tag{15}$$

$$MSE = \frac{1}{N}\sum_{i=1}^{N}(f_i(x) - y_i)^2 \tag{16}$$

where $\alpha = 1$, N is the total number of testing samples for y_{test}, $f_i(x)$ is the output result by the model, and y_i is the original data.

The trained deep learning model shall be carefully evaluated to examine its performance since only the model with high performance can provide reliable results after optimization. By comparing the prediction from the model $f_i(x)$ and y_{test} over the testing data, mean absolute error (MAE), root mean squared error (RMSE), and coefficient of determination (R^2) are adopted for models' performance evaluation. MAE, RMSE and R^2 can be calculated by Eqs. (17)–(19), respectively.

$$MAE = \frac{1}{n}\sum_{i=1}^{n}|y_i - f_i(x)| \tag{17}$$

$$RMSE = \sqrt{\frac{\sum_{i=1}^{n}(y_i - f_i(x))^2}{n}} \tag{18}$$

$$R^2 = 1 - \frac{\sum_{i=1}^{n}(y_i - f_i(x))^2}{\sum_{i=1}^{n}(y_i - f_i(x))^2} \tag{19}$$

where N is the total number of samples and y_i represents the ground truth which is y_{test} in this case.

3.3 Virtual-to-Physical Control

The established intelligent module can be further extended providing useful guidance for TBM operation. This process can be achieved through online optimization by adopting the NSGA-II method. However, since the NSGA-II method generates a number of Pareto fronts while only the most ideal solution shall be selected to guide TBM operation, the technique for order preference by similarity to an ideal solution (TOPSIS) method is applied for this purpose. The generated ideal solution shall then be the optimized operating parameter, where advanced control can be achieved by applying a virtually generated number to guide the physical object operation. Detailed elaboration on NSGA-II, TOPSIS, and improvement measurements are shown below.

3.3.1 Pareto Fronts Generation by NSGA-II

The general objective of the optimization process is to determine the minimum values for the fitness functions $f_i(x)$ with the parameters bounded by some constraints. This process can be mathematically expressed by Eq. (20).

$$\begin{aligned} \min .f(x) &= \{f_1(x), f_2(x), f_3(x)\} \\ s.t.g(x) &= \{g_1(x), g_2(x), g_3(x)\ldots\} < g_{req} \end{aligned} \tag{20}$$

where $f_i(x)$ is the fitness function for the ith objective, $g(x)$ are the constraint functions, and g_{req} is the required value for the constraint. In this study, the conventional NSGA-II method is proposed to source for the pool of non-dominating solutions named Pareto fronts.

To perform the algorithm, two sets of data, named parent population $P^0 = \{p_1^0, p_2^0, p_3^0, \ldots p_N^0\}$ and offspring population $Q^0 = \{q_1^0, q_2^0, q_3^0, \ldots q_N^0\}$ are randomly generated, where N is the population size. These populations are assessed by the fitness functions and ranked based on a fast non-domination sorting method. That is to say, individuals with non-dominated performances are assigned to the same group, named as $F_1, F_2 \ldots$, etc. Additionally, individuals within the same group are compared according to the crowding distance (CD), which is calculated based on the distance for an individual to its neighbors in the same rank. CD can be calculated by Eq. (21)

$$CD_i = \sum_{m=1}^{M} p_m(i+1) - p_m(i-1) \tag{21}$$

where CD_i is the CD value for the ith individual, M is the total number of objectives, p_m is the normalized absolute distance between individual i's sorted neighbors at a dimension of m. Based on the ranking and CD values, the most suitable N individuals

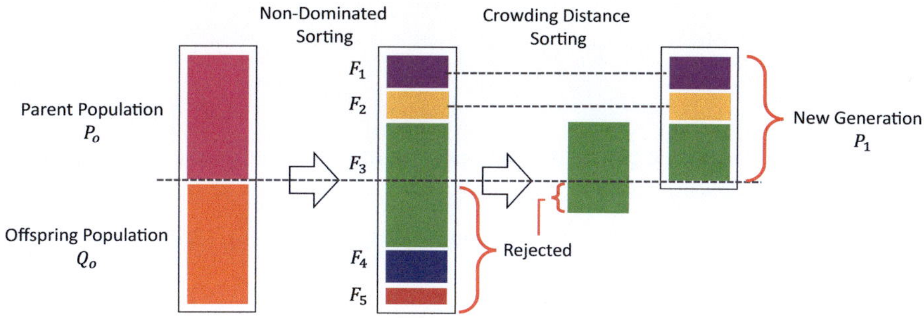

Fig. 4 Individual selection process by NSGA-II

are selected from P^0 and Q^0 and marked as P^1. The next offspring population Q^1 can then be generated based on mutation and crossover operations. The entire process will be repeated for multiple iterations until the termination criteria are met. An illustration of the NSGA-II algorithm is shown in Fig. 4.

3.3.2 Ideal Solution Determination and Online Updating

According to the NSGA-II method, Pareto fronts P_n are generated with N individuals available. However, the best-fit data point must be selected to guide TBM operation. Therefore, this study adopted the TOPSIS method to determine the most ideal solution. TOPSIS method generally calculates and compares the distance of the individuals to the most ideal and most unfavorable points, and the process can be calculated by several steps.

Step 1: Consider the objective j, the most ideal and most unfavorable points to the objective can be derived by Eqs. (22) and (23) accordingly.

$$Z_j^- = \max\{x_{1j}, x_{2j}, \ldots, x_{ij}\} \tag{22}$$

$$Z_j^+ = \min\{x_{1j}, x_{2j}, \ldots, x_{ij}\} \tag{23}$$

where x_{ij} represents the ith individual under objective j.

Step 2: For any individual x_{ij}, the distance to Z_j^+ and Z_j^- can be calculated by Eqs. (24) and (25), respectively.

$$D_i^+ = \sqrt{\sum_{j=1}^{m}\left(x_{ij} - Z_j^+\right)^2} \tag{24}$$

$$D_i^- = \sqrt{\sum_{j=1}^{m} \left(x_{ij} - Z_j^-\right)^2} \tag{25}$$

where D_i^+ and D_i^- are the geometric distance for the ith individual to the most ideal and unfavorable conditions accordingly.

Step 3: TOPSIS value for individual x_{ij} can be evaluated by Eq. (26).

$$S_i = \frac{D_i^-}{D_i^- + D_i^+} \tag{26}$$

Step 4: The best-fit individual will be selected for those with the lowest S_i value and the optimized deviation is determined accordingly.

However, since the developed deep learning model requires both spatial and temporal features as inputs, the optimized results have an impact on the prediction of TBM performances at the next time step. Therefore, online updating is necessary to simulate this condition. In particular, the historical values of performances shall be replaced by the optimized results instead of their original values for the next round of optimization. That is to say, after obtaining the optimized results, O_{i_opt} for time step t, the actual values of the respective objective at t, $O_{i,t}$ shall be replaced by O_{i_opt}, accordingly before performing the optimization for the next time step. The procedure will be repeated with every time step moving forward. The pseudocode of the proposed online optimization method is summarized in Algorithm 1.

Algorithm 1 Online MOO with deep learning model and NSGA-II

Input: X_n: Operational parameters, $Model_i$: Deep learning model for objective O_i, N: Number of populations

Output: O_{i_opt}: Optimized parameters

Reframe Input data $X_k = X_{n,t\sim t-n}, O_{i,t-1\sim t-n}$
Construct Fit functions $O_i = Model_i(X_k)$ using deep learning models
Define optimization problem: $min.f(x) = \{Model_1(X_k), Model_2(X_k), Model_3(X_k)\ldots\}$
For k in range ($len(X_n)$):
    At timestep $t = k$
    Generate initial populations P^0 and offspring Q^0 through mutation and crossover
    while termination criteria $\neq$ True:
        Calculate the rank and CD for individuals in $\{P^0, Q^0\}$
        Select N least crowded individuals from $\{P^0, Q^0\}$, denoted as P^1
        Generate new offspring Q^1 from P^1 through mutation and crossover operation
    return Pareto fronts P^N
    Determine the most ideal solution by TOPSIS with respective $X_{opt,t=i}$
    return $O_{i\_opt} = Model_i(X_{opt,t=k})$

(continued)

(continued)

14:	**Update** $O_{i,t} = O_{i\_opt}$
15:	**end**

3.3.3 Improvement Measurement

The effectiveness of the recommended parameters can be validated by comparing the optimized results with the original performances. By comparison, the improvement achieved by the proposed method for each objective can be calculated by Eq. (27).

$$Imp_i = \left(O_{i,org} - O_{i,opt}\right)/O_{i,org} \tag{27}$$

where Imp_i is the improvement magnitude in performance, $O_{i,org}$ is the original value, and $O_{i,opt}$ is the optimized result for objective i, where $i = 1, 2, 3, 4$. The overall improvement is calculated based on the improvement of a single objective, which can be derived from Eq. (28).

$$Imp_{overall} = \sqrt{\frac{1}{n}\sum_{i=1}^{4}\frac{Imp_i}{|Imp_i|}(Imp_i)^2} \tag{28}$$

where n is the total number of objectives, and $|Imp_i|$ is the absolute value of the improvement for objective i.

4 Experimental Studies

To verify the effectiveness of the proposed optimization method as well as the developed DT-based approach, data collected from Singapore's Circle Line 6, C885 tunnel project is utilized as a case study for demonstration purposes. Detailed elaborations on the project background and digital twin model development, intelligent module establishment, and result analysis are presented in Sects. 4.1–4.3, respectively.

4.1 Digital Twin Model Development in Tunnel Construction

The C885 tunnel project includes the construction of 2 bound tunnels under Singapore's circle line that connects Cantonment Station and Prince Edward Road Station. An EPB-TBM with an outer diameter of 6.68 m and a total length of 9.81 m is utilized for tunnel construction. The encountered geological condition can be categorized into

3 types, namely Marine clay, Jurong formation type III to IV, and Jurong formation type V to VI. To avoid the negative impact of different geological conditions, the data utilized in this study are from Ring No. 20 to 190 where the geological condition is identified as Marine clay. The IoT sensors that are installed on TBM record the working status of TBM in real-time and thus can provide useful information for advanced control.

For better monitoring and control of the key operating parameters, a 3-dimensional (3D) DT-based platform is developed that enables real-time transmission of the information from the physical object into the digital model. The 3D model is developed through the Unity platform, which includes the key TBM components that are critical for TBM operation. Specifically, 6 sensors are installed beside the cutterhead chamber that records the earth pressure at the top, middle, and bottom of the chamber. The thrust force is generated by hydraulic cylinders, which can be controlled under 4 groups. The transportation system involves a screw conveyor (SC) that removes the soil from the cutterhead chamber, a belt conveyor that provides horizontal transportation and a wagon that delivers the collected mucks to the launching shaft. Key operational parameters from these components are displayed in the developed DT-based platform, which will further assist in better monitoring and subsequent intelligent control of TBM operation. An illustration of the developed DT-based platform with TBM's key components and a display of the major operational parameters are shown in Fig. 5a–d, respectively.

The selection of input parameters is critical for achieving reliable performance estimation and effective TBM control. To avoid redundancy and improve the model training efficiency, only the key operational parameters are selected as the inputs. The key operating parameter is generally determined by referring to some related studies [21] and consulting from site experts' experience. As a result, 3 active parameters and 6 passive parameters are chosen as the input parameters to construct the intelligent module for the digital twin model. In particular, TBM's key operational parameters, including thrust force (x_2), cutterhead torque (x_3) are the active parameters that can be actively adjusted during TBM operation; SC's torque (x_3), cutterhead rotational speed (x_4), the volume of foam (x_5), SC's pressure (x_6), and the earth pressure at the middle left and right of the chamber (x_7, x_8), and net stroke for each ring (x_9), are the passive parameters that generally reflect the localized geological conditions and state of tunnel ring building. That is to say, the active parameters can be adjusted and optimized during tunnel construction, while the passive parameters shall be maintained at their original values for each tunnel ring.

According to the literature review and problem statement, TBM's penetration rate (O_1), under excavation ratio (O_2), energy consumption (O_3), and tool wear of the cutter head discs (O_4) are identified as the key performance indicators. Therefore, these are the objectives for estimation and optimization by the intelligent module of the digital twin model. The description of the input parameters is listed in Table 1, and a summary of the TBM operational parameters is presented in Table 2 for reference.

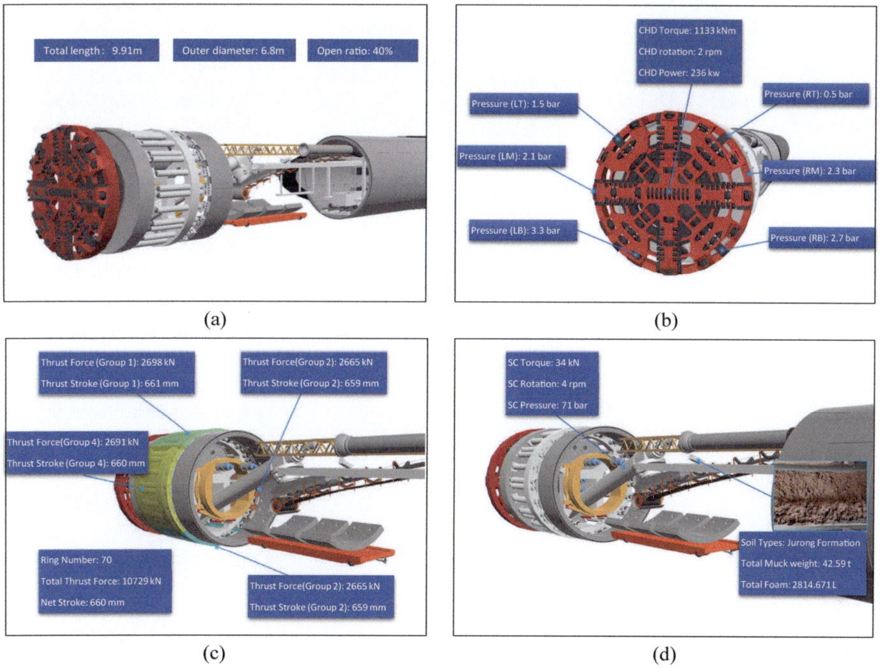

Fig. 5 DT-based platform for physical to virtual presentation for **a** overall model geometry, **b** cutter-head parameters and soil chamber pressure, **c** jacking system parameters, and **d** earth removal system parameters and soil identification

4.2 Intelligent Module Enabled Physical-Cyber Interaction

With the physical TBM being transformed into a digital twin virtual model and real-time data of its working status being recorded and extracted through the sensors and data acquisitions system, an intelligent module can be integrated with the digital twin model which aims to provide accurate estimation for TBM's key performance indicators and useful guidance for the TBM control to enhance its overall performance. As elaborated in the previous section, the proposed intelligent module consists of two major functions, namely TBM performance estimation, and parameter optimization.

To achieve an accurate estimation of performance indicators, four distinctive deep learning models are established to evaluate the four indicators, which are penetration rate, over-excavation ratio, energy consumption, and tool wear (denoted as O_1, O_2, O_3 and O_4, respectively). The pre-processed data are divided into 2 groups with around 80% of the total dataset being used to train the models and the remaining 2000 datasets are utilized for models' performance evaluation. The established deep learning model consists of two layers of GCN neurons stacked with three LSTM

Table 1 Explanation of the key TBM operational parameters

Parameter		Symbol	Unit	Description
Active parameters	Thrust force	x_1	kN	The main driving force that enables the cutterhead to move against the soil along the longitudinal direction of the tunnel
	Cutterhead torque	x_2	kN m	The torque exerted on the cutterhead to maintain its rotation during excavation
	SC torque	x_3	bar	The torque is used to maintain the rotation of the SC that removes the excavated earth from the chamber
Passive parameters	Cutterhead rotational speed	x_4	rpm	The rotational speed of the cutterhead in revolutions per minute (*rpm*)
	SC pressure	x_5	bar	The pressure within the screw conveyor
	Volume of mixed foams	x_6	m^3	The volume of additives consumed during excavation, where additives mainly serve as lubricants for excavation
	Average earth pressure	x_7, x_8	Bar	The pressure in the cutterhead chamber, which is used to counter-balance the earth's pressure and prevent collapse. Sensors are installed on the middle left and right side of the chamber
	Net stroke	x_9	mm	Actual advance distance for each ring

layers and ends with a dense layer for result output. A summary of the structure of the proposed meta-model and hyper-parameter settings is shown in Tables 3 and 4, respectively.

The proposed deep learning meta-model employs a spatio-temporal prediction approach, in which the historical and present data of the key TBM operating parameters, x_1 to x_9 from timesteps $t-2$ to t and historical data of the 4 objectives, O_1 to O_4 from timesteps, $t-2$ to $t-1$ are utilized as the inputs for model training. This arrangement is practical since the actual performance of the TBM can be shown to the operators immediately after the excavation is executed. The training and validation loss function for each epoch for the four meta-models is plotted in Fig. 6. According to the figure, the loss function values drop rapidly in the first few epochs

Table 2 Data description of the input parameters

Factor	Count	Mean	Std	Min	25%	50%	75%	Max
x_1	10,710	8588.469	1749.285	1940	7658	8575.5	9855	13,904
x_2	10,710	885.362	341.135	298	619	796	1104	1995
x_3	10,710	22.874	11.491	− 2	14	21	31	60
x_4	10,710	1.931	0.133	1.4	1.9	2	2	2.5
x_5	10,710	57.198	17.368	12	43	55	69	114
x_6	10,710	1771.154	1395.753	27.685	754.703	1418.587	2369.550	6947.269
x_7	10,710	1.920	0.185	1.4	1.8	1.9	2.1	2.5
x_8	10,710	1.907	0.163	1.5	1.8	1.9	2	2.4
x_9	10,710	694.186	389.821	20	360	681	1021	1580

Table 3 Structure of the proposed GCN-LSTM model

Layer	Type	No. of neurons	Activation function
1	GCN	12	tan*h*
2	GCN	12	tan*h*
3	LSTM	265	*ELU*
4	LSTM	128	*ELU*
5	LSTM	64	*ELU*
6	Dense	1	*ReLU*

Table 4 Hyper-parameters for the GCN-LSTM models

Hyper-parameters	Settings
Epochs	30
Batch size	32
Loss function	*MAE*
Optimizers	*Adam*
Metrics	*MSE*

and stabilize afterward. The loss functions then converge with minimum variances after around 20 epochs, which indicates that 30 epochs sufficiently boost the models' performance.

Succeeding the establishment of deep learning models, these models can be used at the fit function to locate the optimum solutions through the proposed online NSGA-II. In this study, 200 datasets from the number 1000–1200 of the testing data are optimized for demonstration purposes. According to actual TBM operation, active parameters, x_1 to x_3 are adjustable while the rest operating parameters (x_4 to x_9) shall follow their original values to reflect the actual condition of the localized excavation condition. Additionally, the historical value of O_1 to O_4 from timesteps, $t - 2$ to $t - 1$ that are involved in the optimization process are determined based on the

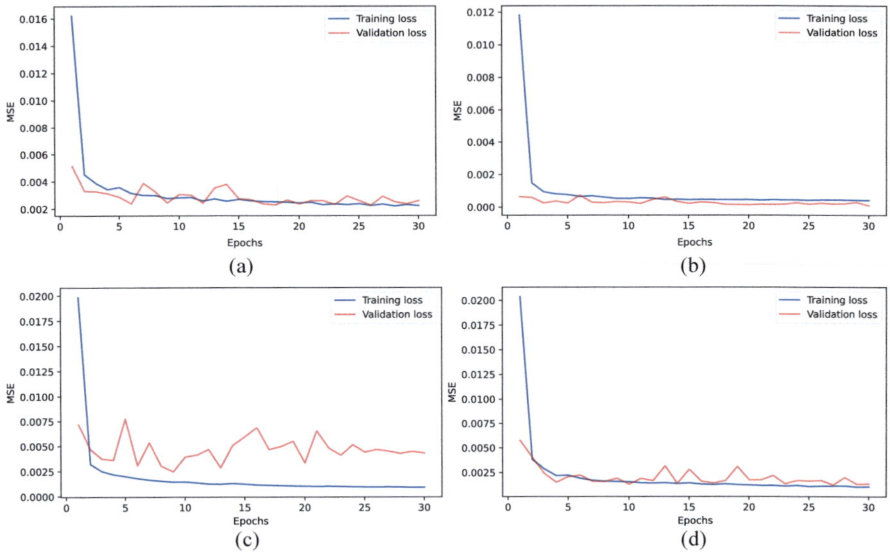

Fig. 6 Line plots of training and validation loss for the model training of **a** O_1, **b** O_2, **c** O_3, and **d** O_4

optimized values of O_1 to O_4 at timestep t. These values will be progressively updated according to the sequence of excavation. Proper boundaries shall be applied to the active parameters to simulate the actual operation condition. This is to prevent the optimized results from being unachievable or causing excessive damage to TBM.

For simulation and comparison, constraints of 20%, 30%, and 40% are applied for these parameters, which allows them to be adjusted in the range of [− 20%, + 20%], [− 30%, + 30%], and [− 40%, + 40%], respectively. To perform the NSGA-II algorithm, 200 populations are chosen with simulated binary crossover and polynomial mutation to generate the same number of offspring. The entire optimization process converges after numerous generations until the termination criteria are met or 50 generations are processed. From the 200 Pareto fronts for each dataset, the TOPSIS method is applied to select the most ideal individual which is the optimized parameter for TBM operational control.

4.3 DT-Enabled Real-Time Advanced TBM Control

The reliability of the proposed meta-models as well as the proposed optimization framework is further analyzed for performance evaluation. Common metrics including MAE, RMSE and R^2 are employed to validate the reliability of meta-models. Estimated performances after optimization are compared with the original values to examine the effectiveness of the proposed optimization framework. A detailed result analysis is presented as follows.

(1) The proposed deep learning method in the intelligent module provides a reliable estimation of TBM's performance indicators. Referring to Table 5 for the evaluation results by the intelligent module for all four indicators, the proposed models can predict the TBM performances with minimum errors. In particular, the MAEs produced by the models are 2.083 mm/min, 0.020, 0.399 kwh, and 3.89×10^{-5} mm for O_1 to O_4, respectively. RMSE results are as low as 2.760 mm/min, 0.035, 0.929 kwh, and 6.81×10^{-5} mm for O_1 to O_4, respectively. Although the numbers are at different scales, they are very small considering the scale of the original data. For instance, the original values of penetration rate vary from around 10 mm/min to 50 mm/min, while an MAE of 2.083 mm/min and RMSE of 2.760 mm/min is considered minimal. In addition, the prediction by the intelligent module generates high R^2 values which are 0.929, 0.978, 0.835, and 0.909 for O_1 to O_4, respectively. These high R^2 values further demonstrate that there are minimal differences between the prediction and ground truth. Line plots and scatter plots of the prediction against the original values for O_1 to O_4 are shown in Fig. 7a–d, respectively. It is clearly shown that the predicted deviations closely adhere to the original values in line plots and the points are close to the center line in scatter plots. These results further prove that the performance estimation by the established intelligent module is reliable. The predicted values and the historical results against the actual record are presented in the digital twin model as shown in Fig. 8, which will be helpful for the TBM operator to evaluate TBM's performance and for decision-making on the operation.
(2) TBM's overall performance can be effectively improved through the proposed online optimization method in the intelligent module. Referring to Table 6 for the mean values of the original performance indicators and values after optimization, the performance in all indicators and the overall performance can be improved regardless of the different constraints. In particular, the original performance indicators of O_1 to O_4 for the selected dataset have average values of 40.230 mm/min, 0.915, 1.574 kwh, and 2.797×10^{-4} mm for O_1 to O_4, respectively. Despite the different constraints implemented to TBM's operating parameters, the average penetration rate can be significantly increased to 50.931 mm/min, while the average over-excavation ratio, energy consumption, and tool wear can be reduced to 0.905, 1.202 kwh, and 2.111×10^{-4} m,

Table 5 Evaluation of meta-models for O_1 to O_4

Metrics	O_1 (mm/min)	O_2	O_3 (kwh)	O_4 (mm)
MAE	2.083	0.020	0.399	3.89×10^{-5}
RMSE	2.760	0.035	0.929	6.81×10^{-5}
R^2	0.929	0.978	0.835	0.909

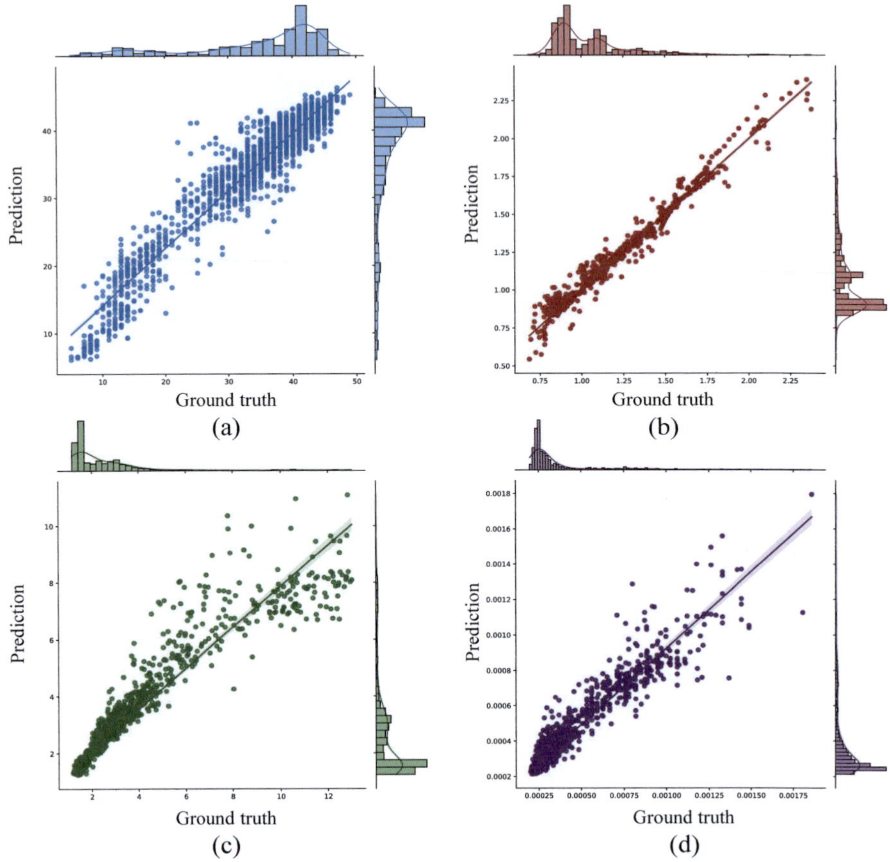

Fig. 7 Joint plots of predicted results against ground truth for **a** O_1, **b** O_2, **c** O_3, and **d** O_4

respectively. The result indicates that the proposed online optimization method creates an average improvement of 25.75%, 1.09%, 22.93%, and 23.46% for O_1 to O_4, respectively. Considering all four deviations together, there is an average overall improvement of 21.12% realized. The results clearly demonstrate the effectiveness of the proposed optimization method and the importance of having an intelligent module.

(3) The overall TBM performance improves with the release of the constraints on the operating parameters. Referring to Table 6 for the optimized performances for O_1 to O_4, under the constraints of 20%, 30%, and 40%, respectively, the indicators perform better with the increment in the allowable adjustment range for the operating parameters. For instance, the penetration rate (O_1) can be increased from 40.230 mm/min to 50.012 mm/min if the operating parameters can be adjusted by 20% from their original values. When the allowable range increases

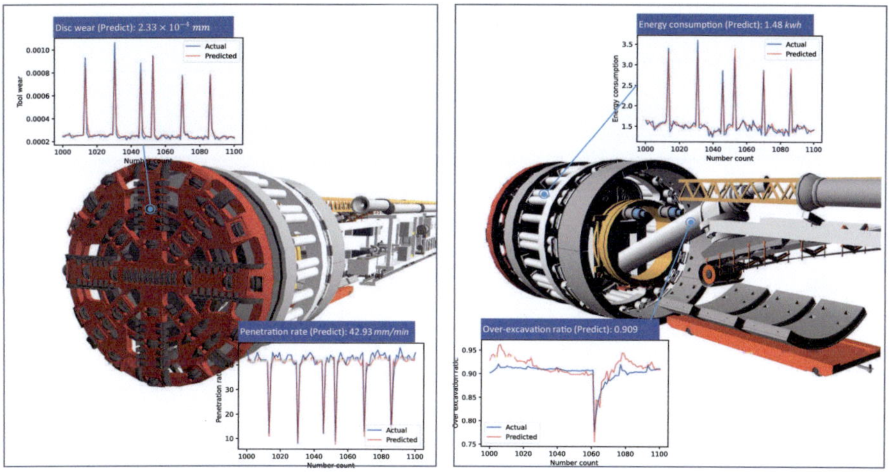

(a) (b)

Fig. 8 Virtual presentation on TBM's performance estimation for **a** penetration rate and disc wear and **b** over-excavation ratio and energy consumption

Table 6 Improvement results under different constraints with respect to 4 objectives

Constraint	Evaluation	O_1(mm/min)	O_2	O_3(kwh)	O_4(mm)	Overall
Before optimization	Mean (original)	40.230	0.915	1.574	2.797×10^{-4}	–
20%	Mean (optimized)	50.012	0.908	1.219	2.198×10^{-4}	
	Improvement (%)	23.92	0.88	21.98	21.53	19.74
30%	Mean (optimized)	51.151	0.904	1.197	2.091×10^{-4}	
	Improvement (%)	26.14	1.13	23.24	23.90	21.43
40%	Mean (optimized)	51.63	0.903	1.190	2.043×10^{-4}	
	Improvement (%)	27.20	1.26	23.56	24.96	22.18
Average	Mean (optimized)	50.931	0.905	1.202	2.111×10^{-4}	
	Improvement (%)	25.75	1.09	22.93	23.46	21.12

Note "–" indicates that it is not applicable

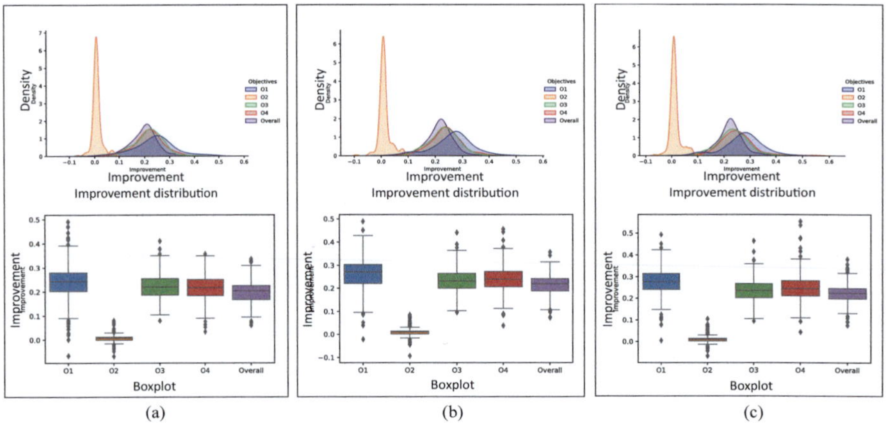

Fig. 9 Box and improvement distribution plots on single and overall objectives under **a** 20% constraints, **b** 30% constraints, and **c** 40% constraints

to 40%, it can be further increased to 51.63 mm/min, which implies an improvement of 27.20%. This conclusion is applicable to all performance indicators such that more improvement can be observed with higher allowable adjustment for the TBM operating parameters. As a result, the overall performance also improves with the release of the constraints. In specific, there is an overall improvement of 19.74%, 21.43%, and 22.18% when the allowable limit increased from [− 20%, + 20%] to [− 40%, + 40%], respectively. The improvements are also reflected in Fig. 9, where the boxplots and distribution for all objectives and overall improvement have higher values in Fig. 8c compared to those in Fig. 8a and b. Although the TBM performance can be further improved with higher allowable limits for the operating parameters, there shall be other constraints and safety concerns to prevent unlimited incremental of the parameters.

(4) The proposed intelligent module provides reliable guidance for advanced control during the TBM operation. Referring to Fig. 10a for an illustration example of the digital twin model, by performing the proposed online optimization method, the digital twin model is able to suggest the most appropriate control values for the active parameters including thrust force (x_1), cutterhead torque (x_2), and screw conveyor torque (x_3) based on the passive parameters of the localized condition. With the suggested values, the performance of TBM is expected to be improved based on the estimation by the trained deep learning models. As shown in Fig. 10b, the estimated performance through the digital twin model is generally improved in all 4 objectives when compared with the original values. The advanced virtual-to-physical control of TBM can be realized through either conventional adjustment or remote control. The conventional adjustment refers to the key operational parameters that can be manually adjusted by the TBM operator according to the suggested values from the DT model. The remote control can be achieved by transmitting the suggested operational parameters

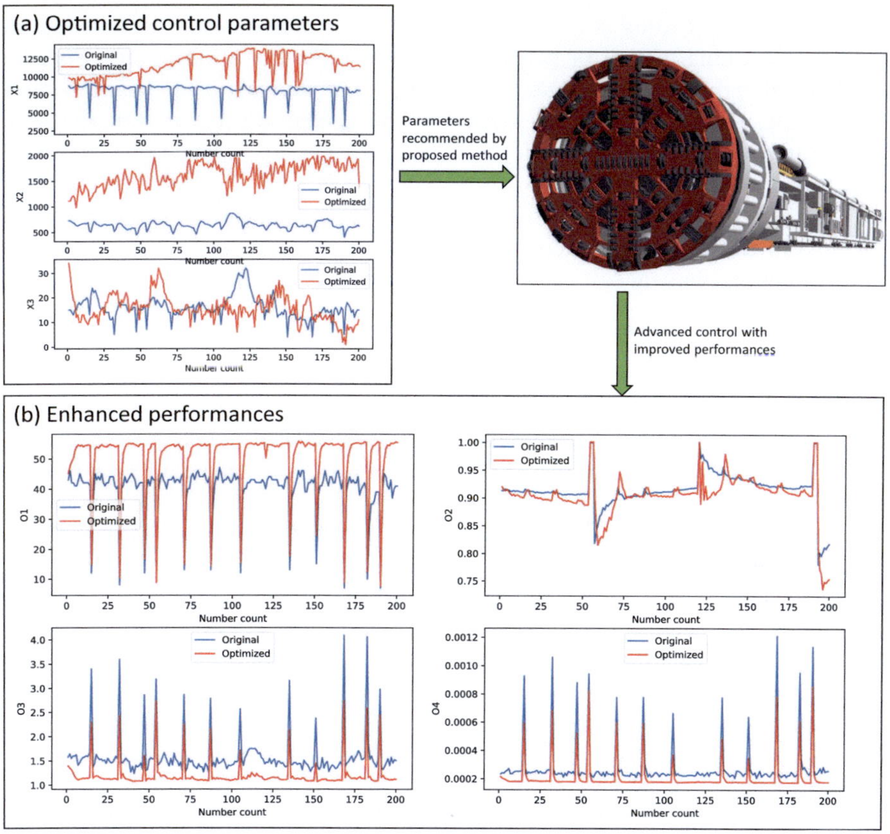

Fig. 10 Realization of virtual-to-physical control through the digital twin model for **a** suggested control parameters after optimization and **b** estimated enhanced performances on the 4 objectives

directly to the TBM through signal transmission. Consequently, the actual TBM operation can be guided by the digital twin model with enhanced performance.

5 Merits Analysis

One of the novelties of the proposed method in the intelligent module is that it applies the technique of online optimization instead of conventional optimization. Instead of only considering the spatial factors of the same timestep, the historical values of TBM performance indicators are derived according to the optimized results instead of their original values. To further validate the importance of online optimization, a comparison study is conducted to evaluate the influence of NSGA-II on online updating.

The comparison study is conducted by using the same dataset while using the same deep learning model for performance estimation. For a fair comparison, the same operation parameters (x_1 to x_9) are utilized for model training and optimization. In particular, x_1 to x_3 are the active parameters that can be adjusted within [+ 40%, – 40%] of their original values, while x_4 to x_9 are the passive parameters that remain at their original values. The major difference in this study is in the optimization part, in which the result of the proposed method is compared with the conventional NSGA-II in earlier studies. A comparison of the optimized performances and box and distribution plots on respective objectives is presented in Table 7 and Fig. 11, respectively.

According to the results, the proposed online optimization method is more practical and creates better improvement than the conventional MOO approach. Referring to Table 7 for the mean optimized TBM performances for O_1 to O_4, the conventional optimization approach can only improve the performance indicators by 14.73%, 0.58%, 14.49%, and 16.26% for O_1 to O_4, respectively, with an overall improvement of 13.40%. In comparison, the proposed online optimization methods managed to improve the performance indicators by 27.20%, 1.26%, 23.56%, and 24.96% for O_1 to O_4, respectively. Despite that same active operating parameters are adjusted with the identical range of allowable limit, the proposed online optimization achieves an overall improvement of 22.18%, which is 65.52% [(22.18%–13.40%)/13.40%] better than the conventional optimization method. Such a result can be clearly reflected in the box and distribution plots in Fig. 11, where the major parts of boxplots for both single and overall objectives are much higher than the reference line in the proposed online optimization method.

Previous results from the case study have demonstrated the effectiveness and practicality of the proposed digital twin model. According to the proposal, big data generated from the physical object will be transmitted into the digital model for real-time monitoring of the operating status. The integrated intelligent module can then estimate the performance and recommend the most optimum parameter under in-situ

Table 7 Comparison of the improvement under 40% constraints

Constraint	Evaluation	O_1(mm/min)	O_2	O_3(kwh)	O_4(mm)	Overall
Before optimization	Mean (original)	40.230	0.915	1.574	2.797×10^{-4}	
Proposed method	Mean (optimized)	51.63	0.903	1.190	2.043×10^{-4}	
	Improvement (%)	27.20	1.26	23.56	24.96	22.18
Without updates	Mean (optimized)	46.129	0.914	1.339	2.291×10^{-4}	
	Improvement (%)	14.73	0.58	14.49	16.26	13.40

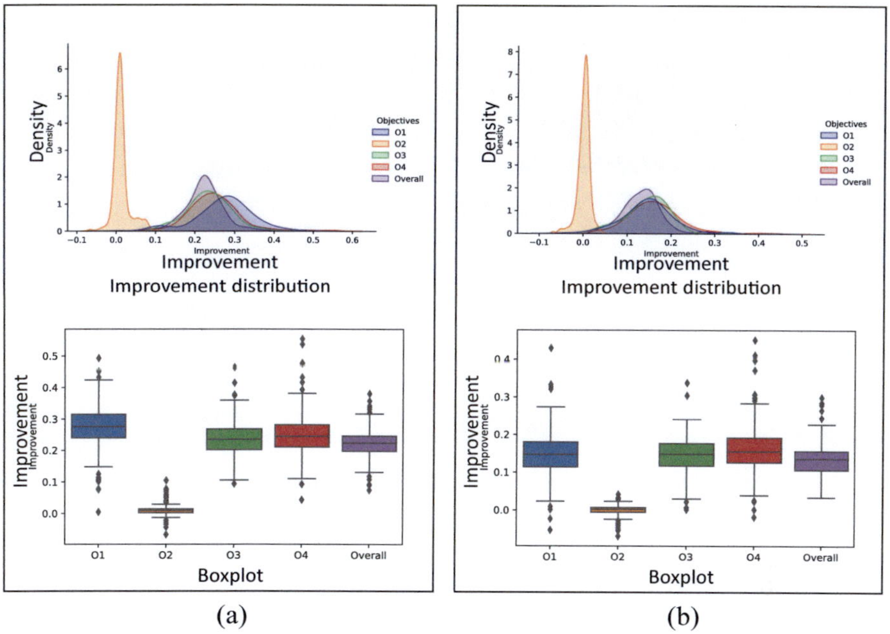

Fig. 11 Box and improvement distribution plots on the improvements for **a** online optimization and **b** conventional optimization

circumstances. With the successful implementation of our proposed idea, the TBM operator can just follow the suggested values to achieve the optimum performance for the TBM. With the continuous advancement of technologies including IoT and the 5th generation of cellular networks (5G), a more efficient and reliable real-time data transmittal can be achieved and the TBM can even be operated by experts from above ground or autonomously. This will further reduce the labor cost, eliminate the potential safety hazards, and improve the overall level of comfort for the tunnel excavation environment. Continuous implementation of the proposed technology will progressively improve the level of intelligence for the digital model since the collection of more data can cover a wider range of geological conditions. An illustrative sketch that prospects the future autonomous TBM is displayed in Fig. 12.

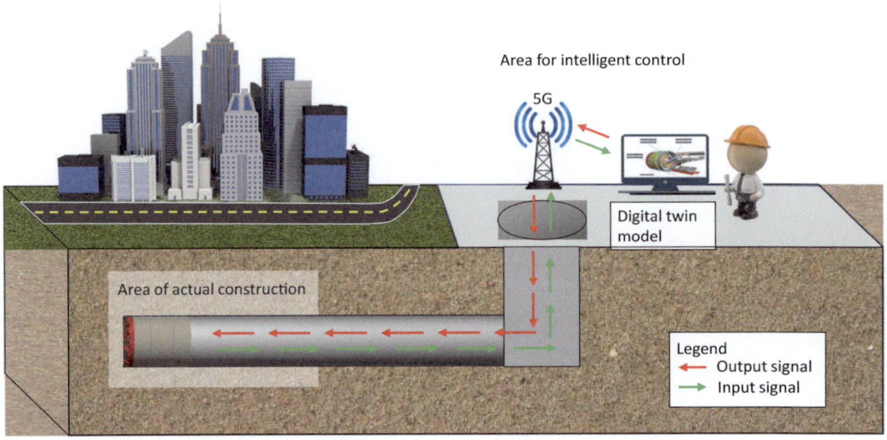

Fig. 12 Conceptual ideal for future tunnel excavation: remote control through digital twin interaction

6 Chapter Summary

This paper establishes a digital twin model that enables real-time physical-cyber interaction for monitoring the operating status and enhances the overall performance of TBM. Through the IoT sensors that are installed on the TBM, the actual operating status of the TBM is transferred from the physical object to the digital model. The collected data is further analyzed for performance estimation by the proposed deep learning method that integrates GCN and LSTM. An online MOO approach is then proposed to determine the optimal operating parameters that can enhance the overall performance of TBM. The developed approach can be used as an intelligent tool to assist tunnel operation and improve TBM performance under varying conditions.

Real data from Singapore's Circle Line 6 C885 project is utilized as a case study to examine the applicability of the developed digital twin model. A total of 9 operational parameters are extracted from the data logger with 3 parameters for optimization and the remaining 6 to represent the localized condition. TBM's penetration rate, over-excavation ratio, energy consumption, and tool wear are selected as the objectives (O_1 to O_4) for optimization. The results indicate that (1) The digital twin model can provide a reliable estimation of TBM's performance indicators, with high R^2 values of 0.929, 0.978, 0.835, and 0.909 for O_1 to O_4, respectively. (2) TBM's overall performance can be effectively improved through the proposed online optimization method, with an average improvement of 25.75%, 1.09%, 22.93%, and 23.46% for O_1 to O_4, respectively, and an overall improvement of 21.12%. (3) The proposed digital twin model provides reliable guidance and advanced control during the TBM operation. (4) The proposed online optimization method is more practical and creates a better improvement than the conventional MOO approach, with an overall improvement of 22.18% against 13.40%.

The major contributions can be summarized below: (1) A novel intelligent approach is proposed that integrates a DT, deep-learning, and MOO methods to guide TBM operation with enhanced performances in real-time; (2) An online-NSGA-II method is developed that considers the impact from both spatial and temporal factors from TBM operating parameters for better practicality, where critical objectives including penetration rate, over-excavation ratio, energy consumption, and tool wear are considered for a more realistic optimization during TBM operation. The developed holistic approach can be used as a decision tool to achieve real-time monitoring and control to enhance the overall performance of tunnel boring machines (TBM) during excavation.

References

1. Pan Y, Zhou X, Qiu S, Zhang L (2023) Time series clustering for TBM performance investigation using spatio-temporal complex networks. Exp Syst Appl 225:120100
2. Zhou C, Kong T, Zhou Y, Zhang H, Ding L (2019) Unsupervised spectral clustering for shield tunneling machine monitoring data with complex network theory. Autom Constr 107:102924
3. Gao X, Shi M, Song X, Zhang C, Zhang H (2019) Recurrent neural networks for real-time prediction of TBM operating parameters. Autom Constr 98:225–235
4. Wang F, Li H, Dong C, Ding L (2019) Knowledge representation using non-parametric Bayesian networks for tunneling risk analysis. Reliab Eng Syst Saf 191:106529
5. Deb K, Agrawal S, Pratap A, Meyarivan T (2000) A fast elitist non-dominated sorting genetic algorithm for multi-objective optimization: NSGA-II. In: Schoenauer M, Deb K, Rudolph G, Yao X, Lutton E, Merelo JJ, Schwefel H-P (eds) Parallel problem solving from nature PPSN VI. Springer, Berlin Heidelberg, Berlin, Heidelberg, pp 849–858
6. Zhang L, Guo J, Fu X, Tiong RLK, Zhang P (2024) Digital twin enabled real-time advanced control of TBM operation using deep learning methods. Autom Constr 158:105240
7. Xie X, Merino J, Moretti N, Pauwels P, Chang JY, Parlikad A (2023) Digital twin enabled fault detection and diagnosis process for building HVAC systems. Autom Constr 146:104695
8. Pregnolato M, Gunner S, Voyagaki E, De Risi R, Carhart N, Gavriel G, Tully P, Tryfonas T, Macdonald J, Taylor C (2022) Towards civil engineering 4.0: concept, workflow and application of digital twins for existing infrastructure. Autom Constr 141:104421
9. Naderi H, Shojaei A (2023) Digital twinning of civil infrastructures: current state of model architectures, interoperability solutions, and future prospects. Autom Constr 149:104785
10. Shim C-S, Dang N-S, Lon S, Jeon C-H (2019) Development of a bridge maintenance system for prestressed concrete bridges using 3D digital twin model. Struct Infrastruct Eng 15:1319–1332
11. Lu Q, Parlikad AK, Woodall P, Ranasinghe GD, Xie X, Liang Z, Konstantinou E, Heaton J, Schooling J (2020) Developing a digital twin at building and city levels: case study of West Cambridge campus. J Manag Eng 36:05020004
12. Jiang Y, Li M, Guo D, Wu W, Zhong RY, Huang GQ (2022) Digital twin-enabled smart modular integrated construction system for on-site assembly. Comput Ind 136:103594
13. Lin K, Xu Y-L, Lu X, Guan Z, Li J (2021) Digital twin-based collapse fragility assessment of a long-span cable-stayed bridge under strong earthquakes. Autom Constr 123:103547
14. Yu G, Wang Y, Mao Z, Hu M, Sugumaran V, Wang YK (2021) A digital twin-based decision analysis framework for operation and maintenance of tunnels. Tunn Undergr Space Technol 116:104125

15. Park J, Kim K, Cho YK (2017) Framework of automated construction-safety monitoring using cloud-enabled BIM and BLE mobile tracking sensors. J Constr Eng Manag 143:05016019
16. Cheng JCP, Chen W, Chen K, Wang Q (2020) Data-driven predictive maintenance planning framework for MEP components based on BIM and IoT using machine learning algorithms. Autom Constr 112:103087
17. Lee D, Lee S, Masoud N, Krishnan MS, Li VC (2022) Digital twin-driven deep reinforcement learning for adaptive task allocation in robotic construction. Adv Eng Inf 53:101710
18. Hamledari H, Davari S, Azar ER, McCabe B, Flager F, Fischer M, UAV-Enabled Site-to-BIM Automation: Aerial Robotic- and Computer Vision-Based Development of As-Built/As-Is BIMs and Quality Control, in: Construction Research Congress 2018, 2018, pp. 336–346.
19. Fu X, Wu M, Ponnarasu S, Zhang L (2023) A hybrid deep learning approach for dynamic attitude and position prediction in tunnel construction considering spatio-temporal patterns. Exp Syst Appl 212:118721
20. Zhou C, Ding L, Zhou Y, Zhang H, Skibniewski Miroslaw J (2019) Hybrid support vector machine optimization model for prediction of energy consumption of cutter head drives in shield tunneling. J Comput Civil Eng 33:04019019
21. Pan Y, Fu X, Zhang L (2022) Data-driven multi-output prediction for TBM performance during tunnel excavation: an attention-based graph convolutional network approach. Autom Constr 141:104386

Conclusions and Future Trends

1 Main Conclusions

The construction industry, including tunneling projects, is entering the next generation to boost the demand for intelligent development. The goal of intelligent construction, which extends to tunneling engineering, is to deliver automatic, smart, reliable, and sustainable construction engineering products and services by leveraging emerging information technologies [1]. Applications of intelligent construction are gaining more attention and becoming increasingly extensive with the purpose of maximizing the potential value in full-lifecycle project management. To capture the valuable opportunities of intelligent construction, this chapter summarizes the body of critical knowledge concerning the theoretical, methodological, and technological advancements.

The development of intelligent construction, including tunneling projects, relies heavily on various advanced artificial intelligence methods. AI technologies, such as Building Information Modeling (BIM) and cloud platforms, facilitate the standardization of design, construction, and operation, including tunnel design and excavation processes. The development of BIM, DT, and other technologies has greatly promoted the digital transformation of tunneling projects by allowing real-time data collection and decision-making throughout the tunnel construction process. Intelligent algorithms, such as machine learning and deep learning, can significantly enhance the intelligence and automation of tunneling operations, improving efficiency and safety. Technologies such as the IoT, cloud platforms, and blockchain can effectively promote collaboration across different projects. Additionally, methods such as generative design and automatic design further expedite the development of tailored tunneling solutions, enhancing the adaptability and precision of tunnel designs. These AI-driven strategies collectively advance the construction and tunneling sectors, providing solutions to modern project management challenges, including tunnel

L. Zhang et al., *Intelligent Construction in Tunnels*, Lecture Notes in Civil Engineering 679, https://doi.org/10.1007/978-981-96-8785-5_15

design, excavation, and long-term maintenance. Figure 1 illustrates the development strategies of intelligent construction, and Table 1 presents their advantages. The main benefits of intelligent construction practices are elaborated below.

(1) Standardization. Standardization is one of the key development strategies for intelligent construction, including tunneling projects, with the purpose of improving construction efficiency and enhancing process transparency. However, achieving standardization in tunneling is complex due to the diverse construction environments, processes, and equipment involved, making it more challenging than standardizing factory production [2]. Additionally, differing preferences and working habits among personnel complicate the establishment of universally accepted standards. Importantly, standardization in intelligent buildings encompasses the entire construction cycle, including processes, components, organization, and product quality, and extends to the specific requirements of tunneling engineering, such as tunnel lining, segment casting, and excavation techniques.

This approach offers significant benefits, particularly in tunneling, where standardization facilitates the coordination of the construction of tunnel segments, shields, and other components through a centralized database. It reduces complexity and uncertainty while fostering engineering standardization in the challenging environments of tunnel construction. Standardized components, such as tunnel segments, segmental lining systems, and pre-cast concrete elements, are widely applied in intelligent tunneling. Moreover, established standardized processes, including those for steel reinforcement, welding, and segment installation, contribute to improving efficiency and quality in tunneling projects. The large-scale application of BIM technology in construction, including tunneling, facilitates the integration of various tunnel segment libraries and model interfaces, which has greatly advanced the development of standardization in intelligent construction and tunneling.

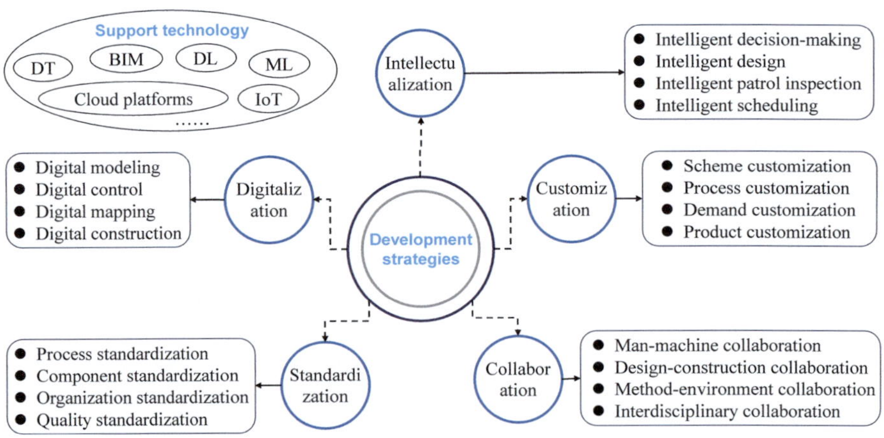

Fig. 1 Development strategies of intelligent construction

Table 1 The benefits of development strategies in intelligent construction

Strategies	Traditional construction	Intelligent construction	References
Standardization	• Lack of unified standards • Rely on specific construction personnel • The project results are not generalized • High complexity and uncertainty	• Implement unified standards • Rely on personnel and intelligence • The project results are generalized • Low complexity and uncertainty	[8]
Digitalization	• Information is recorded through paper documents • Information is transmitted through personnel • Information is calculated through manuals • Information is not real-time	• Information is recorded by the computer • Information is transmitted through the network • Information is calculated through software • Information is real-time	[9]
Intellectualization	• Taking humans as the carriers • Manual or software-aided calculation • Risk control driven by fewer factors • Feasible decision-making	• Taking smart devices as the carriers • Intelligent computing • Risk control driven by massive factors • Optimal decision-making	[10, 11]
Collaboration	• Manual control equipment operation • Weak equipment correlation • Poor environmental adaptability • Collaboration in a single field	• Automatic control equipment operation • Strong equipment relevance • Good environmental adaptability • Interdisciplinary collaboration	[12]
Customization	• Construction product orientation • Regardless of user preferences • Traditional and universal design techniques • Cast-in-situ structure dominated	• User demand orientation • Consider user preferences • Advanced techniques such as generative design, flexible design, and topology optimization • Prefabricated structure dominated	[13]

(2) Digitization. The construction industry, including tunneling projects, is undergoing a rapid digital transformation to integrate new technologies. The procedure for digital transformation involves various digital tools that significantly enhance efficiency, automation, and fault tolerance in construction project management [3]. Similar to the digitalization of industrial products, the digitalization of construction is mainly realized through computer software, enabling the mapping, tracking, and control of everything in a digital world. The main difference between the construction industry and industrial product manufacturing lies in the nature of the digitalized objects. While industrial products typically exhibit high fault tolerance, construction projects—particularly tunneling—cannot tolerate similar levels of risk, as errors may lead to substantial safety and operational issues. Consequently, digitalization serves as a critical enabler for intelligent construction, offering robust solutions to mitigate potential failures.

On the one hand, digitalization has the potential to establish a digital platform for the construction industry, including tunneling, to optimize construction processes and schedules, particularly during pre-construction phases. This facilitates significant time savings in information recording and communication. In tunneling, it enables the optimization of tunnel route planning, excavation strategies, and resource allocation. On the other hand, digitalization can promote the development of engineering software with higher computing power and stronger storage capacity, thereby addressing the advanced technological needs of the construction sector. Technologies such as sensors, 3D modeling, and digital twins support the digitization of tunnel designs, excavation progress monitoring, and predictive maintenance strategies. Digital offices can be implemented across the entire tunnel design, construction, and maintenance cycle, providing real-time monitoring and analysis. As digitalization becomes integral to intelligent construction, its applications in tunneling are exemplified by digital tunnel models, software control systems for TBMs, and advanced digital analysis techniques for risk management, progress tracking, and performance optimization.

(3) Intellectualization. The traditional engineering design and decision-making processes, including those in tunneling projects, are mostly completed by individuals. However, limited professional knowledge and experience cannot always provide optimal solutions, especially in complex tunnel construction scenarios involving unforeseen geological conditions and safety challenges [4]. To handle this issue, there is growing emphasis on integrating an expansive knowledge base, supercomputing capabilities, and AI systems to replicate human-like intelligence and support decision-making. Consequently, intellectualization has become a crucial aspect of intelligent construction, especially in tunneling, where real-time data and intelligent systems are essential for guiding the actions and decisions of computer systems. AI-driven systems in tunneling can enhance planning, risk analysis, and process optimization, allowing for smarter and more adaptable decision-making during tunnel excavation.

Intellectualization offers valuable opportunities to achieve a high degree of intelligence in tunneling, which can effectively address complex problems across various aspects, such as intelligent decision-making for TBM operation, intelligent tunnel design, real-time monitoring of tunnel stability, and optimized scheduling of construction phases. On the one hand, intellectualization not only reduces production costs but also enhances construction efficiency and safety [5]. For example, AI models can predict potential issues such as ground subsidence or water ingress, enabling preemptive action to avoid delays and costly repairs. On the other hand, intellectualization provides practical solutions to the traditionally low-intelligence field of tunneling by automating routine tasks and refining construction strategies. Smart construction sites, including those in tunneling, have fully utilized intelligent technology, with applications in equipment control, environmental sensing, tunnel monitoring, and material transportation. When combined with advanced techniques such as deep learning, reinforcement learning, and evolutionary algorithms, these systems support engineers in making informed decisions, thereby improving the safety, efficiency, and precision of tunneling operations.

(4) Collaboration. Collaboration is essential for advancing intelligent construction, including in tunneling projects, as it streamlines processes and reduces dependency on personnel. Effective coordination is critical in tunneling, where the integration of personnel, equipment, and technology is important due to the complexity and risks of underground work. For instance, effective cooperation between personnel and equipment, such as TBMs and other mechanized tools, can reduce the need for on-site workers while improving efficiency and safety [6]. Real-time communication between the TBM and support teams allows for prompt adjustments in excavation strategy, minimizing downtime and preventing accidents in challenging conditions like soft soil or high groundwater pressure. Similarly, collaboration between designers and construction teams ensures accurate implementation of tunnel designs and feedback regarding construction processes. The interaction between the tunnel and the surrounding geological environment is critical for managing factors such as ground stability, environmental safety, and ventilation. This multidisciplinary approach strengthens the tunneling process, enabling effective responses to unforeseen challenges like rock falls or soil deformation. Furthermore, the integration of tunnel construction entities with virtual twin systems, wherein real-time data from operations is incorporated into digital models, enhances decision-making for both immediate adjustments and long-term planning.

Such collaboration fosters interconnections among various elements of tunneling projects, facilitating interdisciplinary and multidisciplinary advancements. This approach is crucial, as it encourages a collaborative environment that brings together experts from diverse fields such as civil engineering, geotechnical analysis, and tunnel machinery operations. For example, civil and surveying expertise informs the environmental analysis of tunnel sites, while AI optimizes excavation strategies and predicts potential risks, guiding TBM operations. By leveraging collective expertise and effective communication, tunneling projects can achieve holistic problem-solving and continuous improvements. Technologies like the IoT, cloud platforms, and edge computing further enhance collaboration by connecting man–machine-environment units, allowing for real-time monitoring and data sharing across the project. These technological advancements significantly elevate the coordination and decision-making capabilities in tunneling construction, promoting safer, more efficient, and cost-effective tunneling operations [7].

(5) Customization. Customization is a new requirement of the era for intelligent construction, and it presents a major advantage compared with traditional construction practices, including tunneling projects. In the context of tunneling, customization facilitates the design and fabrication of tunnel structures that are specifically adapted to the unique geological conditions, safety considerations, and environmental factors of each project. This tailored approach ensures that tunnel designs and construction strategies are optimized for the specific challenges posed by each site, enhancing both safety and efficiency throughout the project lifecycle.

Furthermore, the increasing demand for individualized customization in tunneling projects facilitates the development of tailored engineering solutions that address the unique requirements of each project. For instance, tunneling in challenging environments, such as beneath rivers, within seismically active zones, or through unstable rock formations, necessitates specialized approaches that conventional tunnel designs cannot accommodate. The application of advanced techniques such as generative design, flexible design, and topology optimization further enhances this customization, enabling more efficient structural configurations and improved integration with adjacent infrastructure. These technologies allow engineers to evaluate multiple design alternatives and select the most suitable configuration based on the specific challenges posed by the tunneling environment. Ultimately, such innovations contribute to safer, more efficient, and cost-effective tunneling construction practices.

2 Future Trends

Supported by massive data and advanced data analysis approaches, the trend of intelligent construction, including tunneling projects, will embrace the incoming Industry 5.0 to meet the ever-growing demand for digitalization, informatization, and green transition. Tunneling projects, which are inherently complex and environmentally sensitive, can greatly benefit from intelligent construction practices. Leveraging AI and machine learning to analyze large-scale geological data, predict risks, and optimize excavation methods can enhance the safety, efficiency, and environmental sustainability of tunneling operations. However, challenges persist in making intelligent construction more practical, human-centered, and sustainable.

Firstly, although the promise of the digital twin in enhancing project lifecycle insights, its practical application in construction is at the initial phase. In tunneling, digital twins can be used to simulate the tunnel's environment, monitor real-time data from sensors embedded within tunnel structures, and predict potential risks such as ground shifts or water ingress. However, a more standardized and integrated engineering platform is necessary to fully realize their potential, enabling seamless data integration and efficient collaboration across tunnel design, construction, and maintenance. Secondly, the traditional data-driven decision-making process in tunneling often overlooks human and environmental factors, which can influence engineering systems. Adaptive learning mechanisms are essential to account for dynamic factors such as soil composition, groundwater levels, and seismic activity, ensuring more effective decision-making. Furthermore, the sector faces significant challenges in meeting carbon peak and neutrality goals, as tunneling is energy-intensive. To achieve sustainability, innovative technologies must optimize energy consumption, incorporate electric and hybrid machinery, and employ sustainable materials for tunnel construction. To tackle these challenges, several emerging trends are shaping the future of intelligent construction, which are provided below.

(1) **Digital twin-enabled engineering cloud platform**

The digital twin, known as a cyber-physical integration, enhances virtual models by incorporating big data for real-time modeling and analysis. In tunneling, this technology enables the creation of dynamic, detailed representations of tunnel construction and operation, integrating real-time sensor data to monitor structural status, geological changes, equipment performance, and potential risks like ground instability or water ingress. Therefore, the robust frame of digital twins can be extended to build the new generation of an engineering cloud platform with the combination of IoT devices, AI, cloud computing, and other advanced informatics techniques for continuous, intelligent monitoring and analysis. As shown in Fig. 2, the digital twin-enabled engineering cloud platform is made up of three important components, including multi-dimensional digital modeling and AI techniques, intelligent construction machinery equipment and robotics, and ubiquitous perception and IoT. These three components interact synergistically and dynamically to create a complete operation system, which can be regarded as the key technology framework of intelligent construction. Moreover, the cloud environment supports the establishment of the digital twin, where information is constantly and automatically accumulated to form a knowledge base for reasoning and decision-making.

In short, the digital twin-enabled engineering cloud platform inherits advantages from both digital twin and cloud services, driving intelligent construction towards a higher level of digitalization, networking, and intelligence. For one thing, it is equipped with the self-evolution and self-learning ability to facilitate the interaction between the physical and virtual space. Under the integration of various information technologies, it can reach digital-physical convergence and high fidelity from data acquisition to information transmission to digital modeling to decision making to practical feedback [14]. That is to say, it can easily provide a holistic view of the complex construction project such as tunnels in a close loop to meet the ever-growing

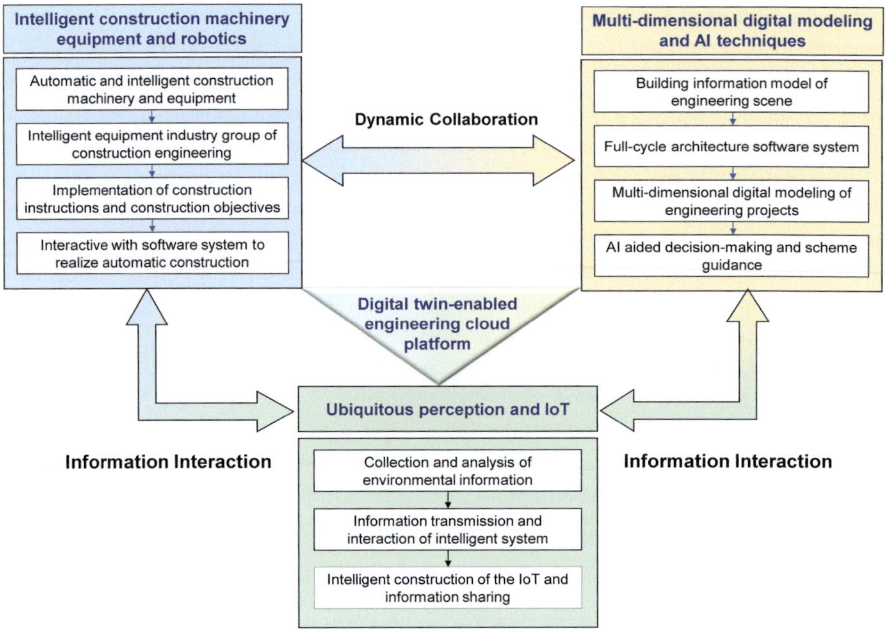

Fig. 2 Overall framework of a digital twin-enabled engineering cloud platform

demand for automation, interoperability, and information cohesion [15]. For another, it streamlines the smart functions across the platform, infrastructure, and project participants to realize high-efficiency delivery and increased resilience throughout the entire lifecycle of a project. In particular, the cloud-based solution serving as an enabler of innovation delivery of emerging techniques can bring benefits of security, flexibility, transparency, cost saving, near real-time collaboration, quality control, and others [16].

Although digital twin can coordinate all aspects of building information and realize the interaction and integration of various technologies, there is still a certain knowledge gap in the development of digital twin. Due to the confidentiality requirements of engineering projects and the protection of intellectual property rights of construction units, a large number of engineering data cannot be shared at present. Especially in the tunnel field, due to the huge differences between tunnel projects, massive data types, and high difficulty in data collection, the open data in tunnel construction is relatively scarce. However, massive data is the basis of intelligent construction technologies, especially for digital twins, and it guarantees the authenticity and reliability of the twin model. A large number of confidential engineering data can easily lead to the fragmentation of the construction industry. For example, since historical data cannot be shared, it is difficult to learn from the experience of previous related projects when facing a new construction project. This could unfortunately bring a lot of cost consumption and resource waste. In general, more open-source data is a necessary condition for the intelligent development of the construction industry, which requires the joint efforts of the government and academia.

(2) Man–machine-environment interaction

For the core of the digital model, it is necessary to take a full account of multi-source information, including monitoring data, expert experience and judgment, project regulations and documents, engineering mechanisms, and environment feedback, all aimed at enhancing the reliability of the model. In tunneling, this means integrating data from sensors embedded in tunnel structures, real-time information from TBMs, geological data, and feedback from environmental conditions such as soil moisture, temperature, and ground stability. Currently, most modeling processes for intelligent algorithms only depend on IoT data, such as sensor readings from the tunnel construction site, but this approach alone cannot always ensure the quality and fidelity of analytical results.

Therefore, a question that is worthy of exploring is to perform the human–machine-environment interaction. It is a new evolving synthesis of frontier science to deeply explore the three elements named human, machine, and environment, and their relationships, leading to a more trustworthy and efficient engineering system. The structure of human–machine-environment interaction is demonstrated in Fig. 3. It can be understood as a proper arrangement of humans and machines interacting with the environment to make the information acquisition and analysis more resilient. As expected, the flow of information can run smoothly, and the engineering system can be optimized continuously under the mutual harmony of the three elements. In the interactive system, both the human and machine can be perceived from the environment, and then the human can fully understand the current state of the environment to inform decision-making for controlling the machine. Moreover, the synergy of the three key elements contributes to efficiently and accurately creating a data-physical-driven model as a more realistic reflection of actual engineering conditions, which can more flexibly change according to the project environment.

In this research direction, how to build the human-in-the-loop system remains a problem worthy of investigation. However, it is not a straightforward task to augment human interaction with the physical world. Difficulties mainly come from the unstructured human knowledge, which is hard to be directly incorporated into

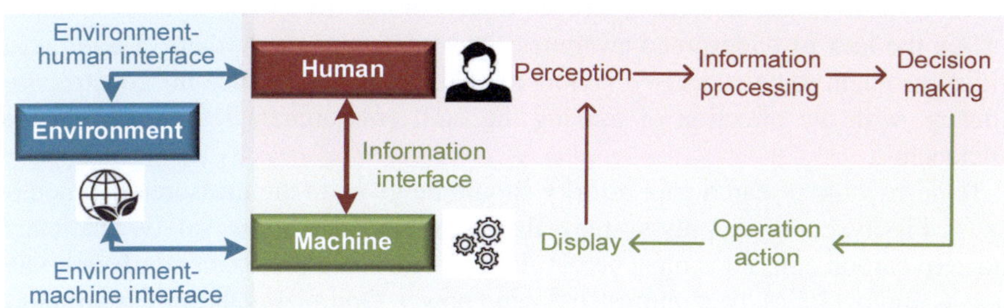

Fig. 3 Implementation process of human–machine-environment interaction

the system. Meanwhile, the incorporation of human knowledge into the modeling process covers an extensive research area, including computer science, cognitive science, and psychology. To bridge the research gap, natural language processing (NLP) plays important roles in text classification and summarization, syntactic and semantic parsing, question answering, sentiment analysis, and others. It can be helpful to build a close loop simulating humans' cognitive thinking and reasoning ability. Apart from human knowledge, the dynamic and complexity of the machine and environment being investigated also cause great difficulties in realizing the man–machine-environment interaction. Since the status of machines and environments in construction is constantly changing, it is required to develop an adaptive mechanism to conduct timely adjustments in the decision-making process. Since the digital twin shows great potential in dynamic environment modeling and real-time decision reasoning, the function of man–machine-environment interaction could be seamlessly incorporated into it. This helps to further empower the digital twin, enabling it to be constantly updated to represent current observations, associated knowledge, and decision-making in an intuitive and easy manner.

(3) **Great Sustainability under carbon peaking and carbon neutrality goals**

For a long period, the construction industry, including tunneling projects, has not seriously taken environmental impact into consideration, which has become a big user of natural resources to burden the ecological environment. The construction industry is responsible for the embodied energy within the building and tunnel materials, as well as the substantial energy consumption during excavation, transportation, and the operation of heavy machinery such as TBMs. With the growing concerns over climate change and finite natural resources, there are great opportunities to bring positive improvements towards sustainable construction. In 2021, China issued a policy document aimed at achieving carbon peaking and carbon neutrality, which outlines a roadmap for transitioning to a low-carbon economy [17]. This initiative provides a framework for balancing long-term environmental benefits with short-term economic goals, thereby advancing the adoption of green and low-carbon construction strategies.

As the key area for peaking carbon dioxide emissions, urban and rural construction is a huge concern. However, realizing carbon peaking and carbon neutrality goals in the construction industry is challenging due to several factors, such as the use of carbon-intensive materials, the nature of project complexity and life cycle assessment (LCA), the lack of widespread awareness and innovation, and others. Therefore, a critical problem to be addressed is how to effectively decarbonize the construction industry, with the intention of making the built environment reach great carbon efficiencies.

To narrow the research gap, priority should be given to the measures outlined in Fig. 4. Firstly, energy supply systems that prioritize cleanliness, safety, efficiency, and zero carbon emissions must be developed, with an emphasis on non-fossil fuels such as photovoltaics, heat pumps, and wind power. Optimizing the energy structure by reducing fossil fuel use is crucial. Secondly, the adoption of renewable, recycled, and repurposed materials should be prioritized to reduce carbon footprints in green

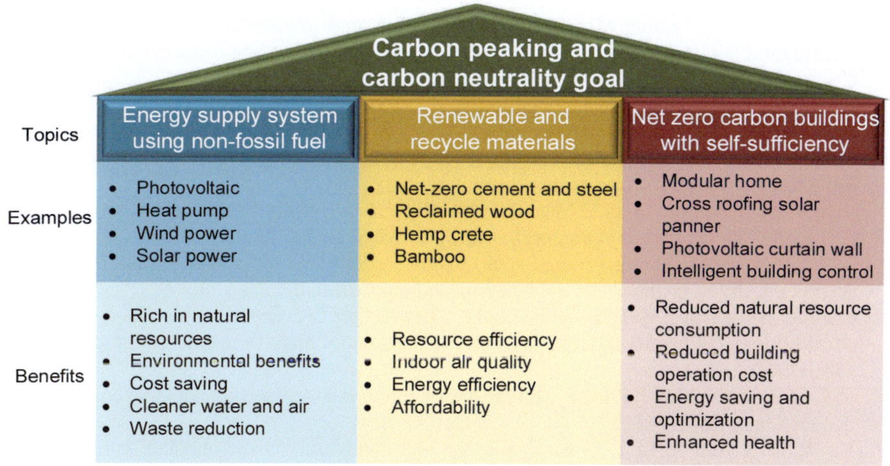

Fig. 4 The supporting measures towards the goal of carbon peaking and carbon neutrality

construction. Notably, cement and steel, which account for 6.5% and 7.0% of global CO_2 emissions, respectively, require cleaner manufacturing processes, such as CO_2 storage in concrete, steel recycling, and advancements in production technologies. Thirdly, transitioning to net-zero carbon buildings, capable of self-sufficient energy production, should be accelerated. While the initial and maintenance costs of these buildings may exceed those of conventional ones, and market penetration remains low, long-term savings and environmental benefits justify the investment. However, the high upfront costs and limited market demand pose challenges to widespread adoption.

Another thing to be mentioned is that the digital twin technology can actually serve as a useful tool to realize carbon emission prediction and control. Herein, digital twin can be understood as the cooperation of physics-based software, real data from the building, and machine learning algorithms. By integrating the intelligent services of building decarbonization and optimization into the digital twin-based platform, data-driven decisions can be more flexibly informed to prioritize sustainability goals throughout the full lifecycle of buildings.

References

1. Zhang L, Li Y, Pan Y, Ding L (2024) Advanced informatic technologies for intelligent construction: a review. Eng Appl Artif Intell 137:109104
2. Chen FY, Wei YC, Ji HC, Xu GY (2024) A multiplex network based analytical framework for safety management standardization in construction engineering. Adv Eng Inf 59:102278
3. Drobnyi V, Li SY, Brilakis I (2024) Connectivity detection for automatic construction of building geometric digital twins. Autom Constr 159:105281

4. Ye XL, Zeng NS, Konig M (2022) Systematic literature review on smart contracts in the construction industry: potentials, benefits, and challenges. Front Eng Manag 9:196–213
5. Chen H, Yang S, Feng Z, Liu Y, Qin Y (2023) Safety evaluation of buildings adjacent to shield construction in karst areas: an improved extension cloud approach. Eng Appl Artif Intell 124:106386
6. Bakhshi E, Golsanami N, Chen LJ (2021) Numerical modeling and lattice method for characterizing hydraulic fracture propagation: a review of the numerical, experimental, and field studies. Arch Comput Method Eng 28:3329–3360
7. Baduge SK, Thilakarathna S, Perera JS, Arashpour M, Sharafi P, Teodosio B, Shringi A, Mendis P (2022) Artificial intelligence and smart vision for building and construction 4.0: machine and deep learning methods and applications. Autom Constr 141:26
8. Chen X, He Q, Qiu J, Wang L, Su D, Liu M, Chen K, Qiu T (2025) Simultaneous construction of subways and utility tunnels: a case project in Shenzhen, China. Tunn Undergr Space Technol 158:106421
9. Li T, Li X, Rui Y, Ling J, Zhao S, Zhu H (2024) Digital twin for intelligent tunnel construction. Autom Constr 158:105210
10. Xu W, Li X, Wu X, Li T, Chen H, Liu Y (2025) Research on the application of a hybrid intelligent algorithm in reducing the negative response of shield tunneling construction. Eng Fail Anal 169:109044
11. Mostafa S, Sousa RL, Einstein HH (2024) Toward the automation of mechanized tunneling "exploring the use of big data analytics for ground forecast in TBM tunnels." Tunn Undergr Space Technol 146:105643
12. Liu L, Song Z, Li X (2024) Artificial intelligence in tunnel construction: a comprehensive review of hotspots and frontier topics. Geohaz Mech 2:1–12
13. Xu Q, Qiu L, Peng J, Bai C, Chang S, Lu Q, Li W, Zhao J, Chen H, Chen F (2025) Suitability analysis of prefabricated underground utility tunnels crossing active ground fissures. Alex Eng J 110:261–276
14. Liu M, Fang S, Dong H, Xu C (2021) Review of digital twin about concepts, technologies, and industrial applications. J Manuf Syst 58:346–361
15. Boje C, Guerriero A, Kubicki S, Rezgui Y (2020) Towards a semantic construction digital twin: directions for future research. Autom Constr 114:103179
16. Bello SA, Oyedele LO, Akinade OO, Bilal M, Delgado JMD, Akanbi LA, Ajayi AO, Owolabi HA (2021) Cloud computing in construction industry: Use cases, benefits and challenges. Autom Constr 122:103441
17. Wei Y-M, Chen K, Kang J-N, Chen W, Wang X-Y, Zhang X (2022) Policy and management of carbon peaking and carbon neutrality: a literature review. Engineering 14:52–63

MIX
Papier aus verantwortungsvollen Quellen
Paper from responsible sources
FSC® C105338

If you have any concerns about our products,
you can contact us on
ProductSafety@springernature.com

In case Publisher is established outside the EU,
the EU authorized representative is:
Springer Nature Customer Service Center GmbH
Europaplatz 3, 69115 Heidelberg, Germany

Printed by Libri Plureos GmbH
in Hamburg, Germany